Edited by
Florian Rudolf-Miklau
Siegfried Sauermoser
Arthur I. Mears

The Technical Avalanche Protection Handbook

Edited by
Florian Rudolf-Miklau
Siegfried Sauermoser
Arthur I. Mears

The Technical Avalanche Protection Handbook

Ernst & Sohn
A Wiley Brand

The Editors

DI Dr. Florian Rudolf-Miklau
Austrian Federal Ministry of Agriculture,
Forestry, Environment and Water Management
Austrian Service for Torrent
and Avalanche Control
Marxergasse 2
1030 Wien
Austria

DI Siegfried Sauermoser
Austrian Service for Torrent
and Avalanche Control, Section Tyrol
Wilhelm-Greil-Str. 9
6020 Innsbruck
Austria

Arthur I. Mears
Arthur I. Mears, P.E., Inc.
555 County Road 16
Gunnison, CO 81230
U.S.A.

This translation is the revised version of the original German edition titled "Handbuch Technischer Lawinenschutz" (ISBN 978-3-433-02947-3), published 2011. Translated by Monica M. Boensch, Mondsee, Austria. Technical-english review by Emily S. Procter, Bolzano, Italy.

Cover: Snow supporting structures in the municipality Warth/Arlberg, (Vorarlberg, Austria)

Photo: © WLV Vorarlberg, Austria

Library of Congress Card No.: applied for

British Library Cataloguing-in-Publication Data
A catalogue record for this book is available from the British Library.

Bibliographic information published by the Deutsche Nationalbibliothek
The Deutsche Nationalbibliothek lists this publication in the Deutsche Nationalbibliografie; detailed bibliographic data are available on the Internet at <http://dnb.d-nb.de>.

Coverdesign: Sophie Bleifuß, Berlin, Germany
Typesetting: Thomson Digital, Noida, India
Printing and Binding: Strauss GmbH, Mörlenbach, Germany

Printed in the Federal Republic of Germany.
Printed on acid-free paper.

Print ISBN: 978-3-433-03034-9
ePDF ISBN: 978-3-433-60386-4
ePub ISBN: 978-3-433-60387-1
mobi ISBN: 978-3-433-60385-7
oBook ISBN: 978-3-433-60384-0

Preface

Large, high-energy snow avalanches can have high destructive consequences in developed areas. Each year, avalanche catastrophes occur in many mountain regions around the globe. This causes a large number of fatalities and severe damage to buildings and infrastructure. In some mountain areas, especially in the European Alps, a high level of safety for settlement areas is attained by application of technical avalanche defense construction. Simultaneously, new risk potentials continue to emerge in mountain regions from building in endangered areas, the establishment of new roads and railway lines across the mountains and development of tourism (skiing, alpine resorts). These are sometimes located partially or entirely outside protected areas. Consequently the demand for technical avalanche protection in these regions is constantly increasing.

During the last decades technical avalanche protection has evolved – especially in the Alpine countries Austria, Switzerland, Italy and France as well as Norway, Iceland, USA and Canada – from a specialist field to a stand-alone engineering branch. Currently avalanche defense structures and protection systems are established in practically all inhabited mountain regions worldwide. With this engineering handbook the editors are able to provide the first comprehensive overview of the field of technical avalanche protection in the English language and establish a common *state-of-the-art*. The book is based on the German edition, which was published in 2011, and comprises all relevant facts on fundamentals of avalanche protection technology as well as of planning, dimensioning, construction and maintenance of defense structures.

Technical avalanche protection denotes structural measures (defense structures), which are predominantly applied to protect inhabited areas. In such areas frequent and/or large avalanches may occur and cause significant risks to humans and material assets. The structures may consist of steel, concrete, earth, rock or wood material. Planning of defense structures is based on an intensive analysis and assessment of avalanche hazards and risks. Structure design usually considers a *design event*, which takes into account avalanches with a certain probability of occurrence and the applicable mass and energy associated with this design event. An unusual aspect of design, construction and maintenance is the enormous force of impact by avalanches and the extreme environmental and climatic conditions (alpine high altitude areas, subarctic climate) to which the structures are exposed. The extreme terrain and climatic conditions at the construction sites also bring about extraordinary challenges to workers and engineers.

However, several decades of experience in avalanche protection engineering have demonstrated the limits and usefulness of structural avalanche defense systems. Alternatively new technologies were developed in the field of artificial avalanche release, supported by sophisticated methods of avalanche monitoring. One of the starting points for emerging new technologies was the large avalanche cycle in the Alps in 1999. The new methods can be combined with classical defense structures and applied together with other kinds of protection measures (e.g. avalanche warning, closure, evacuation) for the purpose of an integrated avalanche risk management procedure. Temporary avalanche protection systems – in the wider sense of the term *technical avalanche protection* – are also comprehensively presented in this book.

Until recently the state-of-the-art of technical avalanche engineering was available in several normative documents; however most advances in this field result from empirical developments in engineering practice. The highest stage of development and standardization was reached in the field of snow supporting structures in the starting zone. The oldest and best established standard in this field is the Swiss guideline on 'Defense structures in avalanche starting zones' (in its current version 2007) [194], which represents one of the most important sources of this handbook. Recently in several European countries standardization processes took place which lead to the publication of normative documents, partially in order to adapt the Swiss Guideline to national framework conditions: for example France: Norme Française (1992) [219]; Iceland: Jóhannesson und Margreth [148]; Austria: ÖNORM-Regeln 24805 ff. [244–246]. In other countries such as Norway, USA, Canada or Japan still no specific national standards are available. One of the most important steps was the adaptation of norms to the regulation of the Eurocode (unified European standardization). This handbook includes a comprehensive overview of the relevant standards and guidelines of technical avalanche protection at the current status. The Eurocode refers to Swiss (SIA), Austrian (ON), German (DIN) and US standards.

In Chapter 1 the reader is introduced to the system of technical avalanche protection and its historical development based on a fundamental classification of protection measures. Chapter 2 deals with the fundamentals of avalanche formation and the criteria for frequency, magnitude and risk assessment. Subsequently Chapter 3 presents the physical principles of avalanche dynamics impact on objects and the numerical avalanche process models best established in engineering practice. Chapter 4 is dedicated to the system of hazard and risk mapping, based on hazard and risk assessment, and shows the planning processes for structural avalanche defense. The most important protection concepts and goals are also provided in Chapter 4 as well as criteria of a sustainable planning according to technical, economic and environmental principles. Chapter 5 provides a comprehensive and systematic overview of defense structures in the avalanche starting zone as well as the avalanche path and runout zone. All relevant, applicable and historic construction types are presented by technical description system sketches and photographs. The construction and dimensioning of avalanche defense structures, with special respect to supporting components, building material and geotechnical fundaments of foundation are dealt with in Chapter 6. This chapter also comprises all relevant information for dimensioning and technical calculation of required in engineering practice. Chapter 7 presents the fundamentals of construction works and maintenance for avalanche defense structures and with special respect to the Alpine environment. Details on construction methods, construction site infrastructure, transportation systems and construction equipment is included as well as the system of monitoring (inspection) and maintenance for avalanche defense structures over their useful life. Chapter 8 gives a comprehensive overview of the methods of building protection (object protection) in areas endangered by avalanches. Finally Chapter 9 comprises the fundamentals and technology of temporary avalanche protection by artificial release, avalanche warning and monitoring. In this chapter current developments and best practice examples of artificial avalanche release technology from Switzerland and Austria were added (referring to the chapter in the German edition). Chapter 10 finally presents an international overview (table) of avalanche protection in the most endangered countries (based on the German edition).

During the writing of this handbook the editors were able to bring together an international team of leading experts in technical avalanche protection. Authors from Austria, Switzerland, USA, Norway, Canada, Iceland, Japan, France and Italy have directly contributed to this book or supported it with essential information. The book represents a sequel of publication in the field of natural hazard engineering in the framework of Wiley/Ernst & Sohn Berlin publishing house. The main purpose of this publication is to share specialized engineering knowledge and experience in avalanche protection among experts worldwide and contribute to more safety in mountain regions exposed to avalanche risks.

Special thanks go to the Federal Ministry of Agriculture, Forestry, Environment and Water Management in Vienna, the Austrian Service for Torrent and Avalanche Control, the Austrian Standards Institute, the WSL Institute for Snow and Avalanche Research SLF in Davos, the Tyrolean Avalanche Warning Service in Innsbruck, the Austrian Research Centre for Forests, the Austrian Meteorological Service, the Icelandic Meteorological Office (Reykjavík), the American Avalanche Association (AAA), the South East Alaska Avalanche Center (AAC) and the Canadian Avalanche Association CAA (Revelstoke), who have actively supported the creation and elaboration of this handbook. The publication of this handbook would not have been possible without the intensive translation work by DeAnn Cougler (Munich; MB eurocom international languages Vienna) and the critical review by Emily Procter (Bolzano) as well as the design work of Andreas Herbert (Innsbruck). We also thank the legion of colleagues, who have given technical advice and the companies in the field of avalanche protection, who have supported us by latest information on new technologies. Finally special appreciation goes to the team of Ernst & Sohn in Berlin, especially Claudia Ozimek and Ute-Marlen Günther, for the support, patience and engagement to bring avalanche protection technology to the global engineering community.

Vienna,
Innsbruck and Gunnison,
October 2014

Florian Rudolf-Miklau,
Siegfried Sauermoser, and
Art Mears

List of contributors

The editors

DI Dr. Florian Rudolf-Miklau
Austrian Federal Ministry of Agriculture, Forestry, Environment and Water Management
Austrian Service for Torrent and Avalanche Control
Marxergasse 2
1030 Vienna
Austria

DI Siegfried Sauermoser
Austrian Service for Torrent and Avalanche Control, Section Tyrol
Wilhelm-Greil-Str. 9
6020 Innsbruck
Austria

Arthur I. Mears
Arthur I. Mears, P.E., Inc.
555 County Road 16
Gunnison, CO 81230
U.S.A.

The authors

Dr. Karl Gabl
Austrian Central Institute for Meteorology and Geodynamics ZAMG
Branch office Innsbruck
Fürstenweg 180
6020 Innsbruck
Austria

PhD. Peter Gauer
Norwegian Geotechnical Institute NGI
Sognsveien 72
0806 Oslo
Norway

DI Matthias Granig
Austrian Service for Torrent and Avalanche Control
Staff Unit for Snow and Avalanches
Wilhelm-Greil-Str. 9
6020 Innsbruck
Austria

Dr. Robert Hofmann
State authorised and certified chartered engineer
Consulting engineer for structural engineering
Hochstraße 17/2
2380 Perchtoldsdorf
Austria

Dr. Karl Kleemayr
Federal Research and Training Centre for Forests, Natural Hazards and Landscape BFW
Institute for Natural Hazards
Rennweg 1
6020 Innsbruck
Austria

Dipl.-Bauing. ETH Stefan Margreth
WSL Institute for Snow and Avalanche Research SLF
Organisational Unit Snow Avalanches and Preventiontion Measures
Flüelastr. 11
7260 Davos Village
Switzerland

Mag. Michael Mölk
Austrian Service for Torrent and Avalanche Control
Staff Unit for Geology
Wilhelm-Greil-Str. 9
6020 Innsbruck
Austria

DI Patrick Nairz
Avalanche Warning Service of Tyrol
Tyrolean Provincial Government,
Department for Civil Protection and
Disaster Prevention
Eduard-Wallnöfer-Platz 3 (Landhaus 1)
A-6020 Innsbruck
Austria

DI Wolfgang Schilcher
Austrian Service for Torrent and
Avalanche Control, Section
Vorarlberg
Oberfeldweg 6
6700 Bludenz
Austria

DI Christoph Skolaut
Consulting Engineers Skolaut Naturraum
Herzog-Odilo-Straße 1/1
5310 Mondsee
Austria

DDI Dr. Jürgen Suda
alpinfra, consulting + engineering gmbh
Kuefsteingasse 15-19
1140 Vienna
Austria

Dipl.-Bauing. ETH Lukas Stoffel
WSL Institute for Snow and Avalanche
Research SLF
Organisational Unit Snow Avalanches
and Preventiontion Measures
Flüelastr. 11
7260 Davos Village
Switzerland

Dr. Markus Stoffel
University of Bern, Institute of
Geological Sciences
Dendrolab.ch
Baltzerstr. 1-3
3012 Bern
Switzerland

DI Gebhard Walter
Austrian Service for Torrent and
Avalanche Control
Section Tyrol
Wilhelm-Greil-Str. 9
6020 Innsbruck
Austria

MSc Emily Simone Procter
European Academy of Bolzano EURAC,
Institute of Mountain Emergency
Medicine
Viale Druso, 1
39100 Bolzano
Italy

DI Arnold Kogelnig
Wyssen avalanche control AG
Reimmichlgasse 5
6020 Innsbruck
Austria

Mag. Roderich Urschler
SUFAG Snowbusiness GmbH
Hans-Maier-Strasse 9
6020 Innsbruck
Austria

Marco Larghi
Inauen Schätti
Tschachen 1
8762 Schwanden
Switzerland

The Contributors to the survey in chapter 10

Francois Rapin
National Research Institute of Science and Technology for Environment and Agriculture IRSTEA
Research Unit ADRET
BP 76
Domaine Universitaire
38402 St. Martin D'Heres Cedex
France

Bernhard Zenke
Bavarian Environment Agency
Unit 87 - Avalanche Warning Center, Avalanche Protection
Heßstr. 128
80797 Munich
Germany

Yasuo Ishii
Public Works Research Institute
Erosion and Sediment Research Group
Landslide Research Team
2-6-8, Nishiki-cho, Myoko-shi, Niigata-ken, 944-0051
Japan

Dr. Rudolf Pollinger
Autonomous Province of Bolzano-Bozen
Department 30 – Flood Control
Cesare-Battisti-Straße 23
39100 Bozen
Italy

Krister Kristensen
Norwegian Geotechnical Institute NGI
P.O. Box. 3930 Ullevål Stadion
806 Oslo
Norway

Tomas Johannesson
Iceland Met Office
Veðurstofa Íslands
Bústaðavegi 7- 9
108 Reykjavík
Iceland

Pere Oller
Institut Cartogràfic i Geològic de Catalunya
ICGC Territorial Support Center (CST) Pyrenees
Passeig Pompeu Fabra, 21
25620 Tremp
Spain

Chris Stethem
Stethem & Associates Ltd.
409, 8 Avenue
Canmore, AB T1W2E6
Canada

Contents

1 Introduction

Siegfried Sauermoser, Florian Rudolf-Miklau and Stefan Margreth

1.1 Avalanche hazards

1.1.1 Overview and terminology

Avalanches are defined as large masses of snow or ice that move rapidly down a mountainside or over a precipice. The term *snow avalanche* is more accurate to make the conceptual demarcation from other types of avalanches such as rock avalanches or mud flows. According to ONR 24 805, 3.34 [202], snow avalanches are characterized by rapid movement of snow masses that were triggered from the snow cover. Snow avalanches that cause human losses as well as severe property and environmental damage are classified as *natural catastrophes*.

Throughout history, avalanches have had a major impact on the development of settlements in mountain regions (Figure 1.1). This influence is obvious from the location and structure of historical villages and traffic routes. Typical toponyms like *Lähn* or *Lavin* indicate old avalanche paths and are probably derived from the Latin terms *labi* (gliding down) and *labes* (falling) [7]. For many centuries, humans were not able to protect themselves effectively from avalanche hazards and resorted to simplistic solutions such as avoiding areas at risk. Despite the sparse population in Alpine regions, major avalanche disasters with numerous victims occurred repeatedly in history, as people were not able to assess the risk of these infrequent but catastrophic events.

In the last century, increasing populations in the Alps (1870: 7.8 million; 2010: 13.6 million) in combination with growing demands for mobility and leisure activities in Alpine terrain have increased avalanche risk significantly. Traditionally, Alpine valleys were scarcely populated apart from mountain farms, whereas today there are a wide range of competing interests in land use such as settlement developments, traffic, trade and industry, tourism and recreation facilities. This has created progressive consumption of land and use of higher risk areas for building. Some Alpine valleys in well-developed regions are subject to urban sprawl and in areas where tourism is the only profitable economic branch, intensive development of higher elevation areas has occurred, especially for skiing. Though depopulation has been reported in infrastructure-poor mountain regions (*mountain escape*), the Alps will be subject to intensive land use in the future as well since mountains are a sustainable source of natural resources (timber, water, renewable energy and mining).

Increasing traffic density and volume of transportation have resulted in a growing demand for efficient and safe transit corridors across the Alps (e.g. Tenda, Fréjus, Mont Blanc tunnel, Simplon pass, Lötschberg tunnel, St. Gotthard, San Bernadino, Arlberg, Reschen pass, Brenner, Felbertauern, Tauern and Katschberg tunnel, Tauern railway Böckstein/Mallnitz, Gesäuse railway). Outdoor leisure activities and sports (mountaineering, mountain biking, skiing, hunting) have increased human activity in higher elevation areas. In the last decades, the majority of avalanche victims have been skiers off marked slopes as well as ski tourers and free riders.

The Technical Avalanche Protection Handbook, First Edition. Edited by Florian Rudolf-Miklau, Siegfried Sauermoser and Arthur I. Mears.

Fig. 1.1 Alpine living space, shaped by avalanches (© *Sauermoser*)

Increased human impact is noticeable in the European Alps and can be expected in the future in other mountain regions around the world. Avalanche risk and safety expectations have increased significantly while the risk acceptance of a modern society is constantly decreasing. Consequently, the demand for technical avalanche protection in the Alps increased within a short time and prompted rapid development in defense technology. The diverse technological innovations included both new types of avalanche defense structures with permanent protection effects and high-tech systems with temporary protection effects, especially for monitoring and detection of descent or artificial release of avalanches. The establishment of the field of technical avalanche defense as a stand-alone engineering discipline shows the central role avalanches play in mountain regions.

1.1.2 Avalanche hazards: historical and geographical relevance

An avalanche hazard refers simply to a source of potential harm, and is a function of the likelihood of triggering and the destructive size of an avalanche. The different dimensions of avalanche hazards are expressed in the five-point *European Avalanche Hazard Scale* [79] (Table 4.1). Avalanche risk must relate to a specific element at risk, for example people, buildings, vehicles, or infrastructure. Avalanche risk is determined by the exposure of that element and its vulnerability to the avalanche hazard. Avalanche hazards are not necessarily related to catastrophic events. Most of the avalanche accidents causing loss of human life occur in unsecured areas where the people involved actually triggered the avalanche. These so-called *tourist avalanches* happen frequently but generally do not affect settlement areas, traffic routes or infrastructure and thus are not considered target areas for permanent technical defense structures (also for economic reasons). As avalanche size increases, the probability of occurrence decreases but settlements and traffic routes may also be affected. For example, a so-called *hundred-year avalanche* represents an event that occurs – from a statistical point of view – on average once every 100 years.

Fig. 1.2 The Icelandic village Seydisfjördur is a high-risk area for avalanches (© Sauermoser)

Snow avalanches can occur anywhere where sufficient snowfall occurs within a short time on slopes with an inclination of more than 30 degrees. Avalanches occur throughout the Alps and many other mountain ranges in the world including the Pyrenees, Apennines, Norwegian Fjordland, Iceland (Figure 1.2), Rocky Mountains, Andes, Japanese and New Zealand Alps, Elbrus mountains, Hindu Kusch, Pamir mountain range, Russian Altai and Baikal mountains, Chinese Tianshan or Himalayas (Figure 1.3). In ancient times, the Greek geographer Strabon (63 BC to 23 AC) documented avalanche events in the Caucasus Mountains in his scriptures 'Geographica'. In Austria, more than 6000 avalanche paths have a potential impact on settlement areas [35] and countless other avalanches occur in undeveloped mountain areas or remote, seasonally used regions. In Switzerland, more than 20 000 dangerous avalanches are known. The capital of Alaska, Juneau, is an example of an urban area at high-risk from avalanches [60] (Figure 4.5).

1.2 Technical avalanche defense: classification

1.2.1 Classification scheme of defense measures and their effects

An avalanche hazard is not absolute, but is relative to an element at risk. *Avalanche defense measures* are also designed relative to a specific scenario, and several such measures are presented in this book. In countries where avalanche risk is considered substantial, avalanche defense should use a holistic approach that considers various relevant protection goals and possible measures.

Avalanche defense refers to any measure in the catchment area of an avalanche used to achieve the targeted protection goal [202], and is classified as follows [161]:

- *Active* defense measures prevent avalanches from starting or act directly on the flow process, and
- *Passive* defense measures mitigate the consequences of a potential avalanche hazard.

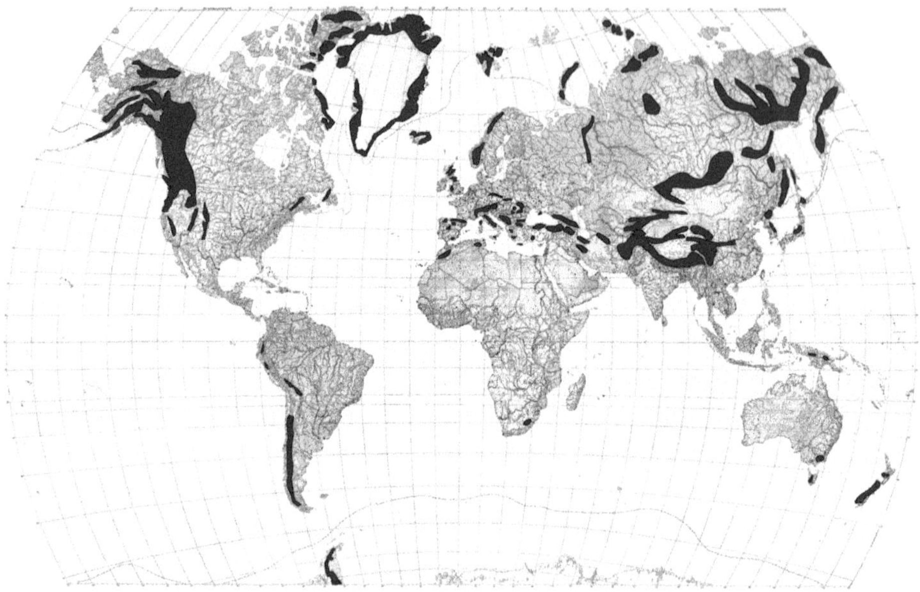

Fig. 1.3 Global overview of mountain regions with potential avalanche hazards (originally elaborated by Glazovszkaya [78]) (The map is only a rough presentation, as no exact survey was carried out)

Active measures are appropriate to reduce the frequency of hazardous avalanches or directly decrease the intensity of the avalanche process. In contrast, passive measures reduce either the damage potential or the vulnerability of objects at risk.

Avalanche defense measures provide either *permanent* (constantly effective) or *temporary* (time-limited effect, adjusted to a specific situation) protection [222]. Table 1.1 gives an overview of the classification scheme of avalanche defense measures.

Another classification of avalanche defense measures uses the *risk cycle of the natural hazard management* [209] (Figure 1.4). According to [222], the hemisphere of precaution comprises prevention, preparation and preparedness; the hemisphere of response (to catastrophes) integrates intervention, assistance and restoration. Most of the measures presented in this book are among the sectors of prevention and preparation.

Holistic systems for avalanche defense have been established in most Alpine countries (Austria, Switzerland, France, Italy, Germany, Slovenia), as well as in other European countries (Norway, Iceland), furthermore in Canada, USA, Japan and New Zealand. Avalanche defense is generally a public service (task of the state), though the degree of responsibility and actual duties varies substantially. This holds true especially for the organization, financing and execution of technical avalanche defense. Furthermore, in other mountainous countries in Europe and around the globe, such as in Poland, Slovakia, Romania, Bulgaria, Spain, Great Britain, Russia, Turkey, China, Andean states, Himalaya and the Caucasus region, avalanche defense has gained in importance due to major events.

Table 1.1 Classification scheme of avalanche defense measures

Defense measure			Permanent effect	Temporary effect
Active	Precautionary effect	Reducing the disposition for an event	Forest and bioengineering measures (protection forest, high-altitude afforestation) Avalanche defense structures: snow supporting structures, snowdrift control structures	Artificial release of avalanches
		Acting directly on the avalanche process	Avalanche defense structures: dams, breakers, tunnels, galleries	Closure for roads Evacuation (of buildings at acute risk)
	Reaction to an event			Emergency measures (after an event) Catastrophe management
Passive	Precautionary effect		Legal measures (regulations, prohibitions) Hazard mapping Planning measures (land use planning) Administrative measures (building permission, relocation of buildings at risk) Structural building (object) protection Catastrophe management plans	Information (risk communication) Avalanche monitoring and prediction Avalanche commissions Avalanche warning service
	Reaction to an event			Preparedness Catastrophe management

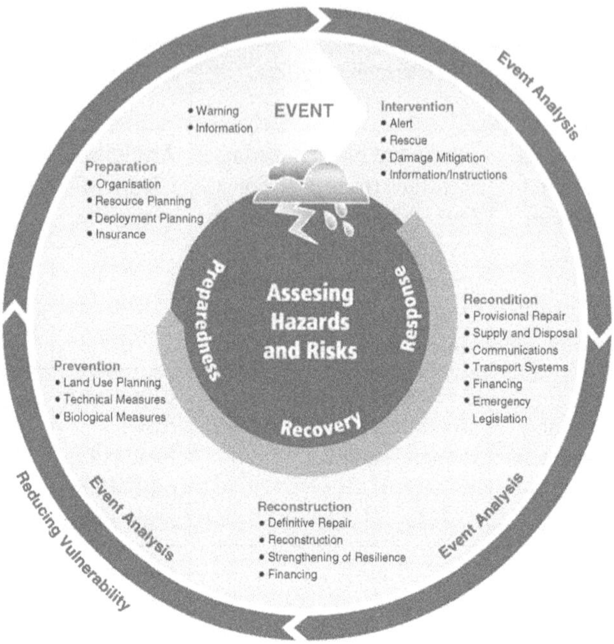

Fig. 1.4 Risk cycle for natural hazard management (© AdaptAlp)

1.2.2 Permanent technical avalanche protection (defense structures)

In the relevant technical standard literature (e.g. *Margreth* [165], *ONR 24805* [202]) the term *technical avalanche defense* is equated with structural (constructional) defense measures with permanent effects – in contrast to the technical avalanche defense measures with temporary effects (Section 1.2.3 and chapter 9). The protection effect of these measures is constant, that is independent of the actual avalanche risk or season.

Technical defense measures typically refer to avalanche defense structures, meaning constructed works (sometimes including mechanical and electronic components) and are termed *avalanche defense structures* in the engineering field (Figure 1.5 a and b).

According to [165], structural avalanche defense is based on one of two strategies:

- hinder initiation or propagation of an avalanche by stabilizing (support) the snow pack in the starting zone or by reducing snow drift (snow displacement by wind), or
- break, decelerate, retard, deflect or retain avalanches in motion (deflection or retarding structures).

Measures based on the first strategy are used in the starting zone of avalanches (Figure 1.5a), whereas measures based on the second are constructed in the avalanche path or runout zone (Figure 1.5b). Table 1.2 gives an overview of the classification and function of structural avalanche defense structures. A third group of measures includes structural building (object) protection, whereby the protection effect is defined for a single

(a)

(b)

Fig. 1.5 Examples of structural avalanche defense structures: (a) snow nets in a starting zone (© Sauermoser); (b) avalanche retarding dam in the municipality of Galtür (Tyrol) (© Rudolf-Miklau)

object (e.g. residential house, towers of a cable car, electricity pole) (Chapter 8). Object protection measures are amongst the avalanche defense structures.

1.2.3 Technical avalanche defense with temporary effects

In this book the term *technical avalanche defense* is used in a broader sense and also comprises active and passive measures with temporary protection effects. These are

Table 1.2 Overview of avalanche defense structures classified by the function and location in the catchment area, according to [161]

Structural avalanche defense	Avalanche defense structures				Object protection
Category of defense measure	Snow drift control structures	Snow supporting structures	Avalanche catching and retarding structures	Avalanche deflection structures, snow sheds and tunnels	Structural building protection
Function (protection effect)	Structures that control the snow drift and snow accumulation in the starting zone.	Structures that stabilize and sustain the snowpack in the starting zone and prevent the release of avalanches.	Structures that stop or decelerate the motion of avalanches or dissipate the energy in order to reduce the run out distance.	Structures that deflect avalanches in motion from objects at risk or to by-pass them from traffic routes (roads, railway lines).	The building at risk is enforced in a way that it is able to withstand the impact (stress) of avalanches with little damage.
Type of defense structure	Snow drift fence Wind baffle Wind roof (Jet roof)	Snow bridge/rake/ net Combined snow bridge (steel/wood) Terrace	Avalanche catching or retarding wall (dam) Avalanche mound Avalanche breaker	Avalanche deflecting dam (wall) Gallery (shed) Tunnel	Avalanche splitting wedge Roof terrace Impact wall
Location in catchment area	Starting zone	Starting zone	Avalanche path Runout zone	Avalanche path Runout zone	Avalanche path Runout zone

measures with effects that are limited in time and that require additional assessment of the actual avalanche danger (e.g. by an avalanche commission).

Technology for the artificial release of avalanches is amongst these temporary defense measures (Section 9.2) and includes structures or facilities adapted to specific avalanche hazard situations. These usually supplement other protection structures but may in a few special cases substitute them. To initiate avalanche release, additional loads are applied to the snowpack and fracturing occurs at natural weak zones/layers. According to [263],

Fig. 1.6 Artificial avalanche release facilities with the system Gaz.Ex (© Interfab Snowbusiness GmbH)

a wide range of new technologies for artificial release of avalanches is available on the market (example system Gaz.Ex; Figure 1.6). Artificial release minimizes the duration of traffic routes closures (roads, railway lines, cableways, ski slopes) or evacuation of buildings (public places). In Europe, this measure is applied predominantly for protection in ski areas and traffic routes, albeit with some limitations, but is rarely used for the protection of settlements and buildings due to legal and safety concerns.

Technical systems (facilities) for avalanche defense are also used for avalanche monitoring, prognosis and warning (alert) (passive defense measures with temporary effect). Avalanche monitoring and prognosis (warning) requires digital measuring technology, remote sensing and computer-based models. According to [111], these technical systems facilitate assessment of avalanche hazards, recording of snow layering and compilation of relevant meteorological data with the aim to create daily updated avalanche reports (warning) with regional relatedness. Computer-based models for prognosis are appropriate to assess the actual avalanche hazards on a local level, if sufficient documents and data on historical avalanches are available (Section 1.3.1). All these technologies represent an indispensable support for the work of regional authorities and avalanche commissions for a specific emergency situation (blockage, evacuation, closure). In comparison, direct remote sensing and detection of moving avalanches (e.g. by high-speed cameras, geophones or radar) is primarily of scientific interest but will gain importance in the future with improvements in technology (Section 9.3).

1.3 Avalanche disasters, development of avalanche defense: historical overview

1.3.1 Chronicle of avalanche catastrophes

1.3.1.1 Avalanche disasters in the Alps

Avalanche events have been recorded throughout the history of settlements in the Alps. Livius [7] reported heavy losses caused by avalanches during Hannibal's crossing of the Alps (loss of 18,000 men and 2,000 horses). Famous Alpine villages such as Heiligenblut or St. Christoph/Arlberg owe their origin to avalanche disasters. One of the most disastrous avalanche winters was in February 1689 with a toll of 265 people including 80 people in the Swiss villages Saas and St. Antönien [117]. In 1667 in Switzerland, the village Anzonico was totally destroyed and 88 people died. Avalanches played a major role during World War I along the Dolomite front. Extraordinary snowfall and low temperatures led to catastrophic avalanches that buried several thousand soldiers on both sides of the front. On the Austrian side of the front alone, some 6000 soldiers were killed on December 16, 1916 on *black Thursday*.

Extreme avalanche events in the twentieth century include (see Figure 1.7):

1909	The most severe avalanche catastrophe of the European Railway history happened during the construction of the Tauern railway when 26 workers were killed [117].
1916	During World War II, Italy and Austria had military bases in the Alps and these troops were soon to find that bombs and enemy fire were not the only threats – heavy snow instigated a series of avalanches in the Tyrol region causing the death of 10 000 soldiers on what became known as *White Friday* 1916.
1924	Avalanches caused 9 deaths in Styria [117].
1935	Heavy snowfall (12.5 m snow depth in Langen am Arlberg, Austria) led to numerous avalanches in Austria, Switzerland and South Tyrol. More than 100 people were killed by avalanches, 50% of them were skiers [117].
1950/51	In Austria in more than 1000 avalanches, 135 people were killed. In Switzerland 98 people were killed.
1954	An avalanche catastrophe in Vorarlberg, Austria, caused 143 deaths [117].
1968/70	There were 24 avalanche victims in the region of Davos, Switzerland [5]. In 1970, several enormous avalanches occurred in Austria (Figure 1.8) and France with 39

deaths in Val d'Isère. An avalanche killed 74 people on the Plateau d'Assy in Savoyen.

1998/99 The most recent avalanche catastrophe occurred in February 1999 and caused 70 deaths in France, Switzerland and Austria; 31 persons died in a large avalanche in Galtür, Austria.

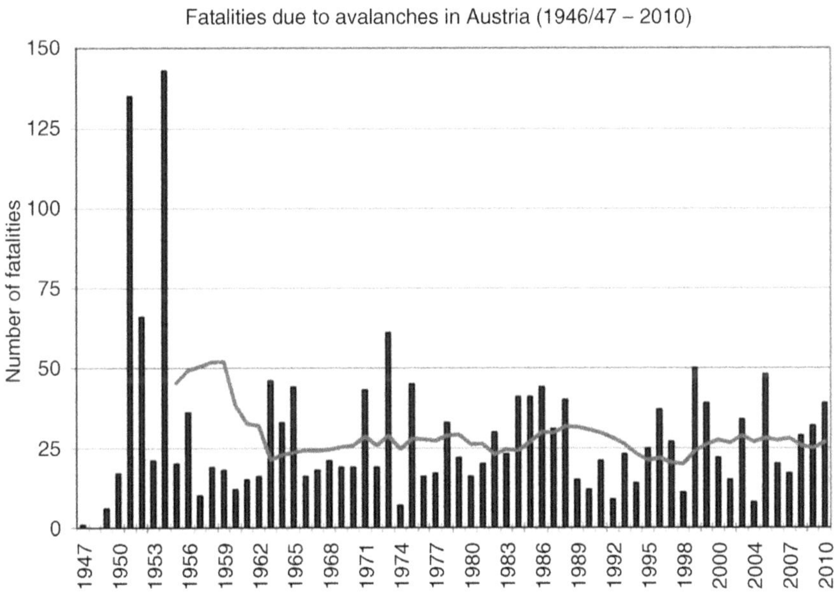

Fig. 1.7 Annual number of avalanche victims in Austria from 1945 to 2010 (© Austrian Board for Alpine Security Surveillance)

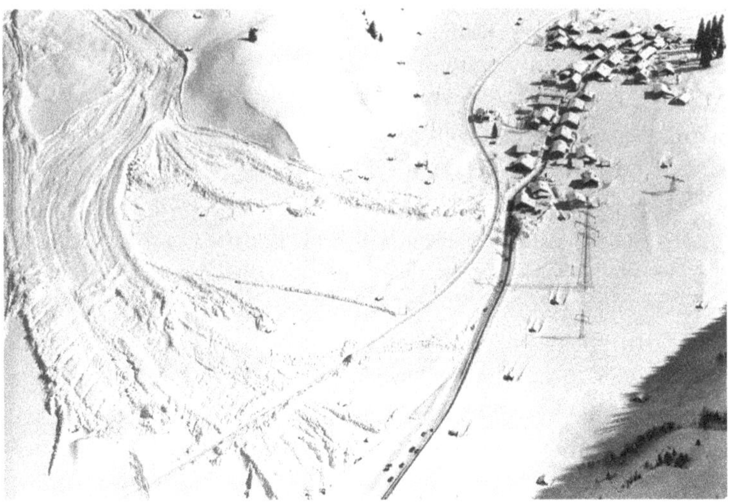

Fig. 1.8 Wiestal Avalanche (municipality Bichlbach, Tyrol) after the avalanche event in 1970 (© WLV Tyrol)

Statistical classification of the recurrence probabilities of historic avalanche cycles is limited because of a lack of long observation periods. For the avalanche event in Galtür in February 1999, the return period has been retrospectively estimated as 100–200 years [169] and the recurrence probability of the snow precipitation as 300 years [68].

1.3.1.2 Avalanche disasters in other regions

Avalanches are observed worldwide. The highest number of fatalities ever recorded in one event was the catastrophic mudflow in Huascaran, Peru, in 1970 where 25 000 people died [88].

Other notable avalanche events include:

Norway	From 1836 to 1998: 1510 people were killed in avalanches (in 1679, there were 130 fatalities in Western Norway; in 1868, there were 161 fatalities) [143]. March 5, 1986: In Vassdalen in Nordland county, a snow avalanche was released from Storebalak. 31 men from the North Norway Brigade were involved, 16 men were killed, 15 survived. March 1909 and March 1956: 51 fatalities on Lofoten islands. February 1928: 45 persons were killed mainly by slush flows in Hordaland, Sogn.
Iceland	1995: 34 avalanche fatalities in two avalanche events in Sudavik and Flateyri [125]. 1974: An avalanche killed 12 inhabitants in Neskaupstadur. 1910: An avalanche killed 20 inhabitants in Hnifsdalur. 1919: An avalanche killed 18 inhabitants in Siglufjördur.
USA/ Canada	1910: Two passenger trains were buried by avalanches at Stevens Pass, Wellington, 97 passengers died [8, 142] 1910: 62 persons died at Rogers pass, Canada. 1965: 40 persons died in the Granduc mine disaster.
Turkey [87]	1975/76: 9 avalanche events with 170 fatalities. 1991/92: 112 avalanche events with 328 fatalities. 1992/93: 31 avalanche events with 135 fatalities.
Chile	The most disastrous avalanche event within the last 100 years happened in August 1944 in the working class district of the copper mine El Teniente with 102 casualties [150].
Pakistan	February 2010: 102 fatalities in Kohistan. April 2012: An avalanche hit a military camp near the Siachen glacier in the Karakoram branch of the Himalaya mountains, 135 soldiers died.
India	1979: A series of avalanches buried the valley leaving at least 200 victims in Lahaul Valley.
Afghanistan	2010: 166 fatalities at Salang Pass. 2012: Several avalanches in the Daspai area killed 201 people.

Corsica	1934: An avalanche from Castagniccia caused 37 fatalities in Ortiporia.
Romania	April 1977: 23 fatalities in Balea lac in the Southern Carpatians.
Slovakia	1924: Vel'ká Fatra avalanche in the Low Tatra mountains destroyed half of the village and killed 18 people. 1956: Vajskorska dolina avalanche caused 16 fatalities.
Japan	1993–1998: 143 avalanche disasters reported, these caused 50 fatalities.

1.4 History of avalanche defense

1.4.1 Historical development in Europe

The first structural avalanche defense structures were built in the form of earth or rock walls and were positioned directly above the endangered object (by hand, since machinery was not available). In Austria, the first known direct defense structure was a rock fill wedge erected in 1613 in the village of Galtür to protect the houses in the area called Birche. In Switzerland, technical defense structures have existed since the sixteenth century. One example is the defense wedge on the church Frauenkirche in Davos (Figure 1.9).

Avalanche defense walls in Austria were known as *Schneearchen*, *Spaltecken* or *Sauköpfe* [122]. Many houses had shed-like roofs to guide avalanches over the house. Although these first technical defense structures were primarily direct defense structures, organizational measures were also common. Houses located in safe areas were designated as meeting spots during evacuations and high-risk periods. It is still possible to find these cellars, called *Lahngrube*, *Lahnkeller* or *Lawinengruften* in old farmhouses today.

Fig. 1.9 Historical avalanche defense structures: Endangered by the avalanche Frauentobellawine, the church Frauenkirche was protected by an avalanche splitting wedge, constructed most probably in the 16th century (© Margreth)

Fig. 1.10 Historical avalanche defense structures: rock fill terraces at the Schafberg mountain in Pontresina (Switzerland), constructed around 1900 (© Margreth)

The need for a more systematic approach to defense structures arose in the 19th century, for example on a larger scale in the starting zones of avalanches. Emphasis was placed on the maintenance and restoration of protection forests and, through this, the frequency of avalanches was effectively reduced. Coaz [42] (the first Swiss forest inspector and pioneer in avalanche defense) reported the first avalanche defense structure in the 18th century; these were earth terraces with a length of 100–200 m, depth of 0.8 m and distance between the rows of approx. 20 m.

The erection of stone or earth terraces was the beginning of systematic protection in starting zones (Figure 1.10). However, Coaz noted already in 1910 that the height of the terraces was too low to support the accumulated snow cover and prevent avalanche release. Newer rock walls were built with a height of up to nine meters, and between 1876 and 1938, approx. 100 km of stone walls had been erected in Switzerland [64].

In Austria, the first systematic avalanche defense structures in a starting zone was built by the Austrian Service in Torrent Control at the Rax Mountain in Lower Austria. In the starting zone of the avalanche Lahngrubenlawine 770 m of rock fill walls and terraces and 23 wooden snow rakes were built [33].

A vast network of technical avalanche defense structures was necessary at the end of the 19th century during the construction of trans-Alp roads and railways. During the construction of the Arlberg railway from 1880 to 1884, serious avalanche accidents occurred on the western part of the pass. The first defense structures were built by Pollack (an employee of the railway company) (Figure 1.11a). In Switzerland, the first avalanche gallery was built on the Splügen and Simplon passes (Figure 1.11b). In

(a) (b)

Fig. 1.11 Pioneer technical avalanche defense structures: (a) avalanche defense structures in the starting zone included rock fill terraces and snow rakes (*Arlbergrechen*) for the Arlberg railroad (© ÖBB); (b) avalanche tunnel with barrel vault, constructed in 1824 at the Splügen pass in Grisons (© Cantonal department of monument preservation Gisons, Switzerland)

Austria, the first avalanche gallery to protect a road was built by Karl Ritter von Geha in 1854. By 1888, 21 avalanche galleries had been constructed with a total length of 1.6 km.

The avalanche disasters of 1951 and 1954 intensified efforts to establish widespread technical avalanche protection in Alpine countries. In comparison to the vertical earth or stone terraces that had been used until this point but were ineffective because of rapid filling with falling or drifting snow, supporting structures made of steel, rope wires or wood erected perpendicular to the slope were thought to be more efficient. Steel snow bridges were developed by the Austrian Alpine Montan Union in 1955 and were used for the first time at the protection site on the mountain Heuberg in Häselgehr, Austria (Figure 1.12).

The first types of steel snow bridges were constructed with concrete foundations. This method carried high transportation costs since the concrete and steel parts had to be transported with a cable crane to the starting zone and from there with a narrow gauge railroad along the hillside to the construction site. To reduce costs, tests were made with anchors and blasting anchors. Nowadays foundations consist of ground plates for the supports and micropiles or anchors for the girders (Section 5.2.3).

Already in 1955, the first edition of the 'Swiss guideline for defense structures in avalanche starting zones' was published [50, 51]. After some revisions, the *Swiss guidelines* [163] have become the technical reference for supporting structures in many countries. Technical measures that influence the distribution of snow in or around the starting zone were also investigated. In Switzerland, the first snowdrift measures were erected in 1908 at the avalanche site Faldumalp in Valais, where rock fill walls were used to influence the snow distribution within the starting zone. In Austria, wind

Fig. 1.12 First avalanche defense structures in the starting zone with prefabricated snow bridges in steel, Heuberg, Häselgehr (© WLV Tyrol).

roofs and wind baffles were erected at the avalanche site Heuberg in Häselgehr (Sections 5.2.4.3 and 5.2.4.4).

If construction of protection measures in the starting zone was not feasible because of costs or unfavourable conditions for foundations, technical defense was implemented along the path or runout zone, usually in the form of deflecting or retarding dams or walls. Historical retarding mounds are visible along the Penzenlehner and Arzleralm avalanches above Innsbruck (Tyrol, Austria). After two severe avalanche cycles in Austria in 1935 and 1951, different kinds of retarding measures (concrete splitting wedges, earth or stone masonry retarding mounds) were erected on the mountain Nordkette near Innsbruck (Tyrol) to reduce the runout distance of the avalanches that endangered the town.

Many of these historical protection measures are still used today, which in part indicates that the basic principles of technical avalanche defense have not deviated far from the original principles introduced in the middle of the 20th century.

2 Avalanches: evolution and impact

Patrick Nairz, Art Mears, Siegfried Sauermoser, Karl Gabl, Markus Stoffel and Stefan Margreth

2.1 Characteristics of avalanches

2.1.1 Definitions and classifications

An *avalanche* is a rapid flow of snow, ice, rock or mud along a slope with a volume of more than 100 m^3 and a path length of more than 50 m. Depending on the material involved, it is classified as a snow, ice, rock or mud avalanche. The term avalanche implies the whole process of movement, from release in the starting zone to displacement (flow) through the path and deposition in the runout zone. The *runout zone* is an area with an average inclination less than 10°, that is where the inclination is insufficient for further movement of snow.

The load on the snowpack may be caused by gravity, in which case failure results either from weakening in the snowpack or from increased load from precipitation, or from other external factors, for example skiers, snowmobilers or explosives. Slow movements of the snow cover (on a scale of mm to m per day) are described as *snow creep* or *snow glide*.

Precise classification of avalanches is necessary for standardized reporting and analysis, however there are several classifications. The first attempts at avalanche classification were simple schemes aimed at alpinists. Flaig [58] developed the *avalanche table* that distinguished dry-snow avalanches, wet-snow avalanches and ice avalanches, and Zsigmondy & Paulke [286] distinguished dry and wet new-snow avalanches and old-snow avalanches.

The first detailed avalanche classification was published by the UNESCO in 1981 (*Avalanche Atlas*) [273]. The purpose of this was to observe and record avalanche characteristics in a condensed form suitable for the evaluation of statistical and physical principles governing these phenomena. A secondary purpose was to enable users to describe an observed or expected avalanche in simple terms, easily understood by others familiar with the classifications. This classification defines *genetic* (i.e. the origin or precipitating conditions) and *morphologic* features of an avalanche (Table 2.1) and this classification is still commonly used for avalanche reporting.

2.1.2 Spatial and temporal occurrence of avalanches

A minimum inclination for snow accumulation and avalanche formation is approx. 30°. At slopes with inclination higher than 50° sufficient snow accumulation is not likely. In the Alps, most avalanche starting zones are situated above the potential timberline, which is between 1800 and 2000 m depending on the position within the Alps (inner or periphery regions). The actual timberline is on average 200–400 m lower because of historical clear-cutting practices in subalpine areas for Alpine pastures [63]. Uniform and trough-shaped slopes promote snow accumulation and avalanche formation. The

The Technical Avalanche Protection Handbook, First Edition. Edited by Florian Rudolf-Miklau, Siegfried Sauermoser and Arthur I. Mears.

Table 2.1 *Morphological avalanche classification* (Avalanche Atlas, 1981 [273], modified in ONR 24805 [202]): the classification is based on the manner of starting, form of movement, form of the path, and manner and form of deposition

Zone	Criterion	Alternative characteristics	
Starting zone	Manner of starting	Starting from a point (loose-snow avalanche)	Starting from a line (slab avalanche) (Figure 2.1a, 2.12)
	Position of sliding surface	Within snow cover (surface-layer avalanche)	On the ground (full-depth avalanche)
	Liquid water in snow	Absent (dry-snow avalanche)	Present (wet-snow avalanche)
Avalanche path	Form of path	Open slope (unconfined avalanche)	Gully or channel (channelled avalanche)
	Form of movement	Snow dust cloud (powder snow avalanche) (Figure 2.1b)	Flowing on ground (flow avalanche)
		Mixed (mixed avalanche)	
Runout zone	Surface roughness of deposit	Coarse (>0.3 m) (coarse deposit)	Fine (<0.3 m) (fine deposit)
	Liquid water in snow debris at time of deposition	Absent (dry avalanche deposit)	Present (wet avalanche deposit)
	Contamination of deposit	None apparent (clean avalanche)	Present (contaminated avalanche)

inclination of slopes above the tree line in typical glacial-formed alpine valleys is generally between 30° and 50°, thus typical for starting zones. Heavier precipitation in periphery areas of the Alps causes higher avalanche activity than in drier, inner alpine regions.

Frequent avalanche activity seems to occur in so-called *avalanche cycles* in which widespread areas are affected concurrently by avalanches due to the spatial extent of precipitation events. Weather conditions conducive for avalanche formation mostly concentrate on the northern or southern regions of the Alps. The avalanche cycle in 1999, for example, covered only the northwestern part of the Austrian mountains, and the avalanche cycle in 1954 covered a small area in the westernmost part of Austria.

(a)

(b)

Fig. 2.1 Types of avalanches: (a) Fracture of a slab avalanche and (b) powder snow avalanche in motion (© WLV Tyrol)

2.2 Meteorological principles of avalanche evolution

2.2.1 Weather conditions forming avalanches in the European Alps

2.2.1.1 General remarks

To a large degree, avalanches occur because of extreme weather phenomena (snowfall, temperature, wind). Strong snowfalls, warm and cold air occurrences, strong winds, on both the northern and southern sides of the Alps, are caused by specific weather conditions.

Depression areas, which determine the weather in Europe, mainly occur over the Atlantic Ocean and travel westward in the direction of the continent. Because of the

Fig. 2.2 Typical avalanche path in Alpine valleys (© Sauermoser)

temperature inversion in the valleys, the temperature in the high mountain areas cannot be calculated directly based on the valley temperatures. If the Alpine area is located at the edge of a high-pressure area over Russia, high-altitude continental cold air flows into the Alpine area. When snow volumes are low, these weather conditions lead to the formation of depth hoar. However, if the Alpine area is being affected by a high-pressure area centred over the Alps or Mediterranean, the eastern Alps will be influenced by tropical warm air.

Especially during the winter, warm fronts carry intensive precipitation with them, because, physically speaking, warmer air also can absorb more moisture than colder air. After a cold front has passed through, the current often turns from northwest to north. This results in strong build ups of precipitation along the northern reaches of the Alps.

The following weather conditions are relevant in the Alps in connection with the formation of avalanches.

2.2.1.2 Northwestern (precipitation build up) area

Between a depression area reaching from Poland to Northern Scandinavia and a high-pressure area over the Atlantic (Azores to the Gulf of Biscay), there forms a northwestern upper airflow in the Alps. While the depression area on the western side guides the Arctic cold air from the Arctic Ocean in the direction of central Europe, the high-pressure area over the Atlantic distributes heat and, especially, moist air from the northwest to the Alps. Cold and warm fronts (Figure 2.3) occur at the border of these air masses.

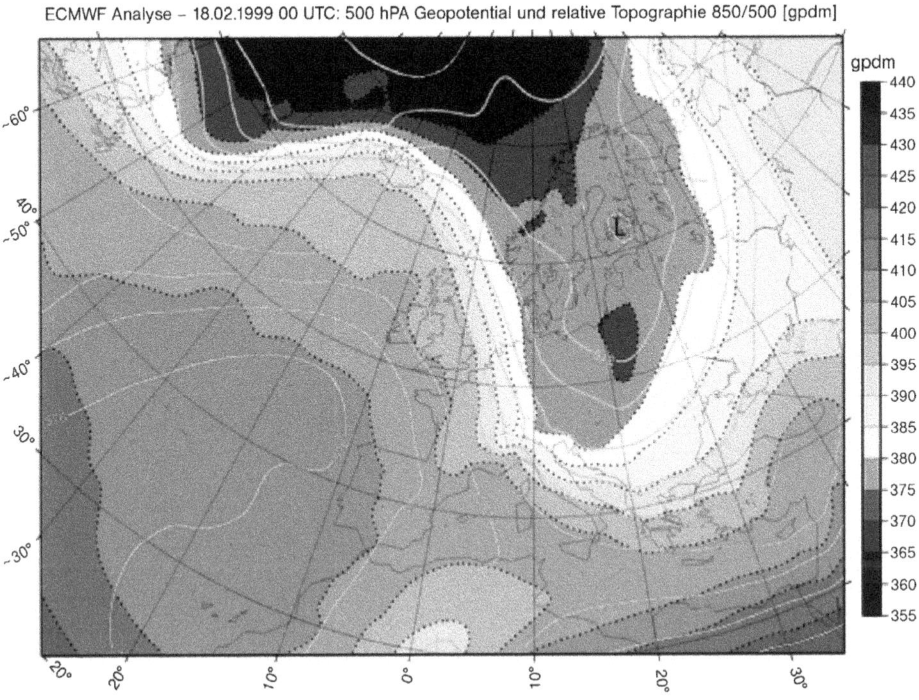

Fig. 2.3 Northwestern area on 20.02.1999 (500 hPa area) (© ZAMG, EZMWF)

The northwestern current results in a lifting on the northern rim of the Alps, and buildup of the air in the Alps. Under these conditions, the northern borders of the Alps and the buildup areas experience large amounts of new snow. The Alpine regions receive smaller amounts, and it is predominantly dry on the southern side of the Alps (the north-foehn side). Under such weather conditions, 222 cm of new snow fell in Rauz in Arlberg within 2 days on January 1954 (125-year event), or in Galtür in February 1999, 254 cm in 10 days (300-year event).

The large amounts of new snow in the Alpine region from the northwestern areas are due to warm fronts and vorticity. Warm air from the northwest glides on the cold air, and, furthermore, turbulences in the northwestern current result in heavy precipitation caused by the strong vertical movement. Strong snow drifting can be deduced from the weather maps (isobars).

2.2.1.3 Western weather conditions

Depression areas with connected fronts pass the Alps from west to east. The quick interchange of warm and cold fronts results in larger temperature fluctuations.

As already mentioned, the strong lifting effects in the atmosphere are responsible for the heavy precipitation. Besides the described lifting in the mountains (damning up), on the

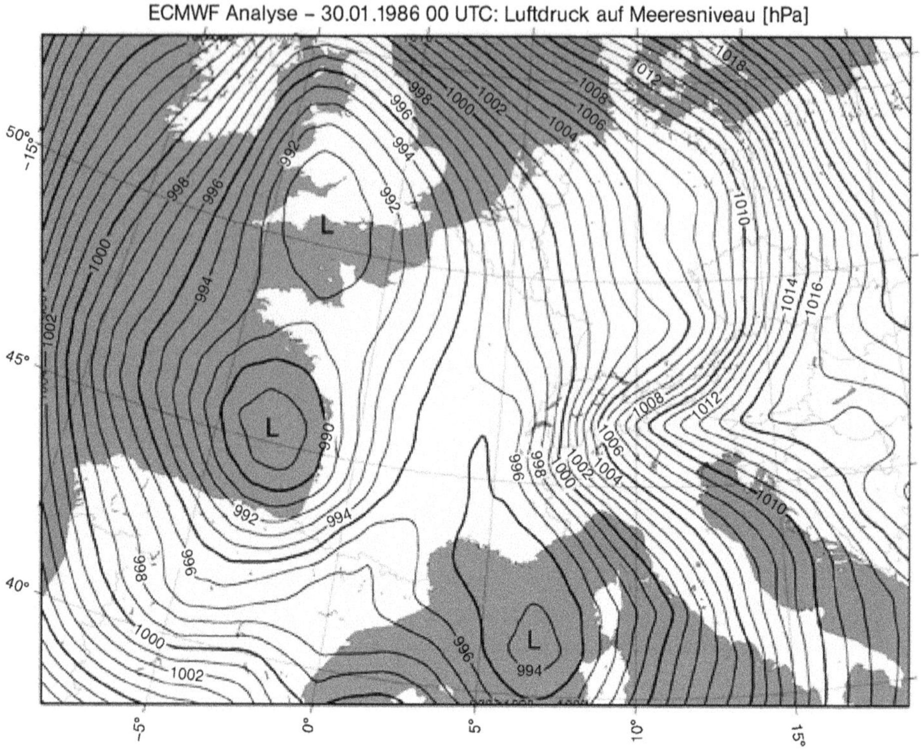

Fig. 2.4 Ground weather map at a south foehn situation (© ZAMG, EZMWF)

fronts (frontal lifting), as well as upgliding warm air, the strong high-altitude current (jet stream) and turbulence formation (vorticity) play a major role. During cases of heavy precipitation during the winter, the presence of the jet stream must be taken into account when seeking meteorological advice. Turbulence in the atmosphere may occur due the velocity differentiation in a large current area, and result in the air mass either lifting or sinking. By evaluating the numerically calculated turbulence size, significant liftings – at least in large areas – can be recognized and used to forecast extreme snowfall.

2.2.1.4 South foehn situation

As opposed to the northwestern area when the depression is located in Eastern Europe, for a southern current the controlling depression area is located above Western Europe. In Figure 2.4, there is a depression area with its centre in the vicinity of the Gulf of Biscay, and close to the British Isles, as well as a secondary depression above the Gulf of Lion. The resulting current transports humid warm area from southern Europe toward the Alps, during which the air can absorb large amounts of moisture above the Mediterranean. These humid air masses concentrate on the southern side of the Alps and on the main Alpine crest, and lead to large-scale and abundant precipitation and snowfall.

The situation on the northern side of the Alps – just a few kilometers to the north – is completely different. The falling air movement leads to disintegration of the clouds. There is only mild precipitation; the southern foehn disperses the clouds. Extensive wind transport takes place on the northern slopes because of the foehn.

2.2.1.5 Occlusion from the North East – north-east location

Under these weather conditions, a depression system forms to the North East of the Alps. The occlusion from this depression reaches the eastern Alps from the north east and carries warm air into upper altitudes which then glides onto the cold air in the Alps or is forced to rise in the Alps. This leads to intensive precipitation. Depending on the location of the occlusion, heavy precipitation is to be expected along the northern rim of the Alps.

2.2.2 Weather conditions forming avalanches in North America (western ranges)

There are two basic types of climate conditions (*snow climate*) forming avalanches in the Rocky Mountains (USA, Canada: western ranges), which are characterized as *maritime* or *continental*, and additionally as *transitional* (in US: *intermountain*) between the two [175]. The climatic conditions in the western part of North America are controlled by the longitudinal extension of this mountain range from north to south. McClung & Shearer [172] describe these types of snow climate as follows:

The *maritime snow climate* is characterized by relatively heavy snowfall and relatively high temperatures. While rain may fall at any time during the winter, cold Arctic air can appear several times per winter. Maritime snow covers are deep and often very unstable with rapidly fluctuating instability. Typical examples of maritime ranges include the Cascade and Sierra Nevada Range in the western United States and the Coast Mountains in British Columbia. The average annual snowfall in the maritime ranges of North America is 15 to 25 m. Avalanche formation in maritime snow climates usually takes place during and immediately following winter storms, with failure occurring in the new snowstorm near the surface. A significant cause of major avalanching can be rain if it immediately follows deep, new snow. Rainfall may also cause formation of ice layers, which can act as future sliding layers when buried by subsequent snowstorms.

The *continental snow climate* is characterized by relatively low snowfall, cold temperatures, and a location considerably inland form coastal areas. Typical examples of continental ranges include the Rocky Mountains in Canada and Colorado, and the Brooks Range of Alaska. The annual snowfall in the continental ranges of North America is usually less than 8 m. As with any snow climate, most avalanches are caused by snow loading (new snow or drifting snow), and most form within new snow. Major avalanches, however, are often related to buried, structural weaknesses, which persist for long periods with a different character than maritime snow climates. Consequently, failures of old snow layers are a distinguishing feature of a continental snow climate. Often in a continental range, it is possible to find nearly the entire snowpack consisting of weak faceted crystals and depth hoar grown under high temperature gradients, at least where the snowpack is thin.

It must be stressed that in the mountain ranges of the American West considerable overlap occurs, with many areas exhibiting transitional *snow climate* features associated with both classes. That is, maritime conditions can occur in a continental range and vice versa. Examples of ranges that usually display transitional features are the Selkirk Range of British Columbia and the Wasatch Range of Utah. The transitional snow climate resembles closest the conditions in most parts of the European Alps. It should also be stressed that persistent heavy snow leading to exceptional avalanches may occur as a result of strong southwest flow driven by the high-altitude jet stream. This has occurred in the Sierra Nevada of California, Mountains of Wyoming, Utah and Colorado. For example, in an 8-day period in February 1986, 700 mm of snow-water equivalent (SWE), all in the form of snow, was deposited in the Sierra Nevada of California, 500 mm of SWE in the Wasatch Mountains of Utah and 350 mm SWE in central Colorado. This produced exceptionally large avalanches with apparent return periods of 100–300 years or more.

2.2.3 Weather conditions forming avalanches in other mountain regions

The types of snow climate conditions described in Section 2.2.2 in general can be transferred to other mountain areas on the globe. The prevalence of a specific snow climate is primarily a function of the position and extension as well as of the distance of the mountain range to the ocean or sea. Additionally the local climatic conditions of an avalanche catchment depend on whether it is situated on the windward side or lee side of the predominant cyclonic flow.

Examples for mountain ranges with maritime snow climate are the Andes in South America, the mountains of the Norwegian fjord land, the mountain areas of Iceland or the west coast of New Zealand. Continental snow climate dominates, for example, the Pamir Mountains, most parts of the Caucasus Mountains and the Altai Range in China.

Snow climate in the Himalayas is predominately governed by the monsoon. Both snow accumulation and snow cover duration differ widely along the Himalaya ranges. Snowfall can occur during the whole year in the western Himalaya, due to both summer and winter monsoon and mid-latitude low-pressure systems bringing moisture into the region. The Central Himalaya and on the Tibetan Plateau, a much more marked dry season occurs from October to March. Snow cover does not have a pronounced seasonal cycle in these regions, since it depends both on the quite variable duration of the monsoon and on the rare but possible occurrence of snowfall during the extra-monsoon period [174].

2.3 Nivological principles of avalanche evolution

2.3.1 Properties of material snow

Snow is characterized as either single or collective crystals in solid form, made up of precipitation or its deposition on the ground [180]. Snow is amongst the most complex, continuously changing material that we know. The primary reason for this is that snow consists of forms of all three aggregate conditions: solid (= snow crystals), fluid (= embedded water), and gas (= air and water vapour in the cavities). Depending on the mixing ratio of the three components and the degree of impurity, snow has very varying physical properties, which have an effect on avalanche events.

New snow reflects a large part of the incoming short-wave radiation. Conversely, with regard to long-waved radiation, snow is a *black beamer* (complete absorption of the radiation). The radiation that penetrates the snowpack decreases exponentially according to depth. Snow is, because of its air voids, a perfect insulator. The thermal conductivity of the snow is, therefore, low; it is structurally and texturally dependent and increases as density increases. Snow is furthermore a viscous medium with regard to material properties, and becomes brittle under high deformation velocity.

2.3.2 Genesis of snow

The entire weather activity takes place in the troposphere. The water vapour contained in the air is decisive for the formation of precipitation. Its proportion of the total air volume is between a few parts per mil and a maximum of 4%. What is also important are the aerosols contained in the air, that are small solid particles (e.g. soot particles, pollen, or salts). They form the condensation nucleus to which the water vapour in the air docks when the saturation level is correct. Very clean (under-cooled) water droplets form, which still can be in a liquid state at temperatures around –40 °C.

In order for snow to form, frozen nuclei are also required. Aerosol particles are also a factor. They, however, must have a similar structure to ice crystals, and they are encountered significantly more rarely than condensation nuclei. If the air temperature drops below –40 °C, the water droplets freeze spontaneously into ice crystals. Further crystal growth takes place through so-called deposition or resublimation, that is the direct deposition of water vapour on the already existing crystals.

In their basic shape, *ice crystals* are always hexagonal. This has to do with the placement of the oxygen and hydrogen atoms at clearly defined locations in a hexagonal lattice (60-degree angle). Depending on the air temperature and humidity, very different types of crystal shapes are formed (Figure 2.5). Typical are stars, needles, pillars, prisms, and flakes.

The crystals are whirled around inside the clouds. In this connection, single crystals may stick or become hooked together; the ensuing aggregates are termed snowflakes. Ice crystals or snowflakes only fall to the ground when the uplift within the clouds is less than their weight. Soft hail constitutes a special variety of snow. It forms when under-cooled water droplets embed in existing ice crystals. The latter, in this case, serve the same purpose as frozen nucleus.

Frost and solid precipitation (also: *rime*), conversely, count as settled precipitation (Figure 2.6). Frost forms when the water vapour in the air settles on the snow surface. Solid precipitation, on the other hand, forms when under-cooled water droplets hit cold, solid objects. This requires somewhat strong winds. Solid precipitation therefore forms against the wind direction.

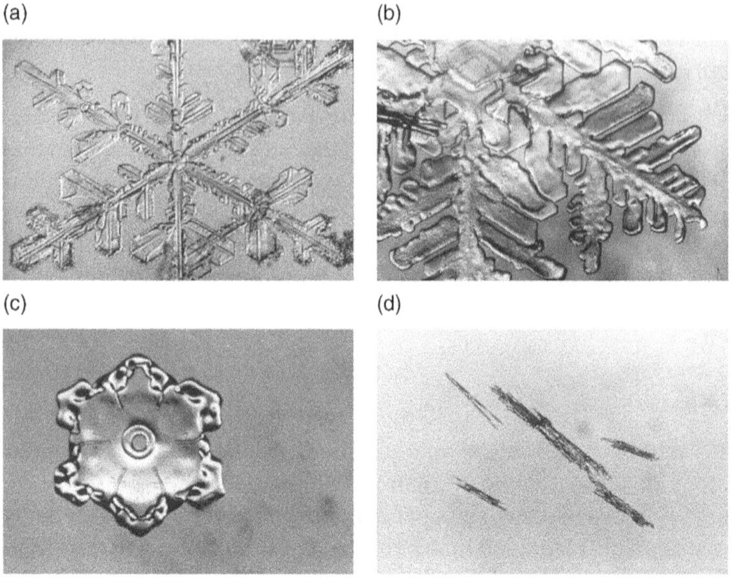

Fig. 2.5 Different types of snow crystals. (a): Star-shaped snow crystal (Dendrit); (b): Transition shape between Dendrit and flake-shaped snow crystal; (c): Leaf-shaped snow crystal; (d): Needle-shaped snow crystal (© EAWS)

Fig. 2.6 Deposition of solid precipitation (rime) on an automatic weather station (© Nairz)

2.3.3 Snow metamorphism

Editor's note: The terminology used is this section results from a translation of the original German text. It is not intended to replace terminology on snow metamorphism currently used in North America.

2.3.3.1 General remarks

The *snow metamorphism* describes the transformation of the lying snow. This is a continuous process and begins in the atmosphere. Already during the snowfall, the temperature, moisture and wind conditions change at different heights, and this affects the crystal growth and destruction. The metamorphosis takes places at the time of the snow deposition on the ground up until the final melting process. The metamorphosis is decisive for avalanche formation because this causes weak layers to form in the snowpack.

2.3.3.2 Principles of snow metamorphosis

During the metamorphosis, the snow's density, structure, texture and strength change. This can happen because of different processes:

- The wind causes a mechanical destruction of the ice crystals.
- Diffusion processes within the snowpack lead to growth and shrinkage of snow crystals.

Furthermore, melting and rainwater influence the snowpack, particularly during spring. Furthermore, the dead weight of the snowpack affects the individual snow crystals and, as a result, leads to densification.

Decisive for the metamorphosis processes, which result from diffusion processes, is the temperature gradient within the snowpack: the steeper the gradient, the more intense the metamorphosis process. A gradient of 25°/m is considered high and leads to a massive exchange of mass. Due to the high vapour pressure of ice and snow, material transport takes place largely through vapour diffusion, that is ice crystals sublimate, while water vapour is deposited on solid parts [245]. Recent research by Schneebeli et al. [244] shows that material transport between the crystals within the snowpack is considerably more dynamic than was previously suspected. It has been shown that – when the temperature gradients are large – most crystals consist of completely new ice encased in the snowpack after only three days.

A basic differentiation is made between the following metamorphosis processes in the snowpack:

- Initial metamorphism
- Equilibrium growth metamorphism
- Faceting
- Melt-freeze metamorphism.

A classification of snow crystals is included in the *International Classification of Seasonal Snow on the Ground* [57].

Fig. 2.7 Redistribution of snow by wind (© Nairz)

2.3.3.3 Initial metamorphism

Initial metamorphosis generally results from wind and gravitation. This mechanically destroys the crystals. Typically, the snow collects on the leeward side, that is on the side away from the wind, where it also forms thick, partly dangerous snowdrift accumulations (Figure 2.7). A prerequisite for this is, however, a poor connection with the old snow layer.

2.3.3.4 Equilibrium growth metamorphism

Equilibrium growth metamorphism begins immediately after the snowfall. At that time, the snow crystals are often arborescent and loose the tips and edges. Because of this, in extremely steep areas individual loose crystals, which were once connected, may become instable. This minimal impulse is often sufficient to pull additional crystals along and cause loose-snow avalanches. Furthermore, the equilibrium growth metamorphism leads to settling of the snowpack. The pore volume decreases, the snow becomes denser. This is usually accompanied by a stabilization of the snowpack, because the contact points between the crystals increase. If, on the other hand, loose new snow falls on a particularly weak layer (e.g. depth hoar), which is subsequently bound, for example through radiation or temperature rise, this may also have a negative effect since tension is transferred suddenly onto larger areas and may cause slab avalanches.

Vapour diffusion processes play a significant role in equilibrium growth metamorphism. This places the snow crystal in a more advantageous position, energy-wise. Furthermore, the dead weight plays a certain role in the metamorphism. The air temperature is decisive. The warmer it is, the faster the equilibrium growth metamorphism takes place.

(a) (b)

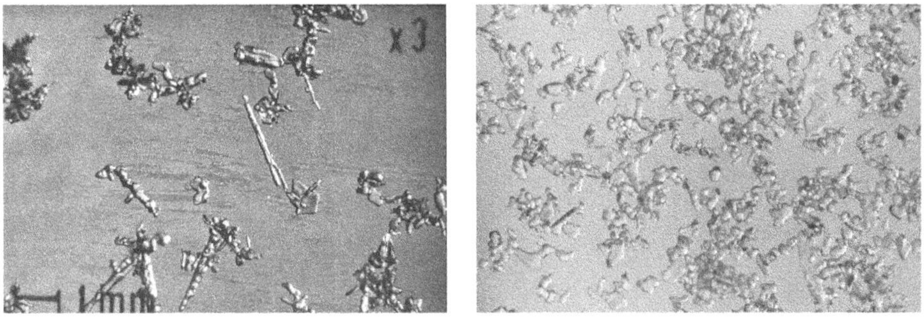

Fig. 2.8 (a): Felted and (b): round snow (© EAWS)

The crystal shapes during equilibrium growth metamorphism include felted and round snow (Figure 2.8 a, b).

2.3.3.5 Faceting

In contrast with the equilibrium growth metamorphosis, the crystals grow during *faceting*. The snow's thermal properties mean that on the ground the temperature measured is usually around 0 °C, while the snow surface temperature often can reach −25 °C. At a snow depth of 1 metre, the decisive temperature gradient for the facting is 25°/m. Since the warm air close to the ground can absorb more water vapour than the cold air above it, a water vapour pressure gradient comes into play, which increases in size proportionally to the temperature difference. Subsequently, water molecules diffuse from the bottom to the top, and embed themselves in the colder crystals. From there, a constant yet smaller mass transport takes place. The crystals continue to grow and become increasingly cohesionless.

Edgy shapes form in the beginning and develop into jar-like *depth hoar* (Figure 2.9). This is called depth hoar because it primarily is formed close to the ground and therefore cannot be recognized from the outside, for example without snow profile recordings or a *stock test*. Furthermore, depth hoar is also observed inside the snowpack, primarily under hard layers. Decisive for this is a high temperature gradient in a small space, which is present particularly when cold snow falls on a moist snowpack and vice versa.

The formation of *surface hoar* is comparable to the formation of depth hoar. The water vapour required for its formation comes from the surrounding air. The surface hoar forms in the vicinity of water streams during clear, cold winter nights but may also form a widespread weak layer on top of an existing snowpack.

Surface hoar and depth hoar are amongst the most critical weak layers. Together with the edgy shapes, they are to be seen at the majority of avalanche accidents. During prolonged cold periods, the entire snowpack often metamorphoses destructively, and is then loose and low in tension. To begin with there is a low avalanche hazard, which

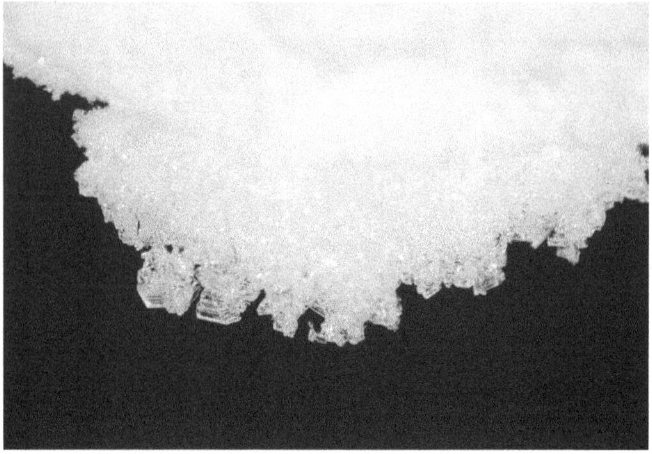

Fig. 2.9 Depth hoar crystals (© Nairz)

quickly increases when snowfall and wind begins, and which within a short time, that is the same day, can become a big hazard.

2.3.3.6 Melt-freeze metamorphism

Melt-freeze metamorphism begins as soon as the snow surface temperature reaches 0 °C. This can happen because of spring-like weather conditions or rains. In this situation, the original ice connections are increasingly replaced by water connections. These results in two opposite processes: the cohesion between the crystals diminishes, while the capillary strength increases. However, free water usually has an adverse effect on the stability of the snowpack.

This is why spring is generally a very active time for avalanches. In just few days, most of the spontaneous avalanches from the winter season are activated as wet-snow avalanches. Decisive factors are a high air temperature in combination with increased humidity, generally higher radiation intensity and little wind, and as well as when the snow surface temperature comes close to the zero degree limit and reaches it. If the snowpack is loose, predominantly loose-snow avalanches occur; if the snowpack is compact and there is at least one embedded weak layer, large, wet slab avalanches will also occur.

On the other hand, the change between humidity during the day and cooling down during the night of the snowpack is generally more advantageous. Under these conditions, *hard-crusts* form predominantly near the surface and with a load carrying capacity. These become increasingly stable the colder the temperature, the lower the humidity, and the clearer the night.

The end product of the melt-freeze metamorphism is spring snow, which is a coarse-grained crystal influenced by water. If the snowpack is completely wet, it is termed isothermal. In this case, the snow temperature is 0 °C everywhere.

2.3.4 Snowpack

2.3.4.1 Formation of snowpack and layering

The *snowpack* presents a picture of the weather event from the time of snowfall, and consists of snow embedded in layers with different characteristics (Figure 2.10). Furthermore, the snowpack metamorphoses in different ways, depending on the exposition and height position. On a small scale, one may observe quite large differences in the snowpack structure. Conversely, on a large scale one finds – if the meteorological conditions are similar (e.g. within the same exposition and elevation) – a very similar snowpack structure. In steep, south side and crest-near areas, the sun radiation may have caused the formation of a compact snowpack with a melting hard crust, while one may find an accident-sensitive snowpack behind the ridge in very steep shadowy areas. On a large scale, the profiles on the south side slopes are similar, as are the profiles on the slopes with shade.

2.3.4.2 Movements and tensions in snow cover

In principle, every snowpack embedded on a sloped area moves down the valley due to gravity. This happens as a result of the *settling* process of the snowpack without noticeable crack formation. The snowpack remains anchored on the soil and only moves in the snow layers embedded above it. The steeper the area, the thicker the snowpack, and the closer to the snow surface, the more intensive is the so-called *creeping* motion and hence the larger is the ensuing tension (Figure 2.11a).

During the *glide* movement, on the other hand, the entire snowpack moves parallel to the slope by a few millimeters to a metre a day. This means that the entire snowpack can tear all the way down to the soil. *Glide snow mouths* are formed (Figure 2.11b) following which creeping snow avalanches may also form. A glide snow mouth may also persist throughout the entire winter without it resulting in a glide snow avalanche.

The effect that the creep and glide movement have on the avalanche formation is discussed in Section 6.2.5.2.

Both the creep and glide movements can result in a decrease or increase in tension within the snowpack. Both tensions, pressure and shear stress are affected by this. Tensile stress typically occurs in extraction areas, that is slope increase, compressive stress in hollow areas, and also at slope decreases. The tension component that is decisive for the formation of avalanches, however, is the shear stress, which becomes increasingly larger proportional to how poorly the layers are connected with each other. Primarily, a snowpack breaks within a weak layer, and not in the tension or pressure zone. However, a weak layer can very well be destroyed in the pressure zone, for example the flat toe of a slope, and the break can spread to the adjacent slope. This has been termed *remote release*. It is only after this that the snowpack is fractured, both in the tension and pressure zones.

The fracture mechanism was simulated experimentally by Van Herwijnen et al. [275]. This research confirmed the conjecture that the fracture within the snowpack is initiated at one point (e.g. through additional load). A fracture of the snow crystals occurs locally within the weak layer, which results in a weak settlement. Sometimes the typical 'wumph' noise can be heard at this point. The fracture now moves on, but an avalanche

(a)

(b)

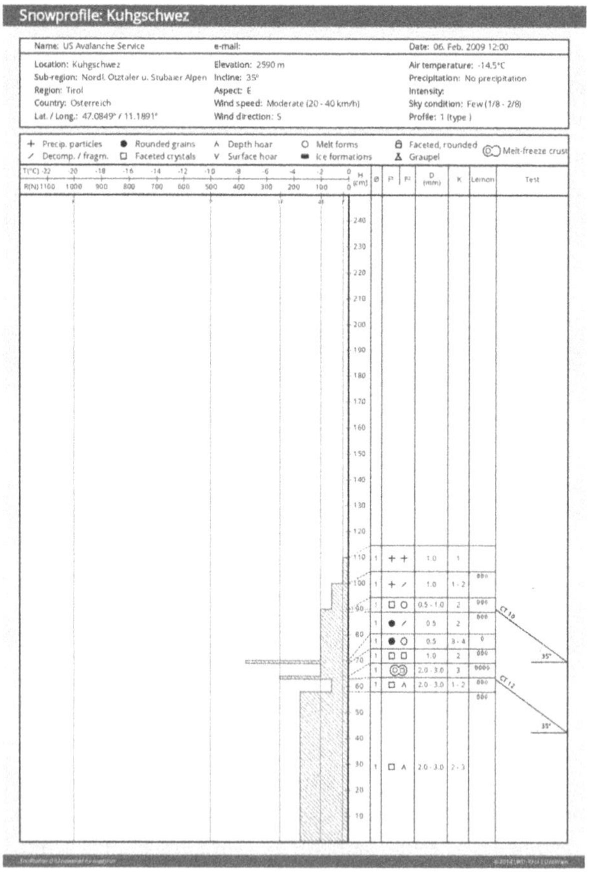

Fig. 2.10 Layer construction of a snowpack: snow profile (a): in the area (© Nairz); (b): survey form (© Nairz: notional sample version)

(a)

(b)

Fig. 2.11 Movements in the snowpack. (a): Creep movements (© Sauermoser); (b): Glide movement: glide mouth (© Nairz)

initiation is, at the end of the day, also dependent on the steepness of the area. The typical angle is more than 30° because at any threshold angle below that, the friction exceeds the downslope stress.

2.3.4.3 Avalanche classification according to evolution

The changeable (not location specific) framework conditions for the formation of avalanches are the basis for the *genetic avalanche classification* (Table 2.2). The criteria for this classification correspond to the previous formation mechanisms for avalanches.

Fig. 2.12 Fracture zone of a slab avalanche (© Nairz)

2.4 Frequency and magnitude of avalanche events

2.4.1 Criteria for frequency, magnitude and risk assessment

Avalanches, similar to other geophysical processes have an inverse relationship between magnitude, M, and frequency or probability, p, for example $M \propto 1/p$. Avalanche magnitude M is measured by mass, volume, energy, runout distance, impact pressure or some other relevant measure and p is the probability. p is usually normalized to a one-year period (e.g. a 1% annual probability). The inverse of p, the return period T is sometimes used, where $T = 1/p$.

For engineering or planning purposes, both M and p (or T) are usually defined at a specific location in the avalanche path where the object or activity (structure, human activity) is exposed. This defines the *hazard* at that location. Avalanche hazard maps (e.g. red and blue or yellow zones) in Switzerland, Austria and sites in the United States and Canada are defined in terms of magnitude and/or frequency (Section 4.1.2.1). The *encounter probability* E is the probability that an avalanche will reach a given location once during a *time period* of L years. If the return period T is known and since avalanches in each year are statistically independent, encounter probability can be calculated.

$$E = 1 - \exp(L/T) \quad (2.1)$$

From Equation 2.1, if the return period T and the observation period L are equal then the probability of observing an avalanche at some point is about 65%. E increases to

Table 2.2 *Genetic avalanche classification* (UNESCO Avalanche Atlas, 1981 [273]): A stringent genetic classification would designate an avalanche according to its formation cause. However, since most avalanches result from various interacting causes, the genetic classification was an attempt to provide an overview of the location-specific and changeable framework conditions

Changeable framework conditions for the formation of avalanches	Effect on avalanche intensity
1 Continuous weather (period approx. 5 days back)	
1.1 Snowfall	Increasing load, increasing mass of low stability; most important factor in the formation of avalanches
– Type of new snow	Fluffy snow: Loose-snow avalanches Cohesive Snow: Slab Avalanches
– Daily new-snow increment	Increased instability med new snow height, new or old-snow fracture
– Intensity of the snowfall	With high intensity progressive instability, furthering the new-snow fracture, expansion of danger on slopes with small inclination
1.2 Rain	Furthering wet loose-snow avalanches or soft slab avalanches. Mixed snow and landslide
1.3 Wind	Two effects: – Increase in snow deposits – Increase in brittleness of the snow
– Wind direction	Increased formation of slab avalanches on leeway slopes; formation of snow cornices
– Wind speed and duration	As speed and duration increase, the local formation of slab avalanches increases; slab avalanche possible
Thermal conditions Normative conditions: – Snow temperature – Snow moisture	Ambivalent effect on strength and tension, that is on formation of avalanches Increase in snow temperature effects instability, but in the end stabilization Increase in moisture leads to formation of avalanches
Air temperature	Similar effect on all slopes
– Sun radiation	Predominant effect on sun exposed slopes

(continued)

Table 2.2 *(Continued)*

Changeable framework conditions for the formation of avalanches	Effect on avalanche intensity
– Temperature radiation	Cooling of the snow surface at night and in the shade, significant when no clouds on the sky. Furthers the formation of surface and depth hoar
2 Old snow conditions *Previous weather influence integrated across the entire previous winter*	
2.1 Total depth of snow cover	Not predominant factor for avalanche hazard. Significant for volume of full-depth avalanches. Important for the fastening and metamorphism of the snowpack
2.2 Layering strength profile	Stability determined by the weakest layer Tension Condition
– Surface layer	Looseness (surface hoar), brittleness, roughness important for the adhesiveness of further depositions
– Inner layers	Old-snow fracture cause by weak middle layers (old surfaces) and depth hoar
3 Release conditions	
3.1 Natural release	Natural avalanche
– Inner influences	Spontaneous avalanche
– Outer (not human) influences	Naturally released avalanche
3.2 Release through humans	
– Unintentional release	Unintentional avalanche
– Intentional release	Artificially released avalanche

about 90% if $L = 2\,T$. For example, if a complete record of avalanches in some path is 200 years long and some location within the potential path has not been reached during that time, there remains a 10% chance that the location is within a 100-year avalanche.

When hazard is known and the exposure of people or facilities at some location in the avalanche path is also known, the *risk* to persons or facilities may also be determined. Appropriate mitigation at that location may then be proposed or designed based on the risk. Risk is defined as the probability of encounter and the consequences of this encounter. (Risk is discussed more fully in Section 4.1.2.2)

Consequences are defined as the amount and cost of property damage and/or the probability of death or injury. A *quantitative risk assessment* (QRA) is sometimes done to quantify the risk and the cost. In a QRA, the risk to a fatality or person may be compared to some acceptable baseline risk that is understood and accepted by society. An example of a baseline risk is the probability of a natural death (e.g. unrelated to avalanches) in the healthiest segment of the general population (10–14 years old). *Acceptable* risk can be defined by an individual, society or a political organization. Individuals have a very wide range of risk tolerance and consequently engage in highly variable activities. In contrast, societal risk is usually related to the least tolerant acceptable risk of an individual within that society. A political body (government, etc.) may have a duty to protect all members of the society it is responsible for, including legal risks.

2.4.2 Frequency (recurrence) of avalanche events

Frequency of avalanches is an essential element of hazard or risk analysis. Avalanche frequency varies substantially with climatic region, elevation, location of a mountain range with respect to storm tracks, location within a particular avalanche path and whether artificial (man-induced) triggering of unstable snow slopes is used to create avalanches deliberately. Some snow climates (Section 2.2.2) receive large storm and wind events much more frequently than others whereas the location and orientation of some avalanche paths within a particular range can strongly affect frequency. Within an avalanche path, frequency is greater in the upper, steeper portions of the terrain as discussed in Section 2.5. When frequency and risk exceed acceptable levels structural mitigation may be required in spite of high costs. Examples of unacceptably high risk levels may occur when frequent avalanches intersect busy highways or where the annual probability of avalanches is 0.3% to 1% at residential buildings, public facilities or areas with concentrated public use.

2.4.3 Magnitude of avalanche events

Avalanche magnitude is classified in North America by two methods: the *Canadian method* related to destructive effects and absolute size and the *United States system*, which is relative to the size of the maximum potential size within a particular avalanche path [254].

The idea of the Canadian system is to estimate the potential destructive effects of avalanches. This is defined in the following table where avalanches are classified by typical mass, path length or impact pressure.

The largest avalanches within the seasonal snowcover probably range up to 10^6 tons mass, path length to 5000 m or more, and maximum velocity from 50 to 80 m/s. Such avalanches are probably limited to large glaciated mountains of Western North America and may also include glacial ice. Because of the remoteness of these areas, little data are available for these very large events.

In the United States, avalanches are classified according to the Canadian system described above or by the US (relative) system described below. The five sizes [205] relative to the particular path are *very small, small, medium, large, and maximum*. This system is useful in describing the degree of instability in the snowcover and has been used for several decades to describe this instability amongst avalanche forecasters. It is useful for communicating a sense of the general hazard within the snowcover at a particular time. It is not a useful system for engineering or land-use planning applications. Currently avalanche sizes are classified in the United States by both the U. S. relative system and the Canadian absolute size system.

Table 2.3 *Canadian Snow Avalanche Size-Classification System* (from McClung and Schaerer, 2006) [254]

Size	Description	Mass	Length	Impact pressure
1	Relatively harmless to people[a)]	<10 t	10 m	1 kPa
2	Could Bury, injure, or kill a person	10^2 t	100 m	10 kPa
3	Could bury a car, destroy a small building, break trees	10^3 t	1000 m	100 kPa
4	Could destroy a railroad car, large truck, several buildings, or forest area up to 4 hectares	10^4 t	2000 m	500 kPa
5	Largest known, could destroy a village or forest area of 40 hectares	10^5 t	3000 m	1000 kPa

a) People have been killed in roof slides or small avalanches of size class 1.

2.4.4 Intensity of avalanche impact

Engineers often consider potential impact pressures in design of structures [254]. Impact pressures range from relatively harmless powder blasts from mixed-motion dry-snow avalanches or small snow sluffs, to high destructive forces of large dry-flowing avalanches capable of destroying reinforced structures (see size classification, Table 2.3). Impact pressure p_f is proportional to flowing density ρ_f and velocity squared v^2 so the pressure on a large, flat surface is

$$p_f = k \cdot \rho_f \cdot v^2 \quad [\mathrm{kN/m^2}] \tag{2.2}$$

where:

k factor [−]
depending on orientation of the surface to avalanche flow direction, whether the avalanche is flowing with a relatively high volume concentration of solids or is a diffuse powder cloud.

In Section 3.3.1, we suggest that a short-duration pressure peak may be 3 times the average pressure p_f. Because of the relative importance of ρ_f and v in Equation 2.2 above, the largest impact pressures are usually associated with dry-flowing avalanches. These avalanches often constitute the design case. Wet-flowing avalanches may have densities 2 or 3 times larger than dry avalanches but usually have smaller velocities, thus smaller impact pressures. A possible exception to this may be the largest slush-flow avalanches where velocities have reached 30 m/s and flowing densities may exceed 700 kg/m^3. Such slush avalanches also often entrain rocks. This complicates impact further.

In structure design, the impact force F on a structure is at least as important as the magnitude of the pressure because it is directly related to overall stability of a structure rather than destruction at a point. Force is calculated

$$F = p_f \cdot A\ [kN] \tag{2.3}$$

where:

A area exposed to impact factor [m^2]

Furthermore, p_f may vary over the surface area, for example ranging from larger pressures at the lower part of the structure that is exposed to denser flowing snow to lower pressures exposed at higher parts of the structure, especially if they are only exposed to low-density powder blast.

In practice, calculation of p_f may be further complicated by impact of solid objects such as trees trunks and rocks that can be entrained into the flowing snow. When impact of solids must be considered, assumptions must be made about the type and size of the solids and their orientation with respect to the impacted surface. (Section 3.3.1.5)

Impact pressure may range (Table 2.3) over three orders of magnitude, from 1 to 1000 kPa. Avalanche dynamics and impact are discussed in more detail in Chapter 3.

2.5 Morphological principles of avalanche evolution

2.5.1 Avalanche catchment area

The total area affected by an avalanche is designated a *catchment area*. It consists of the starting zone (or release zone), the avalanche path, and the runout zone (or deposit zone) [202, 273]. The *morphological avalanche classification* drawn up by the UNESCO [273] bases this on the morphological division of a catchment area, and largely this remains valid to this day. Austria's Forest Act [55] also mentions the *accumulation zone*. This is the area from which wind-transported snow is received, and which therefore is paramount to the placement of snow drift structures. There is an overview of the location-specific framework conditions for the formation of avalanches in Table 2.4.

Table 2.4 Location-specific framework conditions for the formation of avalanches (*UNESCO Avalanche Atlas*, 1981 [273])

Location-specific framework conditions for the formation of avalanches	Effect on avalanche intensity
1. *Terrain conditions*:	
1.1 Relative elevation	
General topographical situation	Effect dependent on the geographic width and the overheight of the surrounding mountains
– Crest locations and high areas	Strong winds, cornices, local slab avalanches
– Area between tree line and crests	Expanded areas for the formation of slab avalanches
– Area beneath the tree line	Decreased wind effect, decreased formation of slab avalanches
1.2 Inclination	
>35°	Formation of loose-snow and slab avalanches possible
>25° (28°–30°)	Formation of slab and loose-snow avalanches possible
>15°	Stationary or accelerated flowing movement
<20° (10°)	Deceleration and/or deposition
1.3 Slope Direction	
– relative to the sun	Increased formation of dry slab avalanches on shaded slopes, on sunny slopes increased formation of wet-snow avalanches
– relative to the wind	On leeward slopes, drifting snow accumulation increase formation of slab avalanches, opposite on windward slopes
1.4 *Terrain type*	
– Open, even slopes	Unconfined avalanches
– Gullies, craters, treads	Gully avalanches, concentrated limitation from the side
– Gradient changes	Slab and loose-snow avalanches on convex slopes

(continued)

Table 2.4 (*Continued*)

Location-specific framework conditions for the formation of avalanches	Effect on avalanche intensity
– Gradient levels	Power avalanches, cascade formation
1.5 *Roughness*	
– Even soil	Snow slipping (on wet soil) full-depth avalanche preferential
– Protruding obstacles (Rocks, transverse ribs)	Surface-layer avalanche over level of roughness
– Vegetation	Grass: snow slipping and full-depth avalanche preferential Thicket: decrease of formation of avalanches when not covered by snow Forest: formation of avalanches prevented when not snow-covered

2.5.2 Avalanche starting zone

The *starting zone* is the area where an avalanche is released. The main characteristic of a starting zone is the slope gradient, which influences the release of an avalanche considerably. Slopes with an inclination >30° are generally categorized as starting zones [7, 71]. Avalanche releases below 30° are rare, and Margreth [161] defines 28° as the lower limit of the inclination, the ONR 24 805 [202] also defines a slope inclination area to be 28°–55°. In UNESCO's Avalanche Atlas [273], it is mentioned that large and dangerous avalanches predominantly break at an inclination between 35° and 40° (inclination area for the most common fracture zone of slab avalanches).

At above 55°, predominantly loose-snow avalanches break [163] (Figure 2.13). The size of an avalanche starting zone varies in accordance with factors such as slope inclination, roughness, altitude and exposition. In [254], the average relation between the length and width of a slab release is given at 0.5, although the width varies from a factor of ten to a factor of several hundred times fracture height. The maximum size of a starting zone is unlimited and is generally dependent on the factors listed in the genetic avalanche classification (Table 2.5) as location-specific framework conditions.

The *absolute elevation* of a starting zone influences the probability of snow accumulation, as does the exposure, the ridge or high area positions, the total shape in profile and lengthwise, the roughness of the substratum, and the position relative to the timberline.

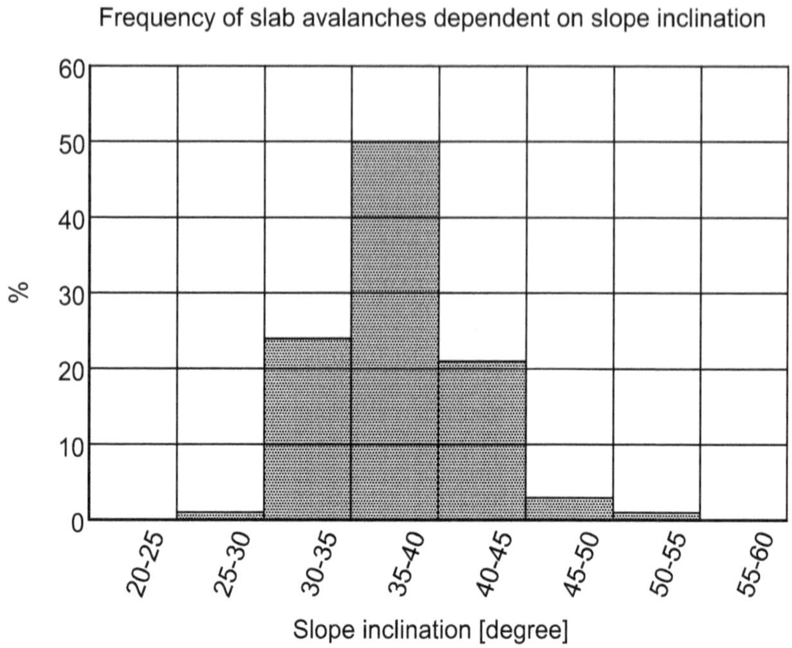

Fig. 2.13 Frequency of slab avalanches dependent on slope inclination, according to [7]

The higher the location of a starting zone, the more snow may accumulate there, the lower the temperatures are, and the more the distribution of the snow is influenced by the wind. In ridge areas, particularly on the leeward side of mountains, experience massive snow accumulations, which far surpass regular precipitation (Figure 2.15). These accumulations are further promoted by the morphology of a starting zone, if it is slightly hollow-shaped, so that drifts are possible from several directions. In contrast, starting zones with a convex profile have a lower possibility of snow accumulation. Snow accumulation on leeward slopes takes place along the northern border of the Alps, predominantly on slopes that face

Table 2.5 Avalanche types by different inclinations according to Schaerer & McClung [254]

Slope gradient	Avalanche type
60°–90°	Avalanches rare, snow glide common in small portions
30°–60°	Dry loose-snow avalanches
45°–55°	Frequent slab avalanches
35°–45°	Slab avalanches of all sizes
25°–35°	Rare (often large) slab avalanches, wet-snow avalanches
10°–25°	Rare wet-snow avalanches and slush flows

Fig. 2.14 Typical hollow-shaped Alpine avalanche starting zone above the tree line (© Sauermoser)

SW–SE, since the weather conditions leading to intensive snowfalls are NW–W weather conditions (Sections 2.2.1.2 and 2.2.1.3), which are associated with strong winds from W–N direction. The situation is the opposite in the southern rim of the Alps where predominantly depression areas (Section 2.2.1.4) are associated with strong southern winds, which drift in toward NE–NW exposed starting zones.

Typical potential starting zones in large avalanches are frequently located in steep ridges, in incline areas that are concave both in profile and lengthwise with low roughness, or significant terrain disruptions (Figures 2.14 and 2.15). Under extreme

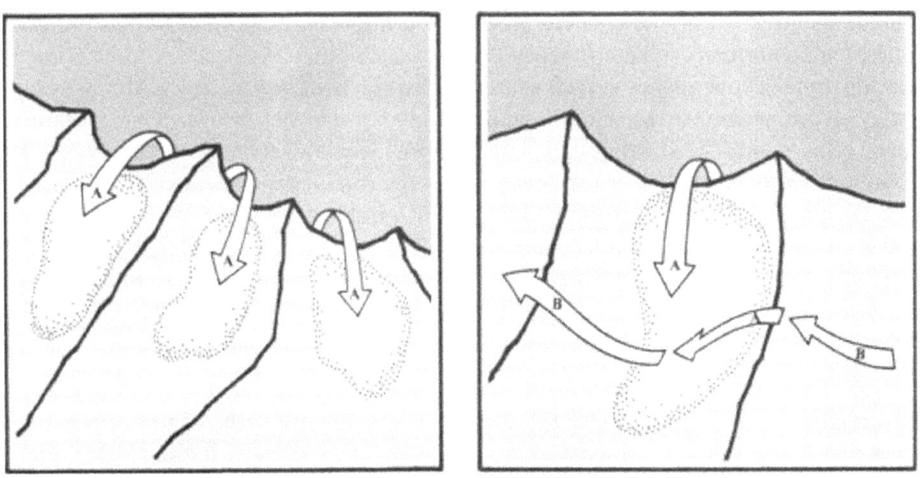

Fig. 2.15 Influence of the ridge location on the snow deposition: (a, left): snow drifting over the ridge (b, right): snow distribution along the slope, according to [7]

Fig. 2.16 Avalanche starting zone above the potential tree line (© Sauermoser)

conditions, fractures may spread via prominent brows, which cannot be forecast beforehand under normal conditions. For this reason, it is often difficult to assess the width of a potential starting zone. The maximum fracture width at the avalanche event in Blatten/Naterns (Switzerland) in February 1999 was 4 km [246]. Fracture widths of 2 km to 5 km have also occurred at various locations in North America. Several terrain ridges, which under normal conditions are considered adjacent, independent starting zones, can grow together into one common starting zone under extreme conditions. This is also the case for potential starting zones that lie above/below each other, which normally release independently of each other. Avalanches from *primary* starting zones above planes – such as trough shoulders in typical inner Alpine trough valleys – can, under extreme circumstances, trigger avalanches from *secondary* starting zones lying within the steep trough. This happens when an increase in the avalanche volume from the primary starting zones causes the runout distance to become so large that the planes above the trough shoulders or cirque threshold [254] are overrun, which does not happen under normal conditions, and the avalanches then are deposited there. The large mass of the destructive avalanche in Galtür in the winter of 1998/99 happened for this reason [234]. What is problematic is the extent of a potential release zone downwards, when the slope inclination in the avalanche path is also more than 30°.

Besides the location relative to the ridges, the position relative to the wind-exposed high areas may also result in significant snow accumulation in potential starting zones. From these accumulation zones, extreme amounts of snow in the upper slope areas can blow in, while the windward areas themselves are almost completely free of snow.

The exposure in a starting zone is not important with regards to the position relative to the main winds, but the position relative to the incidence of the sun's radiation is a decisive factor for the formation of avalanches. Because of the higher infraction on the south slopes, the snow accumulates faster [7], which means that the snowpack gets harder quicker. However, the more intense radiation on the south slopes can lead to the formation of wet-snow avalanches.

A significant factor for the characteristics of potential avalanche starting zones is the *roughness of the substratum*, which is expressed in different soil classes and roughness subvalues [52]. Large, uniform areas without significant terrain roughness, particularly smooth, grassy areas with smooth rocky surfaces and slope-parallel layering, are assigned soil class four and exhibit minimal roughness. The effect of the roughness, however, should not be overestimated. Snowfalls in the early winter can also level out large degrees of roughness, and therefore glide horizons with large expansion may also occur in rough and uneven terrain.

The majority of the *potential* avalanche starting zones are located in the Alps above the *natural tree line* (Figure 2.16). According to estimates by Sauermoser [232], this concerns approx. 2/3 of all potential starting zones. According to [167], the natural tree line is a general climatic limit with significant deviations based on local climate and relief conditions as well as orthographical and edaphical deviations, and a harmonious value can therefore only be indicated roughly. For the Alpine area, 2100–2300 m is indicated, for border Alpine areas SH 1800 to 1900 m. After hundreds of years of human influence, natural tree lines only partially exist and are very difficult to reconstruct. According to [63], the average elevation difference between the natural and actual tree line in Tyrol is 208 meters within the Alps. The average elevation difference from the natural tree line to the summit region is therefore several hundred meters in the eastern Alps, and significantly higher in the western Alps.

2.5.3 Avalanche path

The morphological avalanche classification (Table 2.4) differentiates between a planar (Figure 2.17a) and a confined avalanche path (Figure 2.17b). The planar avalanche path has approx. the same width as the starting zone, while with the channelled avalanche, the avalanche path narrows according to morphological terrain shapes, such as channels or terrain hollows.

Small avalanche catchment areas have avalanche path lengths of less than 100 meters, while large avalanche catchment areas may have avalanche paths that are several kilometers long. It is not only the shape, but also the relief, the inclination, the vegetation and the roughness of the avalanche path that make up the characteristic of an avalanche. Steep avalanche paths lead to high avalanche velocities, relief releases (e.g. rock fall) lead to an additional loosening of the avalanche snow, and therefore an increase in the avalanche's powder component. Because of the inertia of the air-snow-mixture, the powder component follows the fall line rather than the flowing component, which may result in a separation of these two components when the avalanche paths are very curvy.

(a)

(b)

Fig. 2.17 Avalanche path shapes: (a): flat avalanche path (© Sauermoser) and (b): channelled avalanche path (© WLV Tyrol)

Fig. 2.18 Traces of powder snow avalanches and flowing avalanche effect along an avalanche path (© WLV Tyrol)

In the Alpine countries, the avalanche paths are often characterized by tree-free zones within the forest areas (Figure 1.1), and this means that the forest density strongly affects the frequency of avalanches (Section 2.6.1).

2.5.4 Avalanche runout zone (deposition zone)

Slope areas with an inclination of less than approx. 10° are designated as *potential runout zones* or *deposition zones* (Figure 2.19), although avalanches also deposit when the inclination is greater. This means that slope areas, when the avalanches are smaller, can be sedimentation areas while they are being overrun by large avalanches. It is therefore not possible to make a precise attribution.

A slope inclination of 10° marks the inclination area where large dry-flowing avalanches decelerate [85]. The runout zones for large avalanches in the Alpine valleys are usually located in the valley bed areas and alluvial cones, where the flanks in the steep trough valleys transition to the valley ground (Figure 2.20a).

The transition of the avalanche path to the runout zone is abrupt when the avalanche path leading to the valley ground is steep, but generally flat due to an alluvial back filling, which frequently is the case the valley areas of the limestone Alps. Parable-shaped profile of the avalanche path and the runout zone, with a continuous slope decrease from the avalanche path to the end of the runout zone, results in a long runout length for the avalanches. The delineation of the area, which will be coupled with the avalanche path,

Fig. 2.19 Runout zone for avalanches in the valley ground area in an Alpine valley (© Sauermoser)

from the area designated as the runout zone is difficult in these cases. Potential deposit zones from large avalanches frequently are to be found on convex alluvial cones from torrents, which can lead to an expansion of the avalanche deposit (Figure 2.20b). The convex shape of an alluvial cone also leads to the fact that the avalanche may be deflected laterally. It is difficult to evaluate how much a deposit zone will expand, and this evaluation is an object of scenario training during hazard zoning. The greater the volume of an avalanche, the further the runout distance and hence the greater the potential runout zone reaches. This can be very dependent on the type of avalanche. The powder component runs in the avalanche's main direction of impact, while the slower flowing component deposits itself along terrain hollows or alluvial cone edges. Humid flowing avalanches or wet-snow avalanches, which are characterized by low flow velocities, deposit themselves in a finger-shape; however, the reach and especially direction of the individual fingers cannot be predicted. Directional changes of 90° on the alluvial cone are frequently observed.

2.6 Avalanche protection forest

2.6.1 Effects of vegetation and forest on avalanche formation

In mountain areas, the forest stands fulfil important protective functions and may significantly influence the formation, size, and even the reach of avalanches. The formation of avalanches, their frequency, and intensity (magnitude) are also dependent on the structure of the forest stands [18, 171].

(a)

(b)

Fig. 2.20 Avalanche runout zone: (a): different deposition zones, powder and flowing component in a dry-snow avalanche (© WLV Tirol); (b): finger-shaped deposit from a wet-snow avalanche (© Sauermoser)

The ground vegetation in forest stand areas – for example tree trunks, sticks, lying trunks – and other surface structures increase the friction coefficients between the snowpack and the substratum, and hence effect an irregular and disrupted deposition of the snow. Due to falling snow from the tree crowns, as well as scours around the tree butts, the snow structure also changes, and the formation of mechanically weaker layers is destroyed [25]. Because of these characteristics, an intact growing stock and the existence of vegetation in starting zones (Figure 2.21) form an excellent basis for

Fig. 2.21 Forest stands in avalanche starting zone (© Rudolf-Miklau)

preventing large, homogeneous snow areas, and can therefore significantly reduce the risk of the formation of high-frequency avalanches on slopes with an inclination above 30° [242]. In extreme conditions, when trees have a high tree crown line and missing depth load, it is possible however for the falling snow to destroy weak layers, which can trigger an avalanche [61].

A retention of snow with tree trunks, and preventing slab formation in the forest is generally possible, however the support effect of stands only becomes apparent when the number of trunks is relatively high (diameter at breast height [DBH> 8 cm]). A guideline is 500 trunks per hectare at 30° slope inclination and 1000 trunks per hectare at slope inclination of 40° [61, 226]. Besides the topographical and meteorological conditions, it is predominantly the crown density, the size of the forest cleanings, as well as the presence of winter-bare tree species (mainly non-coniferous species, but for instance also larch) that are decisive when it comes to avalanches being triggered on forested slopes. On such slopes, the interception of snow – compared to non-wooded areas – can be between 30% and 70% depending on snowfall, snow volume, tree species, air temperature, and wind conditions. Similarly, the forest climate in the evergreen forests (e.g. in silver fir or spruce stands) causes a light warming of the snowpack during the day and a smaller radiation of the snowpack during the night, and thereby influences the transformation of the snow. Due to the microclimate in the forest, surface hoar and depth hoar occur more rarely than on free surfaces, and the warmer temperatures result in a more compact and stabile snowpack [61].

The protective effect of the forest stands can last for decades on wind throw surfaces, if trunks, root plates, dead wood stands, and stalks and are not removed [243]. It is, however, a prerequisite that the position of the avalanche release under normal conditions and average snow depth do not exceed a slope inclination of 40° [61], and that the entanglement is a minimum of 2 meters. In significantly stepper terrain, and in snow heavy situations, however, it is possible that the tree debris will not be able to resist the snow pressure and may start to move [170]. As a rule of thumb, trees must be at least twice the size of the snow depth in order to contribute to avalanche control [61].

Conversely, beech foliage, unused pastureland, and the existence of shrublets have a negative effect on the stability of the snowpack since they form ideal surfaces for snow glide [182]. When slopes are systematically cleared – as is done in Canada – new avalanche starting zones form [170].

Forest protection initiatives affect the frequency of avalanches and can therefore significantly increase the protection of lower settlements and infrastructure units [23, 61]. For this reason, the uppermost areas in a stand, as well as the timberline region of a forest take on a central role with regard to the stability of the entire forest below. However, since the extreme conditions at the treeline only allow growth when protected by old trees and on ridges, it is necessary to supplement with regular technical initiatives in gullies and if there is no existing adult stand [61, 232].

A *protective forest in the avalanche formation zone* (subalpine and high-montane level) should consist of evergreen coniferous wood, and at a coverage degree of at least 50% and a slope inclination of ≥30° should not have gaps in the fall line that are more than 50 meters. An increasing slope inclination reduces the gap length, which should be less than 25 meters when the angle is ≥45° [61]. Similar indicative values are in place for high and low montane non-coniferous and mixed forests. However, the potential contribution of the forest with regard to protection against avalanches is, as opposed to the coniferous forests, only moderate. At sunny exposures, one can regularly observe glide snow and moist loose-snow avalanches [61].

Under extreme weather conditions, the protective effect of the forest usually decreases; with prolonged snowfalls at low temperatures and a lack of radiation, or strong snowfall with subsequent rains, the trees cannot influence the snow interception, radiation or terrain roughness sufficiently [19, 246]. Similarly, protective forests cannot stop large avalanches that have released above the forest limit, and they only affect the avalanche velocity in an insignificant manner [47]. According to [16], the brake effect of a protective forest is very low on those types of avalanches, and furthermore extensive trunk breakage does not significantly contribute to diminishing the avalanche energy (Figure 2.23). For small events with lower energy, formation zones near the tree line, as well as flowing depths from 1 to 2 meters, one may assume that the forest stands can have a retarding effect on the avalanche [162]. In the runout zone, the avalanche velocity is often low, due to the topographical conditions, so that the forest may have a better retarding effect and may reduce the reach of avalanches [61].

Using the Forest Reserve in Andermatt (Uri/Switzerland, Figure 2.22) as an example, it was possible to demonstrate in [268] that the yearly costs of a 300-year event without forest stands and supporting protection mechanisms – as they often exist today – would amount to anything from € 50 000 to approx. € 2 million. This means that the use of this forest reserve amounts to a benefit of € 1.8 million annually.

Fig. 2.22 Avalanche Forest Reserve in Andermatt (Switzerland) (© Margreth)

2.6.2 Effects of avalanches on trees and forests

Just as the forest has an effect on the formation and movement of avalanches, avalanches also have an effect on the forest (Figure 2.23). In areas with high intensity and frequency, avalanches exercise a certain amount of control on the survival, growth rates, and growth forms of the trees [19]. On avalanche slopes and in gullies on subalpine and mountainous levels with high-frequency avalanche activity, adult and

Fig. 2.23 Uprooted and broken larches caused by the avalanche event, 1999, Lareit Avalanche in Val Poschiavo (Switzerland) (© Reto Tiri, WSL/SLF)

large old-growth trees can only establish themselves poorly because of repeated mechanical strain. Similarly, it turns out that wintergreen tree species in the border areas of avalanche passages are practically non-existent since they – compared to winter-bare tree species – cannot survive because of the larger air resistance [61]. Therefore, larch often grow in these areas of the central high Alps, and in pre-Alp areas, mountain maple and beech grow. Much more frequently, at sites with high-frequency avalanche activity, a natural vegetation (forest) is completely prevented [24, 53], and the vegetation consists of low-nutrient grasses, dwarf shrubs and single conifer seedlings. If, in spite of this, winter-bare tree species manage to establish themselves, the trees have a proportionally small diameter, a smaller size, and narrower growth rings [39, 129, 147]. Furthermore, tree density is far below average [19, 39].

The impact force and frequency of the avalanches cause different mechanical damage to the tree, which generally depends on the flexibility of the tree (dependent on its diameter and height), as well as the position of the tree in the avalanche path [129]. In zones with frequent activity, trees are eliminated as soon as they reach a critical size or a critical diameter [248]. The maximum diameter of flexible trunks for coniferous species, depending on type, is between 6 and 14 cm [129, 173], for non-coniferous species such as maple, birch, and willow, as well as shrub-like trees such as green alder (*Alnus viridis*) or dwarf mountain pine (*Pinus mugo*), a certain amount of flexibility also remains in the thicker trunks [39, 62]. However, the trees lose their flexibility as diameter and height increase. In zones with irregular activity and avalanche starting zones more than 150 meters above the tree line, avalanches can also destroy entire stands of trees with high trunks [244] (Figure 2.18, 2.24). For flowing avalanches, it is assumed that, in order to break a trunk, snow pressure between 10 and 50 kN/m^2 is required [166], for powder snow avalanches, the values are approx. 3 to 5 kN/m^2 lower.

As opposed to other processes, such as fire or mudslides, the destruction of the forest stands by avalanches does not have a considerable effect on new seedlings. This is

Fig. 2.24 Vegetation in the avalanche path (© Sauermoser)

because single events only very rarely lead to a complete destruction of the stands all the way down to the soil [131]. This is also why an evaluation of previous events, which are based on batches in the age-class distribution, are difficult [276] and not suitable. Young trees, which survive an avalanche event more or less unscathed, find themselves in improved light and growth conditions, and contribute to the regeneration of the forest stands by the surviving trees [19, 39].

3 Avalanche dynamics: models and impact

Siegfried Sauermoser, Matthias Granig, Karl Kleemayr and Stefan Margreth

3.1 Principles of avalanche dynamics

3.1.1 Physical principles

3.1.1.1 General remarks

Physical criteria and mathematical models to describe avalanche movement are amongst the most important foundations in the planning of technical avalanche defense measures. The following section summarizes the most important characteristics of avalanche movement and provides a brief overview of basic physical concepts to describe the avalanche motion.

From the physical point of view, *avalanches* can be described as the gravitational flow process of snow (Section 2.1.1). A physically accurate and mathematically complete description of the avalanche movement does not yet exist; this is due to the large variability of the material snow (Section 2.3.1).

3.1.1.2 Characteristic criteria of avalanche movement

For practical purposes, the movement of an avalanche can be characterized by the *velocity v*, and the *flow height h*, and the *mass density* ρ of the avalanche snow. In addition, the maximum *runout distance* of the avalanche and the effect of *pressure* exerted by this are relevant as criteria of the avalanche movement in risk analysis. These measurement parameters can – in designated test areas [119, 120] – be determined experimentally, whereby the flow height can also be determined using optical methods (video grammetry, terrestrial laser scanning) over the surface of the avalanche [274]. Similar processes apply to the measurement of the distribution of the front velocities using radar [69, 213]. The determination of flow velocities or impact pressures within the avalanche body relies on point measurements, such as on a mast or an obstacle in the avalanche path [136, 137, 260, 271]. The technology for measuring the avalanche movement is discussed in Section 9.3.7. Avalanche dynamic nature measurements provide reference values for the characteristics of various avalanche types. It should be noted, however, that such experiments in the field are limited to specially equipped measurement avalanche paths.

3.1.1.3 Avalanche velocity

Each avalanche starts as a laminar flow or as a sliding block, from a velocity of 10 m/s, a smashing of the snowpack and the transition into a turbulent movement form are to be expected [277].

Time-dependent and depth-triggered velocity measurements, together with the flow height, provide information about basic movement patterns and the 'physiological' structure of the avalanche [136]. For flow avalanches, one must differentiate between the head, the body and the tail of the avalanche (Figure 3.1a–d); there are different velocity profiles in each section.

The Technical Avalanche Protection Handbook, First Edition. Edited by Florian Rudolf-Miklau, Siegfried Sauermoser and Arthur I. Mears.

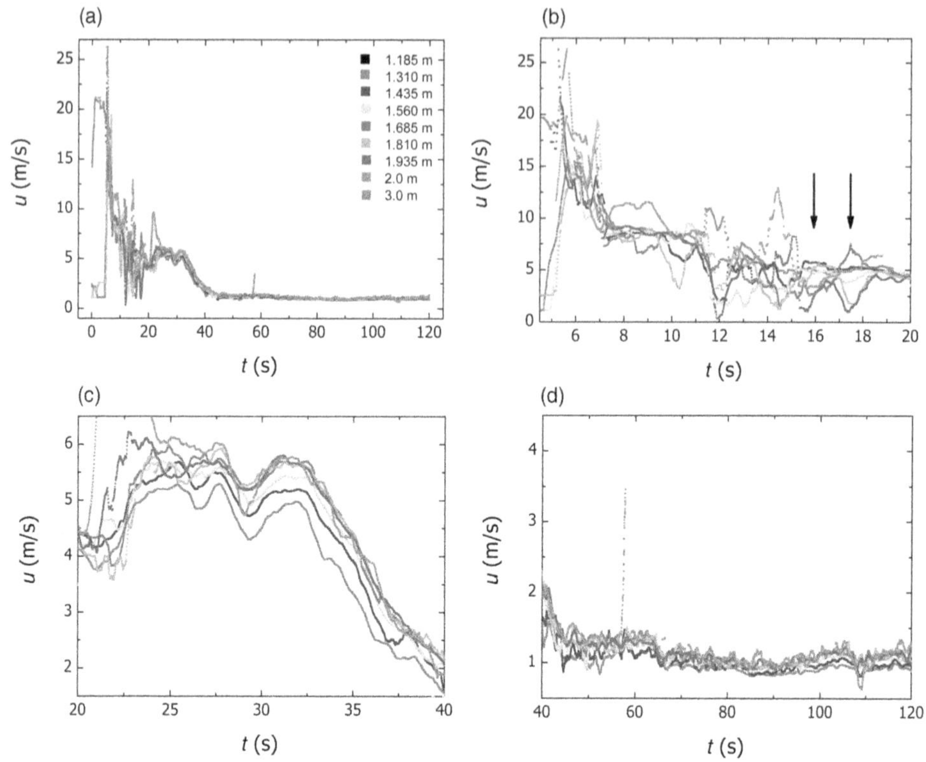

Fig. 3.1 Velocity time series of an avalanche in the Vallée de la Sionne (Switzerland) measurement field: (a): entire avalanche; (b): head of the avalanche; (c): body of the avalanche; (d): tail of the avalanche (u . . . front velocity of the avalanche, t . . . time), according to [137]

In the *avalanche front* (head), the highest velocities (Table 3.1) of the flow section can be observed, whereby the normal flow (i.e. recorded perpendicular to the topography) velocity profile is shown as disordered; the velocities fluctuate heavily (Figure 3.1b).

The *avalanche body* is characterized by a typical shear velocity profile, whereby the velocity increases with the increasing height above the ground. Whether a glide layer exists on the boundary surface between the avalanche and the unmoved snow and the basal velocity of $v \neq 0$ often cannot be clearly determined from the measurements.

Table 3.1 Magnitudes of avalanche velocities according to [7]

Avalanche type	Velocity [m/s]
Wet flow avalanche	10–20
Dry flow avalanche	20–40
Powder snow avalanche	30–70

Model tests on the snow glide path by the WSL Institute for Snow and Avalanche Research SLF in Davos, however, indicate that basal sliding and a basal shear layer can arise in parallel [135, 269]. Often, the shear layer is superimposed by an area with only weak or no shearing motion (plug). This condition is often stable over a certain period of time (Figure 3.1c).

The *avalanche tail*, however, is characterized in many cases mainly by weak or missing shear in the flow, the snow flows in a block on a very thin shear layer or glides directly on the substratum (Figure 3.1d). There are transitional phases between each of the individual avalanche sections.

Measuring the velocity profiles, for example with the help of optical velocity sensors (Section 9.3.7.4) placed in various heights on a mast in an avalanche path, provides insight into the internal dynamic structure of avalanches. Such *invasive measurement methods* affect the avalanche flow, and the measuring equipment must be situated to minimize the disruption of the flow. *Non-invasive measurement methods,* such as video grammetry or radar measurements (Section 9.3.7.5), do not provide depth resolved velocity information but do provide important information on the spatio-temporal development of surface and front velocities of avalanches.

The measured velocity profiles allow conclusions on the friction processes or energy dissipation processes that determine the dynamics of the flowing avalanche in terms that are consistent with model concepts [17, 269].

3.1.1.4 Model laws of avalanche movement

The great variability of the avalanche movement and the considerable uncertainties in measurements performed on real avalanches represent fundamental obstacles for the derivation of *physical model laws*. Therefore, as a substitute, scaled model tests are used (e.g. snow glide path (SLF), laboratory slide with glass beads or polystyrene as a model material).

In scaled tests, one speaks of a similarity to real avalanches when the model avalanche corresponds to real avalanches with regard to its dimensionless ratios, which are usually derived from the Buckingham Π theorem [15]. Trials with scaled model avalanches have so far been used mainly for the design of technical avalanche defense structures (avalanche-retarding mounds, collecting dams) [14, 15, 92–94, 127]. The following *Froude number* describes the kinematic similarity between the real and the model avalanches:

$$F_r = \frac{v}{(g \cdot h)^{1/2}} \tag{3.1}$$

where:

F_r Froude number [–]
v Flow velocity, in [m/s]
g Gravitational acceleration, constant [= 9.81 m/s^2]
h Flow height of the avalanche, in [m]

According to Issler in [127], the Froude numbers for the flow component in dry flow avalanches are between 5 and 10, thus defining avalanches as supercritical flows ($F_r > 1$).

In the interpretation and use of test results with scaled avalanches, note that the similarity to real avalanches is limited. First, the processes that are involved in the movement of real avalanches are not fully known and even if they were, it would not guarantee that a scaled experiment could be realized that would meet all the relevant similar criteria. For example, the dimensioning (design) rules for avalanche dams (Section 6.4) derived from laboratory experiments are based on purely kinematic similarities such as the Froude criterion. Dynamic similarities were only taken into account very roughly since the dynamics of real avalanches are still not sufficiently known to be able to specify appropriate similarity criteria.

3.1.1.5 Mathematical models for avalanche dynamics

Although the complex interaction of the physical processes involved in the movement of avalanches is still not known in detail, avalanches can be formally be regarded as a continuous medium whose motion is described by the *mass* and *momentum balance equations*.

The mass balance equation can be expressed as follows:

$$\partial_t \rho + \nabla \cdot (\rho\, v) = 0 \tag{3.2}$$

where:

ρ Snow density, in [kg/m^3]
v Velocity of the avalanche, in [m/s]

The momentum balance equations are:

$$\partial_t (\rho\, v) + \nabla \cdot \prod = \rho \cdot k \tag{3.3}$$

$$\prod = \rho\, vv - T \tag{3.4}$$

$$T = \sigma - 1\, p \tag{3.5}$$

where:

ρv Momentum density
Π Momentum flow density
v Velocity of the avalanche
vv Dyadic product of v and v
k Forces acting on the flowing snow, such as the downhill force, for example
T Deviatoric stress tensor
σ Stress tensor
p Hydrostatic stress tensor

By choosing appropriate initial and boundary conditions, taking into account characteristic parameters – such as starting zone and released snow mass, topography of the avalanche path, and so on – the avalanche motion can generally be calculated by solving the mass balance and momentum balance equations. This happens – with appropriate simplifications – usually by means of numerical solution methods, such as in the avalanche models AVAL-1 d (Section 3.2.5.2), RAMMS (Section 3.2.5.3) or SamosAT (Section 3.2.5.4). If you average the equations over the flow depth by integrating the vertical dynamic and assuming these as a 2-dimensional surface over the topography (shallow water approximation), the equation system provides the characteristic parameters of the avalanche movement (velocities, flow heights, runout distances and impact pressures).

3.1.1.6 Constitutive law

From a mathematical point of view, the equation system of volume and movement conservation equations is under-determined, that means additional information is necessary to complete the equation system and to be able to determine the solutions for the description of the avalanche movement.

The equation system of mass and momentum balance can be found from a relationship between the stress tensor σ and the deformation tensor D:

$$D_{ij} = \frac{1}{2} \cdot \left(\partial_i v_j + \partial_j v_i \right) \tag{3.6}$$

The following relation is called a *constitutive law* or, in the practice of avalanche dynamics, 'rheology':

$$\sigma = \sigma(D) \tag{3.7}$$

The constitutive law for a material describes the deformation D as a reaction to the applied stress σ.

The simplest case of a constitutive law is a linear relationship between σ and D, which describes the behaviour of an incompressible *Newtonian fluid* (e.g. water). The real rheological behaviour of the medium snow is much more complex. The primary issue of the mathematical description of the snow flow is that the tensorial parameters σ and D in an avalanche have to be determined by point measurements in order to be able to derive a constitutive law from experiments. Even if this could be technically resolved, the strong variability of the material behaviour of the avalanche snow makes it very difficult to derive a universally valid constitutive law from the measurements. The mathematical and numerical avalanche models are therefore invariably based on grossly simplified models for the constitutive behaviour of flowing snow. They generally record the avalanche as a flowing continuum or granules, or as a hybrid of both. In the latter case, the parameterized granular properties are included in a continuum mechanical description. Examples of such highly simplified empirical models are the *Voellmy-like 2-paramerized friction laws* (Section 3.2.5.1) or the *Savage-Hutter equations* (Section 3.2.5.4).

3.1.2 Dynamics of flow and powder snow avalanches

3.1.2.1 Avalanche release

The *avalanche release* (emergence and spread of an avalanche break) for snow slab avalanches is an independent area of research. An overview of the current state of knowledge can be found in [250].

An initial fracture, first starts in a weak layer in the snowpack, when the forces on the weak layer exceed the compression or shear strength. Additional loads can trigger such a fracture. The propagation velocity of the avalanche release can reach several 100 m/s, whereby the weak layer collapses over a wider area (cracking noise). This expansion, called the *primary fracture*, is supported by a spatially homogeneous snowpack with a correspondingly spread weak layer. However, the small-scale variability of the snowpack and the weak layers within are a hindrance for the spread of the primary fracture. An avalanche is released when, in addition to the primary fracture of the weak layer, so-called *secondary fractures* occur and the snowpack can no longer bear the tension trapped by the collapse of the weak layer. The secondary fractures mark the perimeter of the snow slab. Detailed discussions of break propagation can be found in [96, 97, 255].

3.1.2.2 Movement of flow avalanches

Flow avalanches generally evolve from a crack in the snow slab, whereby the sliding snow slab breaks up into smaller blocks as a result of surface roughness. The surface roughness and inner friction caused by shearing and deformations in the flowing avalanche snow act as a retarding force. The essence of the floating section of an avalanche or a flow avalanche is the permanent contact between the individual ice particles over which the movement grows. Typical densities of flow avalanches are in the range of 100–300 kg/m^3 may assumed, where cold, dry sow is expected and typical flow heights of unconfined runoffs are at 1–3 m [127].

The *internal friction* is often combined with ground friction in current avalanche models in which the friction processes occurring in the body of the avalanche are added parametrically to the ground friction.

3.1.2.3 Movement of powder snow avalanches

Powder snow avalanches arise in any case from flow avalanches, but not every flow avalanche becomes a powder snow avalanche. For the formation of an avalanche, the tendency of the avalanche path (>30°) and the properties of snow in particular are crucial.

First, the avalanche moves in more or less dormant surrounding air so that there is a velocity gradient between the snow and air. The shear forces between the snow and air increase with the increasing velocity of the flow avalanche until they reach a certain limiting value and snow particles penetrate into the air as a result of transverse acceleration; a *powder snow cloud* (suspension layer) is formed (Figure 3.2a). From a velocity of approx. 15–20 m/s, the vertical component of the turbulent movement exceeds the fall velocity of the particles and snow-air mixture (suspension) forms, whereby the air penetrated by the particles velocities up in the flow direction. In a fully-fledged powder snow avalanche, air and snow particles move at about the same average

(a)

(b)

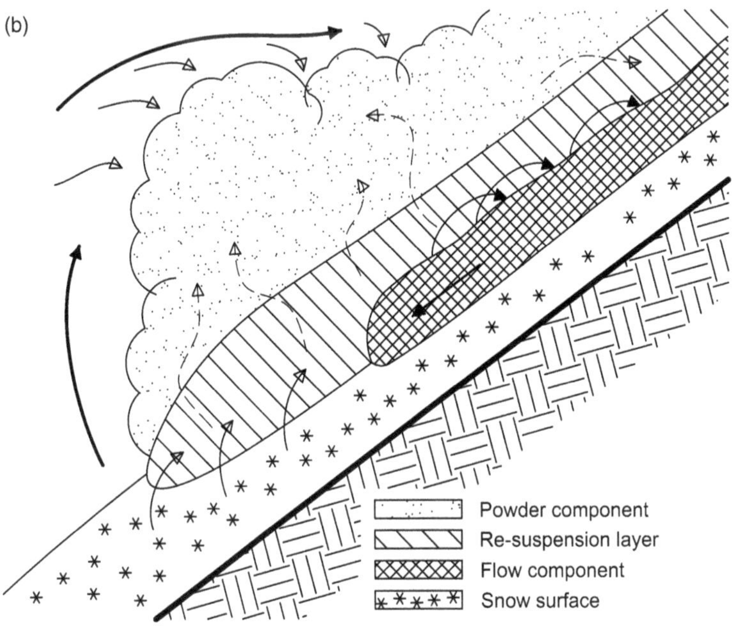

Fig. 3.2 Mixed and powder snow avalanches: (a): powder snow avalanche in motion (© WLV Tyrol); (b): schematic consistance (movement) of a mixed avalanche (flow and powder components) (according to ONR 24805 [202])

velocity. The retarding forces during a powder snow avalanche are much smaller than in a flowing avalanche. They arise during the acceleration of the invading outside air and from turbulent friction.

The powder cloud can be modelled as a turbulent, flowing medium that is characterized by an *energy dissipation rate* dependent on the Reynolds number. Because this approach neglects the fact that powder snow avalanches are two-phase flows and it is not clear how the particles dissolved in the air affect the turbulence structure of the powder cloud, such estimates of turbulent friction are subject to fundamental uncertainties. Newer approaches therefore model powder snow avalanches as *two-phase flows* (e.g. SamosAT, Section 3.2.5.4).

The front velocities of powder snow avalanches are less than the core velocities due to the air resistance. The mass that comes from behind to hit the front is severely decelerated there and therefore slower than the avalanche core; the material is literally swirled back. This process causes the characteristic instabilities of the powder snow avalanche front. Often, large, solid snow chunks of the avalanche core are thrown far out of the front of a powder snow avalanche, like a shot.

The average densities ρ of powder snow avalanches are in the range of 2–15 kg/m^3 and generally increase against the ground, whereby the densities of air (1.25 kg/m^3) or the lightest snow (30 kg/m^3) are estimates for the upper and lower limit of the density of the powder cloud.

3.1.2.4 Movement of mixed-motion avalanches

Since powder snow avalanches always develop from flow avalanches, both avalanche types generally occur together as so-called *mixed-motion avalanches*. This is also the most common type of major disaster avalanches. Observations suggest that the motion of the powder snow avalanche depends on permanent energy and mass additions from the flow of the underlying flowing avalanche. Actually, it can often be observed that the powder cloud starves when the underlying flow avalanche stops, that is it slows greatly and dissolves. Although there are mathematical models for pure powder snow avalanches [271, 272], it is currently disputed whether and under what circumstances a powder snow avalanche can separate and be damaging alone. The powder content of an avalanche follows less the direction of gullies so that in highly structured terrain a separation of flowing and powder components of an avalanche is possible when the flowing component changes direction in an avalanche path and the powder component does not follow this change of direction.

Following the principle of mixed-motion avalanches, powder snow avalanches in the operational avalanche model SamosAT are modelled as a two-phase flow linked to the flowing avalanche with a flow phase as a peripheral condition for the powder component (Section 3.2.5.4). Hereby, the type of transition (*re-suspension layer*) and the physically correct description of the internal turbulent friction in the powder cloud still contain great uncertainties. While the most commonly used avalanche models assume a two-phase flow (Figure 3.2b), there is also experimental evidence [121] that there is a *fluidized transition layer* between the flowing and powder snow avalanches. This fluidized layer can be formed in steep terrain from a fast-moving flow layer. It moves mostly at high velocities at the tip of the avalanche [71] and thus significantly contributes to the dynamics of the whole avalanche. Occasionally this transition layer – also called the *saltation layer* – is seen as the lowest layer of the powder component and an exact separation of the layers is not possible. Physically, this layer is only characterized by individual cases of direct particle contact and the distances between the individual ice particles are already several particle diameters whereas in the suspension layer, direct particle contact in the movement transmission no longer plays a role. In accordance with [127], the density of the transition layer is approx. 10–100 kg and the typical height is 2–5 m.

3.1.2.5 Movement of wet snow avalanches and slush flows

One speaks of *wet snow avalanches* when liquid water is present in the entire sliding snowpack. Wet snow avalanches are triggered by a water-permeable sliding surface, normally immediately after snowmelt starts or after periods of heavy rains. The stability of this is reduced by the free water in the pores of the snowpack. Typically, wet snow avalanches have a low flow velocity – with the exception of a very steep avalanche path – and a clear structure of clumps (Figure 4.3a). Wet snow avalanches occur mostly in the spring but may also be associated with mid-winter rain or thaws and are subject to a strong fluctuation depending on the time of day. Even if the moisture content is higher than for dry snow avalanches, the avalanche mass is formed almost entirely of snow and ice clumps. The content of free water is usually low. Wet snow avalanches can occur both in snow slab breaks as well as in loose snow avalanches. A special form of snow slab breakage is the *ground* or glide avalanche in which the snowpack separates all the way to the ground. Due to their high density, these avalanches can develop very high pressures. For the calculation of avalanche pressures or runout distances for wet snow avalanches, there are still no appropriate calculation models. (The models developed for the calculation of dry flow avalanches cannot be applied to the phenomenon of the wet snow avalanche.) Wet snow avalanches show special deposit patterns, extremely long and unpredictable directions of runout distances. One phenomenon is the spontaneous change in direction of wet snow flows if these encounter small obstacles or terrain unevenness.

Slush flows represent a special form of avalanche related to the wet snow avalanche [101, 102], which occur mainly in Arctic and sub-Arctic areas (Alaska, Canada, Iceland, Russia, Norway, Sweden) [188] and represent a suspension of snow dissolved in water (Figure 3.3b). Contrary to wet snow avalanches, slush flows consist of a mixture of water and ice that has considerable potential for damage due to the high density and low internal friction. A high proportion of free water in the snowpack is a prerequisite for the formation of slush streams. In contrast to wet snow avalanches, the soil roughness and terrain inclination have little effect on the triggering. Slush flows have even been observed at terrain inclinations of 12° or less. The rheology of a slush flow can be described as a hybrid of a liquid and granular constitutive law [70]. In experiments on the snow sliding path, the very high impact pressures could be measured and the liquid content of the material behaviour could be demonstrated [123].

(a)

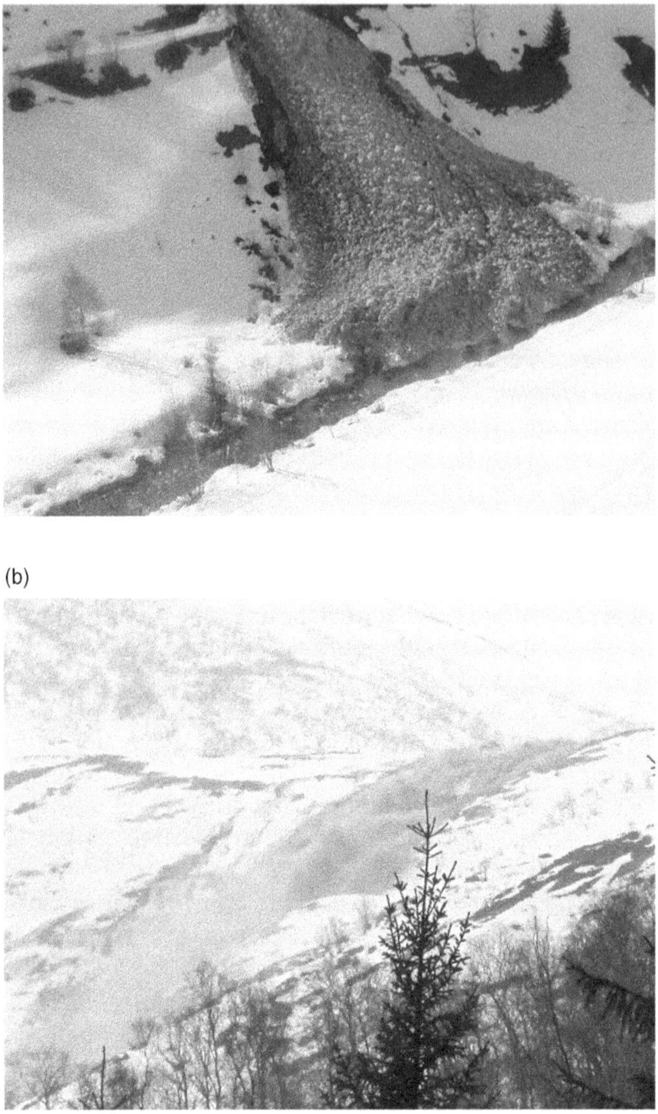

(b)

Fig. 3.3 Wet snow avalanches: (a): deposit of a wet snow avalanche (© Sauermoser); (b): slush flow in motion (© Hestnes)

3.2 Numerical avalanche models and simulation

3.2.1 Application of avalanche models

As was shown in the previous section, the movement of avalanches is a highly complex transfer process that can only be approximated with simplified physical-mathematical models. The avalanche models available differ with regard to their complexity both in terms of the physical methods used and the required initial and boundary conditions as well as regarding the degree of detail in the calculation results. The required input parameters were calibrated with data from recorded (documented) avalanche events and results of avalanche dynamic tests. The use of computer-assisted models makes it possible to calculate avalanches realistically, parameter studies can be used to record the sensitivity of the input parameters. The results of simulation models however represent only a rough approximation of the real processes and require an examination for plausibility and interpretation.

Avalanche simulation models are undoubtedly the most objective tool to determine the characteristic parameters of the avalanche movement (velocity, flow heights, runout lengths, impact pressures) for planning tasks. Calculation models have been used routinely since the 1970s in the engineering practice of avalanche defense. The model development in Central Europe was stepped up after the avalanche winter of 1999; great progress has been made particularly in the multidimensional dynamic models. In the meantime, simulation models are now state-of-the-art in hazard zone mapping and planning of avalanche defense works.

3.2.2 Principles and data for avalanche modelling

An essential basis for the use of avalanche calculations is provided by data collection on location. In the terrain, numerous factors must be determined, such as key points for the runout, 'silent witnesses', the location and size of avalanche starting zones, possible accumulation zones, and areas with additional snow entrainment or surface roughness (block dumps, high forest). The avalanche history, the analysis of 'silent witnesses' and surveys of the local population can provide additional information on the process of avalanche events. The determination of realistic input data is extremely important since the results of the calculations are highly dependent upon these input data.

When modelling an avalanche event with certain recurrence interval, in particular the *size of the avalanche starting zones* and the *potential height of the fracture* must be determined. *Salm* et al. [228] propose using the snowpack growth of the last three days as the initial value for the fracture height at a certain recurrence interval (3-day precipitation) and to correct this value dependent on the elevation and the slope inclination in the area of the crack. In some North America mountains, a snowpack-height growth in up to 10 days is appropriate, depending of snow structure.

Maps, orthophotos and digital terrain models, which are now generated from airborne laser scanning evaluations, serve as a topographic calculation basis. The corresponding friction values for avalanches and the individual simulation parameters can be found in the respective application directives and model manuals. Digital 3D-terrain models (DTM) serve as a calculation basis for multidimensional avalanche models. Airborne laser scanners enable high-resolution DTM in an order of magnitude of normally one data point per square metre. These are then thinned out in the application on a grid with spacing of 5 m for simulation purposes. Studies according to [270] have shown that a higher resolution under 5 m grid spacing causes no significant improvement in the model results and does not justify the considerably higher processing and computational requirements.

3.2.3 Avalanche model overview: classification

An overview of the developments in the area of the avalanche simulation is given by Harbitz [95] in the framework of the SAME project (snow avalanche modelling, mapping and warning in Europe). A revision of the classification of the most common current models of avalanche was presented by the authors (Granig and Margreth) based on the calculation surface in Figure 3.4. Calculation models can therefore be divided into *statistical-topographic* and *physical-dynamic* models. There are separate models for, for example flow or powder snow avalanches, depending on the type of avalanche [176–178, 183, 184].

Large sections of the physical-dynamic models refer to the *Voellmy-fluid constitutive law*, which is based on a combination of the *Chezy* friction term with the dry friction of

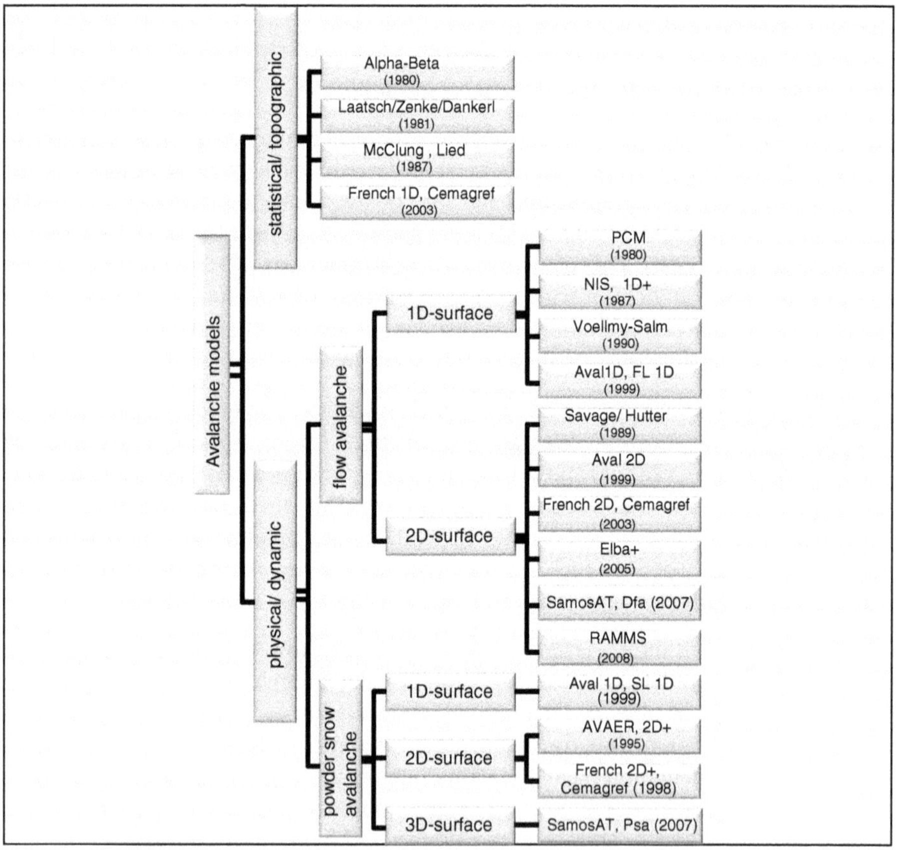

Fig. 3.4 Overview and classification of common avalanche models (classification and dimensioning relative to the corresponding calculation platform) (© Granig, Margreth)

Coulomb [277]. The subsequent developments include the *Voellmy-Salm model* [228] (Section 3.2.5.1), the *center of mass model* according to Perla et al. [206], called the PCM model, which represents avalanches as a one-dimensional granular continuum, the numerical avalanche model AVAL-1D [43] (Section 3.2.5.2), which was widely used by practitioners in many countries, and the two-dimensional simulation model ELBA [278, 279] (currently: ELBA+). The ELBA+-model has the advantage that no specific avalanche path must be defined, because the calculation uses a three-dimensional digital terrain surface. In Austria, after the avalanche disaster in 1999, there was a strong drive to enhance the development of the dynamic 3d-avalanche model SAMOS 99, which could calculate both dense flow and powder snow avalanches [224]. Since 2008, the development *SamosAT* (Section 3.2.5.4) [187, 230] has been available. In 2010, an additional suspension model was calibrated and introduced for SamosAT [130]. The entrainment model of *SamosAT* was studied and implemented in 2013 [84]. At the WSL Institute for Snow and Avalanche Research SLF (Davos, Switzerland), the numerical calculation program RAMMS (Section 3.2.5.3) has been in development since 2005. In this, in addition to avalanches, the next stage of development can also calculate landslides and falling rocks in three-dimensional topographies [44].

Calculation models have been developed for avalanche runout distances not only in Europe. Yoshimatsu [285] developed models for Japan and in Russia; Grigorian et al. (quoted in [95]) developed the flow avalanche model known as *MSU-dense*.

The following section provides an overview of the most common avalanche models in Europe.

3.2.4 Statistical-topographical avalanche models

3.2.4.1 Alpha-beta model

The *alpha-beta model* was developed by Lied and Bakkehoi [152] at the Norwegian Geotechnical Institute (NGI) in Oslo, Norway, to use easily measurable, objectively verifiable criteria to calculate the runout distance. A number of well-known avalanche runout distances were analysed statistically. With multiple regression analysis, it was possible to determine four topographic parameters that are critical for the determination of the maximum runout distance (expressed by the angle α). The four topographic parameters – the vertical fall height H, the inclination of the starting zone ψ, the average inclination of the avalanche path β and the longitudinal profile as the compensation line, expressed by the second derivation y – refer only to the longitudinal profile of the avalanche path (Figure 3.5). The average inclination of the avalanche path β emerges as the dominant factor in the calculation formula. Thus, a statistically validated formula could be set up with only two parameters.

The adaptation of the model for the Alps was carried out on the basis of the evaluation of 80 known and maximum runout distances of avalanches in Austria [153]. The following 2-parameter Equation 3.8 was calculated using regression analysis according

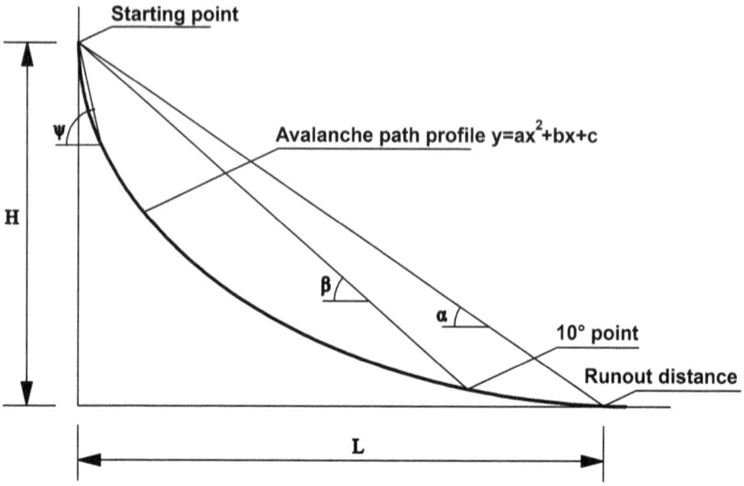

Fig. 3.5 Diagram of the avalanche parameter from the alpha-beta model in the longitudinal section (© WLV, SSL)

to [152, 280] for fast, rough estimates of large dense flow avalanches and flow avalanches with powder components:

$$\alpha = 0,946 \cdot \beta - 0,83^\circ \quad R = 0,96 \quad SD = 1,5^\circ \tag{3.8}$$

where:

α	Angle of maximum runout, in [°]
β	Average inclination of the avalanche path above the 10° point, in [°]
R	Regression coefficient
SD	Standard deviation

To improve the handling, Granig and Luxner [82] presented the model on an easy-to-use calculation platform (*alpha-beta-modell06*). The innovations in the alpha-beta-modell06 are the rapid applicability and the development of a matrix for determining the corresponding standard deviation on the basis of traceable, topological and climatic factors. As part of an update [83], additional regression lines and adaptation possibilities were introduced. The model is suitable for avalanche paths with parabolic avalanche paths to determine in a first approximation the potential maximum runout distance for dense flow avalanches, possibly with powder components.

3.2.4.2 Other statistical models

In the course of time, certain other statistically topographic models were developed, particularly for the *runout ratio* according to McClung (quoted in [254]) as well as for the model of the avalanche according to Laatsch/Zenke/Dankerl (quoted in [280]). This was based on the *slide model for mass movements* according to Heim and Körner [146] and for which the range of the avalanche can only be determined over a generalized slope (1-parameter model). However, these models are no longer used in engineering practice.

3.2.5 Physical-dynamic avalanche models

3.2.5.1 Voellmy-Salm model

After the avalanche disasters of 1951 and 1954, Voellmy [277] developed the first avalanche model in Switzerland (*Voellmy model*). He recalculated the damage caused by the avalanche event in 1954 in Vorarlberg and checked this model based on the laws of hydraulics (combination of Chezy friction terms with the dry friction from Coulomb) against these damage events. In 1966, Salm [227] continued the development of the Voellmy model, where he introduced internal friction and the active and passive flow condition. This is a deterministic-dynamic model in which the avalanche mass is simplified as a centre of mass (block model).

In 1990, WSL Institute for Snow and Avalanche Research SLF issued a manual for the *Voellmy-Salm model* in which the application of the model was presented for practical use [228]. Until the development of the numerical avalanche calculation model AVAL-1D, the Voellmy-Salm model was used worldwide for the determination of runout paths and avalanche pressures (Table 3.2).

The Voellmy-Salm model assumes that the discharge Q from the starting zone is constant along the avalanche path – with the exception of the runout zone where the movement is approximated as a mass point. The velocity v_0 is calculated in the starting zone with the following equation:

$$v_0 = \sqrt{d_0 \cdot \xi \cdot (\sin \psi_0 - \mu \cos \psi_0)} \quad [\mathrm{m/s}] \tag{3.9}$$

$$Q = B_0 . \mathrm{d}_0 . \mathrm{v}_0 \quad [\mathrm{m^3/s}] \tag{3.10}$$

Table 3.2 Overview of the friction parameters in the *Voellmy-Salm model*

Friction parameters		Description
μ	*Dry friction*	Depends on the pressure perpendicular to the ground and the snow properties. For large avalanches (>10^5 m^3) above 1300 m and flow heights of more than 1 to 2 m, a μ of 0.155 is recommended. Wet snow avalanches are calculated with a μ of 0.3. The length of the runout section greatly depends on the friction coefficient μ.
ξ	*Turbulent friction*	Is proportional to the square of the velocity and depends on the roughness of the avalanche path. For flat terrains, a ξ of 1000 m/s^2 is recommended, and for more intense canalizations or in the forest, a ξ of 400 m/s^2. The parameter ξ has a large impact on the velocity.

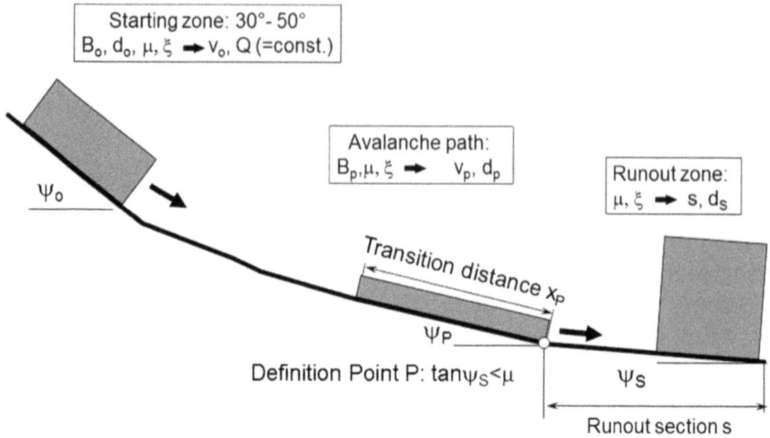

Fig. 3.6 Schematic diagram of the Voellmy-Salm model (© Margreth)

where:

B_0 Fracture width, in [m]
ψ_0 Inclination of the starting zone, in [°]
d_0 Fracture height, in [m]:
v_0 Avalanche velocity in the starting zone, in [m/s]
ξ, μ Friction parameter [–] according to Table 3.2.

Then the velocity v_P and flow height d_P are determined at the start point of the runout zone (Figure 3.6), the point P. Below the point P, ψS must apply to the average slope of the runout section:

$$\tan\psi_s < \mu \tag{3.11}$$

The velocity vP at the point P is calculated for a flow avalanche based on the slope inclination ψP over point P – measured over a transition distance from around 100 to 200 m – and the flow width B_P as follows:

$$v_P = \left[\frac{Q}{B_P}\cdot\xi\cdot(\sin\psi_P - \mu\cdot\cos\psi_P)\right]^{1/3} \quad [\mathrm{m/s}] \tag{3.12}$$

The flow height d_P in point P is:

$$d_P = \frac{Q}{B_P\cdot v_P} \quad [\mathrm{m}] \tag{3.13}$$

The runout distance s of the avalanche from point P is calculated as follows:

$$s = \frac{d_s \cdot \xi}{2 \cdot g} \ln\left[1 + \frac{v_P^2}{V^2}\right] \quad [\mathrm{m}] \tag{3.14}$$

This contains the average deposit height d_s

$$d_S = d_P + \frac{v_P^2}{10 \cdot g} \quad [\mathrm{m}] \tag{3.15}$$

$$V^2 = d_S \cdot \xi \cdot (\mu \cdot \cos\psi_S - \sin\psi_S) \tag{3.16}$$

where:

ψ_S	Average inclination in the runout zone, in [°]
g	Gravitational acceleration, constant [= 9.81 m/s^2]
d_p	Flow depth at point P in the avalanche path, in [m]
v_p	Flow velocity at the point P, in [m/s]
V	Average avalanche velocity in the avalanche runout zone, in [m/s] (supporting variable)

The Voellmy-Salm model is well validated and is suitable for calculating simple avalanches in minimal time. The calculation results can be controlled constantly by entering the parameters step-by-step.

However, the model had a few drawbacks: in the avalanche path, the flow velocities were underestimated and the calculated deposit heights in the runout zone were often not realistic. The position of the point P and the corresponding flow width must be determined by an expert opinion, which could cause difficulties in terraced terrain or terrain slopes near the critical angle. It is therefore rarely used today in practice, but is the basis of most of the following models.

3.2.5.2 AVAL-1D

AVAL-1D [43, 247] is the numerical version of the Voellmy-Salm model. As a significant innovation, the avalanche mass is included in the calculation and the modelling is performed along the realistic topography. It contains two independent calculation modules, FL-1D for flow avalanches and SL-1D for powder snow avalanches. Both modules are based on the solution of differential equations that describe the mass, energy and momentum conservation using the method of finite differences. AVAL-1D was developed by the WSL Institute for Snow and Avalanche Research SLF and is now used worldwide.

FL-1D is a numerical more or less 1-dimensional calculation model based on the Voellmy-fluid material law. The avalanching snow is considered 'quasi liquid' with constant snow density. The flow velocity v is constant over the height and the flow height dP is constant across the width. During the fall, the mass remains constant, a possible snow accumulation along the avalanche path is not considered. The flow resistance is described with three parameters – dry friction μ, turbulent friction ξ, internal

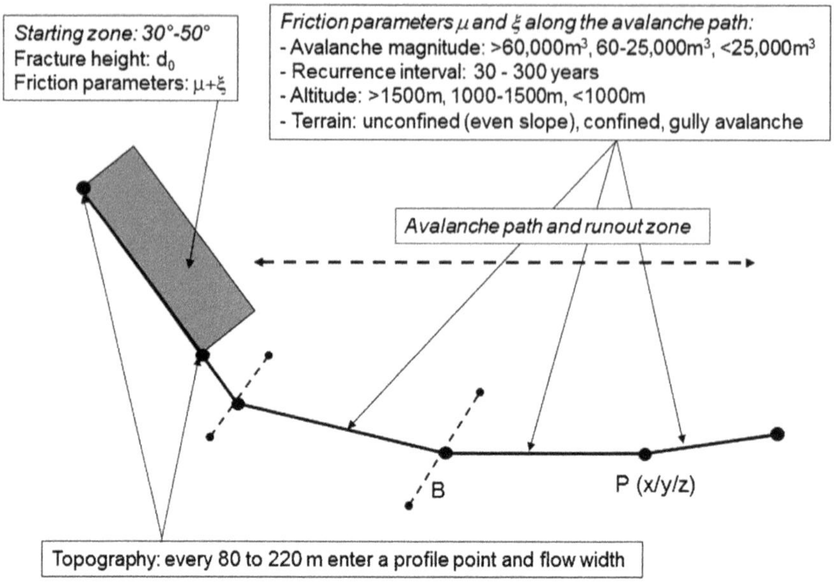

Fig. 3.7 Input parameters AVAL-1D-FL-1D (in accordance with [247])

deformation resistance of snow λ. A schematic representation of the input parameters is provided in Figure 3.7. AVAL-1D provides continuous information about the flow height, velocity and pressure along the entire avalanche path, and the runout distance and mass distribution of the avalanche deposit are calculated.

The avalanche mass has an impact on the length of the runout zone in the modelling, this allows for consideration of, for example, the influence of supporting structures in the starting zone. Since no snow accumulation is taken into account in the model, the calculated deposit heights can be too small. In comparison to the Voellmy-Salm model, the calculated velocities (Figure 3.8) are higher and therefore more realistic. The model does not take centripetal and impact effects into account, in highly meandering avalanche paths the calculation results are therefore conservative. Furthermore, the effect of catching dams cannot be modelled directly.

The powder snow avalanche in model SL-1D consists of a saltation and suspension layer, the flow component is neglected. There is a mass exchange of snow between the snowpack (erosion) and the saltation and suspension layer, taking into account the corresponding friction forces at the layer boundaries. A schematic representation of the input parameters is provided in Figure 3.9.

SL-1D provides continuous information about the flow height, velocity and pressure along the entire path of the avalanche. Furthermore, pressure profiles can be created at any point along the path of the avalanche. The selection of the degree of suspension and the erosion parameters requires a lot of experience and has a great impact on the results.

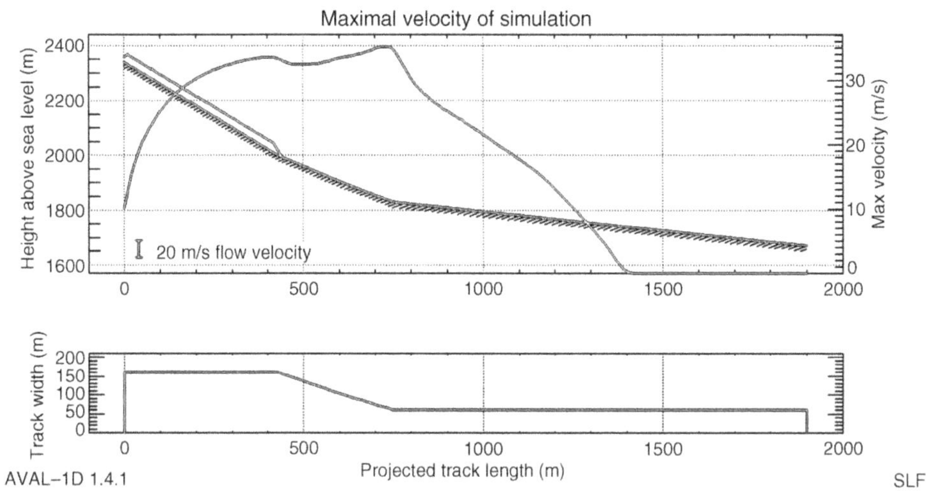

Fig. 3.8 Maximum avalanche velocity along the avalanche path calculated with AVAL-1D-FL-1D (© Margreth)

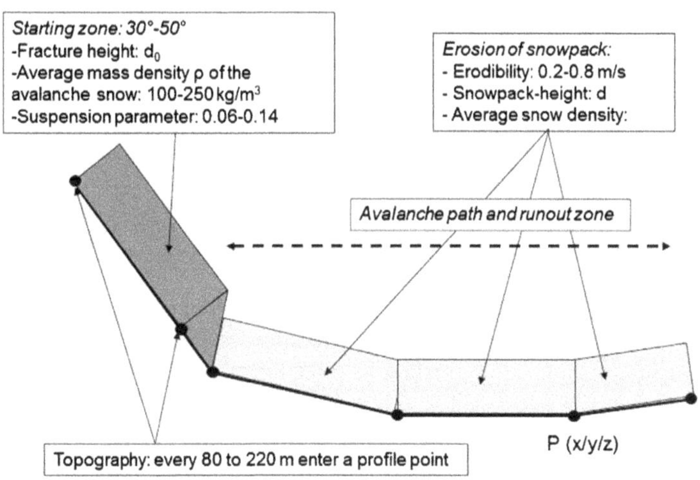

Fig. 3.9 Input parameters AVAL-1D-FL-1D (in accordance with [247])

SL-1D is a purely 1-D model. The influence of the flow width (spreading/channelling) requires expert consideration. The avalanche pressures must be reduced correspondingly, in particular in the runout zone. In the model, the air resistance on the front of the avalanche is neglected; therefore, the delay in the runout zone is too small. Especially in combination with observations, SL-1D allows the realistic calculation of avalanche pressures.

3.2.5.3 RAMMS

The two-dimensional avalanche calculation model, *RAMMS (rapid mass movement system)*, is a direct development of the one-dimensional numerical model AVAL-1D. It has been developed by the WSL Institute for Snow and Avalanche Research SLF (Davos/Switzerland) since 2005 and has been offered commercially since 2010. The velocity vector is calculated in the three-dimensional terrain in two directions and therefore the flow direction and width are determined by the model. The calculation grid is generated from a digital terrain model.

The model is based on the assumption that no internal deformations occur in the body of the avalanche. The friction forces act mostly in a sliding layer located between the avalanche and the substratum. The friction parameters μ and ξ describe the proportion of dry friction from *Coulomb* or the velocity-dependent frictional resistance. Both friction resistances depend on the size of the avalanche. As with the *Voellmy-Salm model,* the basal resistance is indicated by a *Coulomb friction term* proportional to the normal force and a *turbulent friction term* proportional to the square of the velocity. A finite-volume method resolves this in complex terrain with the system related to the shallow water equations. The model is linked to the GRASS GIS system to facilitate the determination of certain input parameters (starting zones and friction parameters) with the help of automatic procedures [44].

RAMMS was calibrated on the basis of many observed large avalanches in the avalanche winter of 1999, also on the basis of avalanches from the SLF avalanche database as well as from the avalanche test site of Vallée de la Sionne (Valais/ Switzerland). To visualize the input parameters (Figure 3.10) and results (Figure 3.11), geo-referenced maps or aerial photographs can be read in as texture data and shown on the topography.

In previous practical experience, RAMMS has proven to be a very useful tool to assess, in particular, flow paths in a complex terrain situation. Furthermore, it has been demonstrated that, in comparison with the Voellmy-Salm model or AVAL-1D, the definition of the location and size of a starting zone is of greater importance. For the calculation, the quality of the terrain model is very important. Small streams, which are filled with snow in winter, can lead to an unrealistic deflection of the avalanche. Fully automated avalanche computations over large areas can be realized with RAMMS.

3.2.5.4 SamosAT

The *SamosAT model* was calibrated on the basis of 22 reference avalanches [187, 231]. This has been used since 2007 in the Austrian Service for Torrent and Avalanche Control in Austria as the successor to Samos99, and in Iceland [76], as well as within the scope of expert assessments in South Tyrol and Russia.

The *model approach of Savage-Hutter* was revised in the SamosAT model and corresponds to an *expanded Voellmy approach* using variable friction coefficients μ and ξ by observing the *shallow water approach* [230]. The friction model for flow avalanche calculations was made dependent on both flow velocities as well as flow

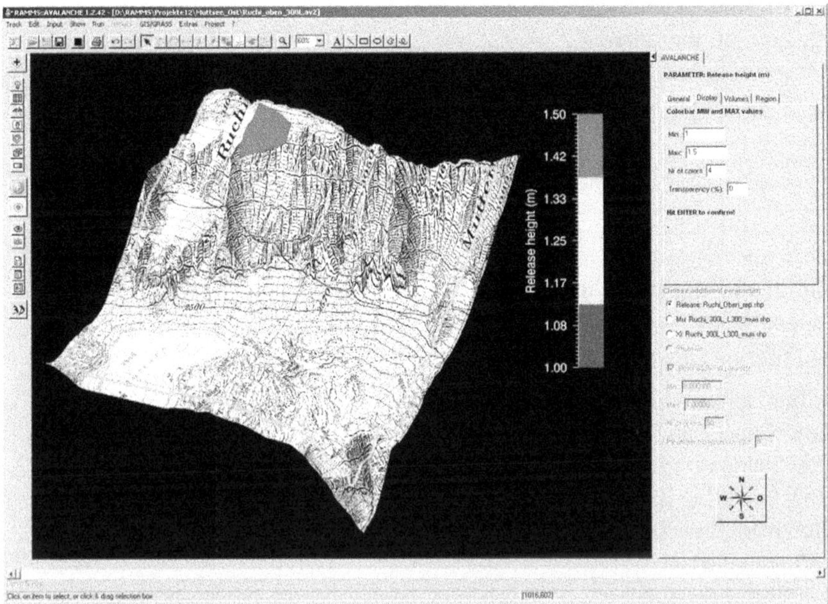

Fig. 3.10 Entering the starting zone and the fracture height in a 3-dimensional representation in RAMMS (© WSL/SLF)

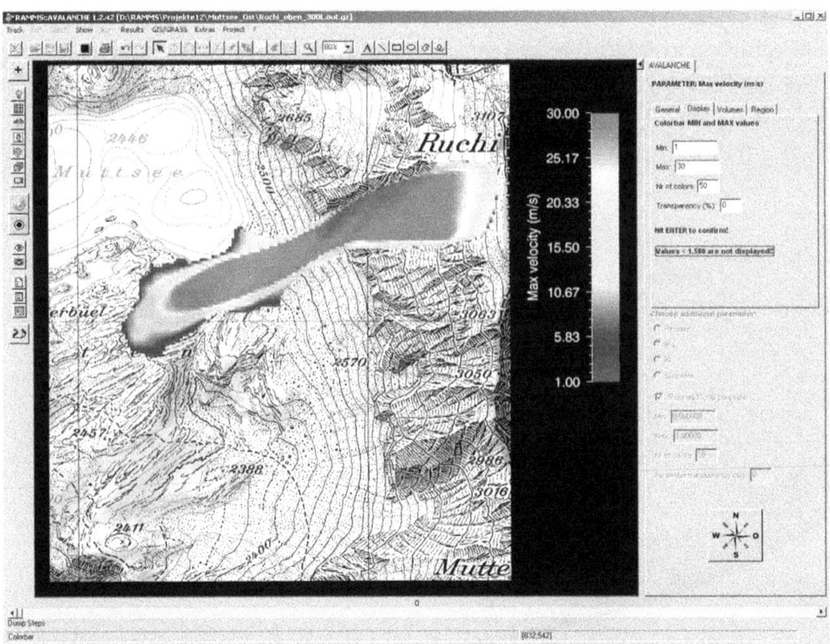

Fig. 3.11 Representation of the maximum velocity of the simulation with RAMMS (© WSL/SLF)

heights. This results in a stronger deceleration of the runout on the one hand, and less subsequent flow on the other.

$$\tau^{(b)} = \tau_0 + \tan\delta \cdot \left(1 + \frac{R_s^0}{R_s^0 + R_s}\right) \cdot \sigma^{(b)} + \frac{\overline{\rho}\,\overline{u}^2}{\left(\frac{1}{\kappa}\ln\frac{\overline{h}}{R}+B\right)^2} \tag{3.17}$$

where:

τ_0	Minimum shear stress, in [N/m^2]
$\tan\delta$	Tangent of the bed friction angle, in [°]
R_S^0	Bed friction elevation
$\sigma^{(b)}$	Normal tension, in [N/m^2]
$\overline{\rho}$	Flow density, in [kg/m^3]
$\overline{u}$	Flow velocity, in [m/s]
$\overline{h}$	Flow height, in [m]
B	Dimensionless constant (Prandl boundary layer)
K	Karman constant
R_s	Roughness constants

The model construction of a powder snow avalanche (mixed avalanche) (Figure 3.12) consists, according to [43, 91, 118], of a dense flow layer, which slides over the surface of the terrain. The flow status without noteworthy temporary air components is referred to as a flowing avalanche (volume component $< 1:10$). The density of such a flow is approx. 300 kg/m^3 on average. The re-suspension layer, which is considered to be very thin, follows the flow layer. A powder layer can develop on top of this. This can be considerably larger than the preceding layers.

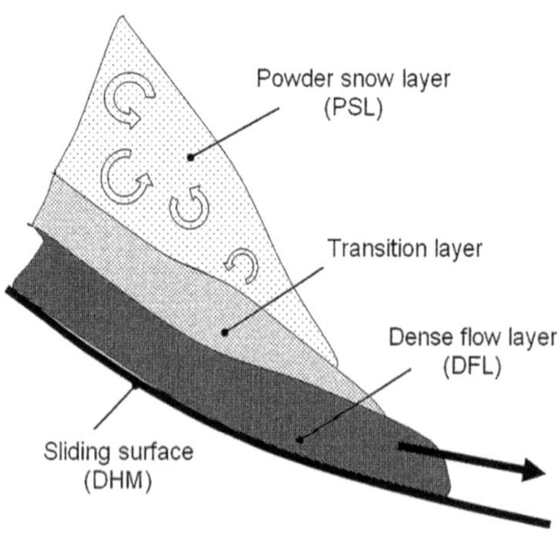

Fig. 3.12 Layer structure of a powder snow avalanche (© AVL)

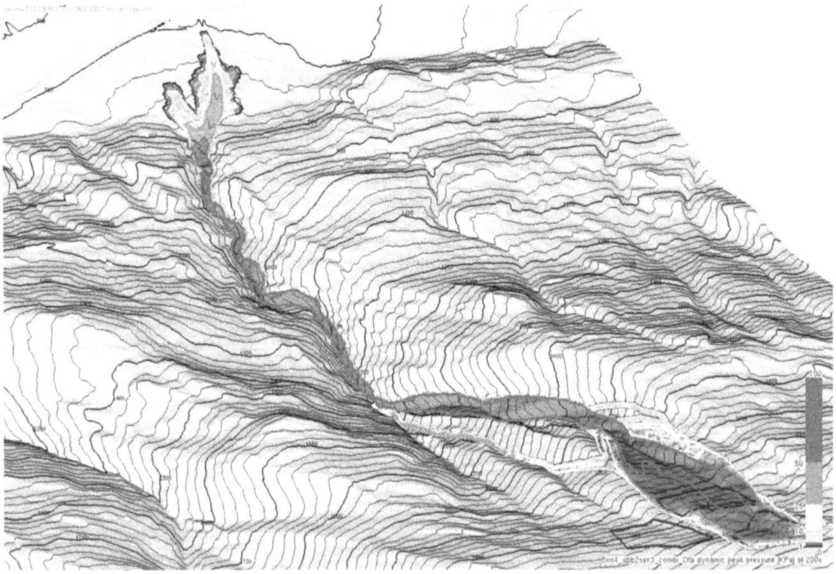

Fig. 3.13 Representation of the maximum flow pressure distribution with SamosAT (© WLV SSL)

In the model, a powder snow avalanche is treated as a turbulent, particle-laden gas flow with a small velocity difference between particles and air. The powder component is calculated in SamosAT in the AVL-FIRE© calculation platform, which uses the re-suspension model to connect the flow component as a boundary condition [230].

SamosAT enables a real two-stage calculation of ice particles and air layers to be able to more realistically represent the gas-dynamic processes in 3-d. In addition to the increase in mass, this method can also show the mass lost by sedimentation along the avalanche path. Thus, this mass loss can also lead to a slowing of the avalanche. In the program, mass increases can be taken into account using frontal or basal entrainment. Additionally, it is possible to activate a direct snow accumulation in the powder module and therefore achieve higher avalanche velocities (Figure 3.13).

3.2.5.5 Application of avalanche models in engineering practice

The classic application of avalanche simulations is in the area of *hazard potential assessment* and *hazard zone mapping*. Here, the maximum runout distance and the pressure distributions within the avalanche are generally necessary for a demarcation of hazard zones in accordance with the relevant guideline criteria [29, 235] (Section 4.2.3). The output of flow heights or flow velocities provides additional information about the potential flow behaviour in a given avalanche. Reports of documented damages by avalanches enable a calibration of the model and its results. The avalanche simulations also allow a continuation of the study of the different avalanche scenarios based on various fracture combinations, drift conditions or friction coefficients.

Simulations are used in the planning of technical avalanche defense measures to improve snow-supporting structures. Often the question of residual risk assessment arises after technical constructions have been built, however this is only partially verifiable with the models. An important area of application is the dimensioning and verification of avalanche dams. Simulations can deliver data on velocities and flow heights at a certain point, which can subsequently be used to calculate the required dam heights (Section 6.4). Most of all, the simulations of the effect of avalanche slide dams provide good data on the effectiveness of these protective structures.

Simulation especially supports the study and analysis of avalanches in complex terrain. The recalculations of documented avalanche events provide insight into the distribution of the mass, the flow paths as well as the pressures, velocities and flow height distributions for the reconstruction of an avalanche. These simulations of reference avalanches in turn provide important information and references to possible improvements in the respective avalanche models and therefore enable feedback for the development of the model.

The numerical modelling is a valuable tool for planning avalanche defense structures and for preparing hazard zone plans, however, the limits of modelling must be taken into account. Wet snow avalanches cannot be calculated with these models. It is important that the user of the model can realistically estimate the possible extent of the process, independent of calculations, and is thus able to validate the results.

3.3 Avalanche action on objects (obstacles)

3.3.1 Dynamic avalanche action

3.3.1.1 Principles

When avalanches hit an object (obstacle), dynamic pressure develops, which is generally considered stationary (i.e. time-independent) [85]. The impact of snow blocks or enclosed solid components (boulders, tree trunks) (Section 3.3.1.5) also causes short-term impact pressures which can be much higher than the stationary pressures.

It is relevant whether the flow is over an obstacle or around it, and whether the obstacle is in the range of the powder or flow component of an avalanche. An obstacle is then considered surrounded, when the majority of the snow particles actually flow around it. In practice, it is not possible to determine whether an obstacle is actually in a potential flow or powder snow avalanche region. Thus, according to a schematically selected procedure, the bottom area of a mixed-snow avalanche is considered the flow component, the next higher is considered the transition layer or saltation layer and the uppermost zone of the avalanche is the suspension layer or powder layer. The actual height of the individual layers can only be calculated partially; in practice, these – with the exception of the height of the flow component – are estimated. It should be noted whether the obstacle hit by an avalanche is located in a channelled section in which the avalanche flow heights are larger due to the constraints in the avalanche path or whether it is an unconfined avalanche. The schematic layer construction of an avalanche is shown in Figure 3.2b. Typical flow heights for the transition layer of an avalanche are 1 to 5 meters and several decameters for the powder layer [127].

The following section discusses the relevant cases of avalanche effects on flow obstacles.

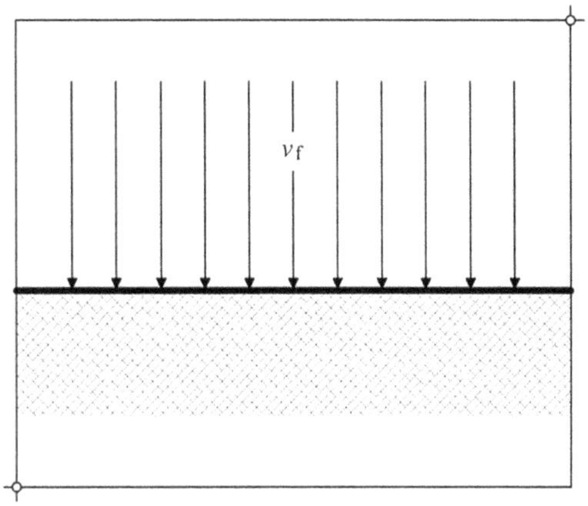

Fig. 3.14 Flow avalanche pressure on an obstacle that has not been surrounded in accordance with [202]

3.3.1.2 Action by flow avalanche on obstacles obstructing the flow

In this case, the flow angle of the flowing avalanche is (more or less) normal on this obstruction and this is so large that the main mass of the avalanche does not flow around the obstacle. This creates a dynamic pressure on the exposure area (Figure 3.14).

According to ONR 24805 [202], the avalanche pressure p_f on a large, obstacle, not surrounded by avalanche flow, is calculated as follows:

$$p_\mathrm{f} = \rho_\mathrm{f} \cdot v_\mathrm{f}^2 \quad [\mathrm{N/m^2}] \tag{3.18}$$

where:

ρ_f Density of the flow avalanche, [recommended for calculation: 300 kg/m^3]
v_f Velocity of the flow avalanche, in [m/s]

According to [127], the widely used method of calculating the avalanche pressure with the Equation 3.18 is too simplified and does not consider the *peak pressure* p_{peak}. This peak pressure can occur during the first milliseconds of the action on an obstacle. The authors suggest calculating the peak pressure as follows as a supplement to the stationary pressure:

$$p_{peak} = 3 \cdot p_f \quad [\mathrm{N/m^2}] \tag{3.19}$$

The section of the contact surface between the avalanche and the obstacle to be subjected to the peak pressure *p*peak has not been specified in detail. However, it is recommended to take into account the expected peak pressure in addition to the base pressure when dimensioning particularly vulnerable buildings.

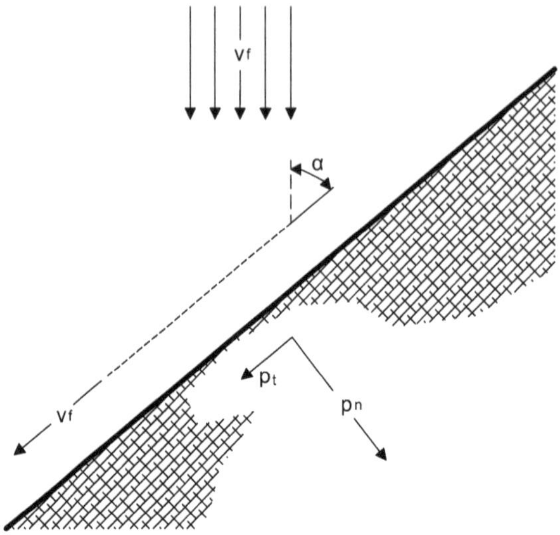

Fig. 3.15 Flow avalanche pressure on a deflecting obstacle, in accordance with [202]

When a flow avalanche is deflected by an obstacle, the normal and tangential pressures pn or pt can be determined as follows:

$$p_{\mathrm{n}} = \rho_f \cdot v_f^2 \cdot \sin^2 \cdot \alpha \quad [\mathrm{N/m^2}] \tag{3.20}$$

$$p_{\mathrm{t}} = \mu \cdot p_n \quad [\mathrm{N/m^2}] \tag{3.21}$$

where:

v_f Velocity of the flow avalanche, in [m/s]
α Deflection angle, in [°]
μ Friction coefficient [–]
0.30 . . . for the contact surface snow/snow or snow/ground.
0.40 . . . for snow on coarse soils or rough walls.

If a flowing avalanche encounters an obstacle (Figure 3.15), the height of the avalanche pressure consists of the flowing height and an additional impact height that is caused by dynamic impact effects. The climbing height h_{dyn} is calculated for an obstacle that has not been surrounded by avalanche flow according to the Equation 3.22, when the avalanche flow has surrounded the obstacle, then according to the Equation 3.25. λ is an empirical parameter intended to reflect the momentum loss when the avalanche hits the obstacle, as well as the effect on friction on the flow of the avalanche during run-up along the upstream side of the obstacle.

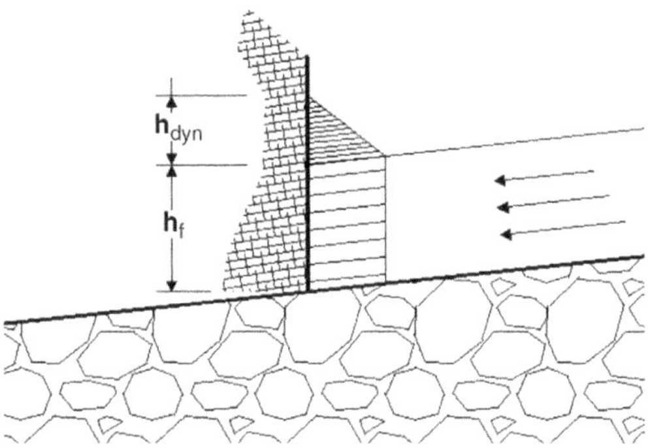

Fig. 3.16 The dynamic pressure reduces linearly within climbing height (*h*dyn) [202]

The climbing height h_{dyn} (Figure 3.16) for an obstacle that has not been surrounded by avalanche flow is calculated as follows:

$$h_{\mathrm{dyn}} = \frac{v_{\mathrm{f}}^2}{2 \cdot g \cdot \lambda} \quad [\mathrm{m}] \tag{3.22}$$

where:

g Gravitational acceleration, constant [= 9.81 m/s^2]
λ Momentum loss factor [−]:
With dry, large avalanches: 1.5; in dense (wet) avalanches: 2 to 3 (in accordance with [85]).

3.3.1.3 Action by flow avalanches on narrow obstacles

An obstacle is considered a *narrow obstacle* if it is so small that the majority of the snow particles flow around the obstacle (Figure 3.17). This applies especially to trees, lift supports or masts for energy supply cables.

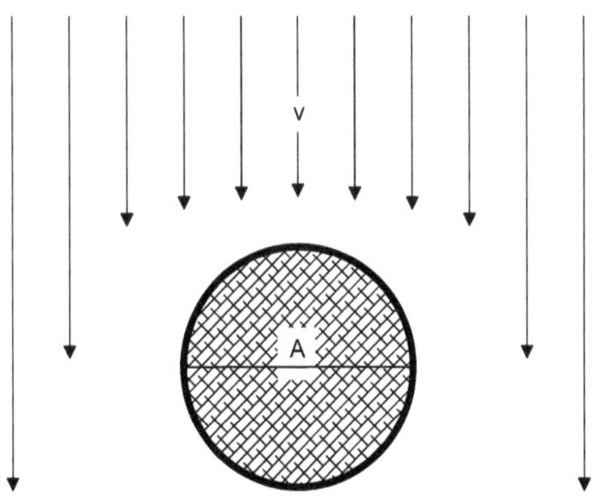

Fig. 3.17 Flow avalanche pressure on an obstacle surrounded by avalanche flow, in accordance with [202]

Table 3.3 Resistance coefficient c_d for flow avalanches according to [127]

Form	Snow type	
	dry	wet
Circular cylinder	1.5	3 to 5
Rectangle	2	4 to 6
Splitting wedge	1.5	3 to 6

The dynamic flow pressure on a narrow obstacle p_f can be calculated according to ONR 24805 [202] as follows:

$$p_\mathrm{f} = c_d \cdot \frac{\rho_f \cdot v_f^2}{2} \quad [\mathrm{N/m^2}] \tag{3.23}$$

where:

c_d Coefficient of resistance, according to Table 3.3

The resulting flow avalanche force on a narrow obstacle is determined as follows:

$$P_\mathrm{f} = A \cdot c_d \frac{\rho_f \cdot v_f^2}{2} \quad [\mathrm{N}] \tag{3.24}$$

where:

A Impact area, in $[\mathrm{m}^2]$

If a rectangular obstacle is surrounded by a flow avalanche, then a friction force parallel to the wall is to be considered on the parallel side of the rectangular obstruction on the avalanche impact side. To increase the security, it is recommended to determine the avalanche flow onto the impact sides at an angle of 20° [73, 74].

The climbing height h_dyn on a surrounded obstacle is calculated as follows:

$$h_\mathrm{dyn} = \frac{v_\mathrm{f}^2}{2 \cdot g \cdot \lambda} \cdot f_{b/df} \quad [\mathrm{m}] \tag{3.25}$$

where:

b Width of an narrow obstacle normal to the flow direction of the avalanche, in [m]
d_f Flow depth of the avalanche, in [m]
f Reduction factor, according to Table 3.4

Table 3.4 Reduction factor f according to [85]

b/d_f	0.1	0.5	1.0	2.0	≥3
f	0.1	0.4	0.7	0.9	1

3.3.1.4 Action by powder snow avalanches

As described in Section 3.1.2.4, large avalanches occur rarely in the form of pure flow or powder snow avalanches; these usually consist of a powder and flow content (mixed-motion avalanches). The transition or saltation layer is located between the two main layers. The transitions are smooth; it is unclear where exactly the suspension or saltation layer is located. The calculation of the occurring avalanche pressures can therefore only follow only a schematic process for which the fictional borders are assumed between these layers. According to the schematic assumption, a pressure image emerges for the calculation of a mixed avalanche, as shown in Figure 3.18.

In the assessment of powder snow avalanche effects and the effect of the saltation layer, consider that these two layers of the avalanche have a considerably lower density; on the other hand, however, they have a higher velocity and therefore flow ahead of the flowing components. Powder snow avalanches are a mixture of ice particles and air, and the snow grains, according to [118], often do not follow the air in the flow around an obstacle and instead hit the surface of the object obstructing the flow. This causes a pressure that is calculated according to the Equation 3.18 or Equation 3.23.

Issler [118] also recommends considering a *dynamic pressure coefficient* (*fdyn*), in which, depending on the velocity of the flow, the particle size and the particle concentration can be assumed to be between 0.5 and 1. In the event of uncertainties or airflow direction normal to the obstacle, assume a dynamic pressure coefficient of 1.

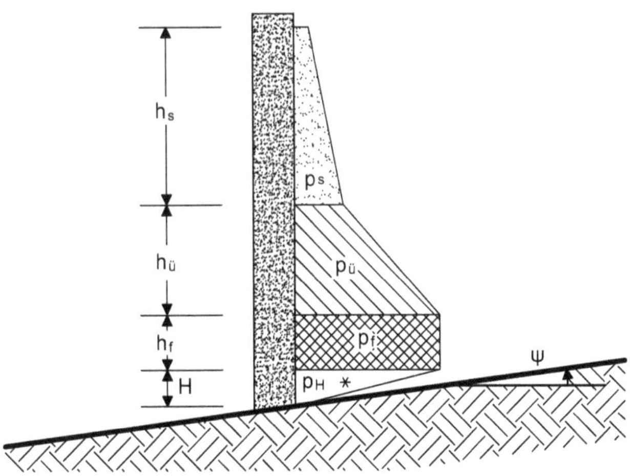

Fig. 3.18 Schematic representation of the pressure of a mixed avalanche on a narrow obstacle, in accordance with [202]

Table 3.5 Resistance coefficients for powder snow avalanches *cd* according to [202]

Form	Values for *cd*	
	Powder layer	Saltation layer
Circular cylinder	1	1
Rectangle	1.5	2
Splitting wedge	1.2	1.5

The density of the suspension layer is to be assumed as 3–15 kg/m^3 and those of the saltation layer as 20–50 kg/m^3.

The *resistance coefficients* c_d for powder snow avalanches are to be selected according to Table 3.5 in Austria according to ONR 24805 [202]. The consideration of a dam pressure coefficient is not foreseen in ONR 24805 due to the high degree of uncertainty in the assumptions and is generally to be assumed with 1.

The pressure distribution in the longitudinal and transverse direction of a powder snow avalanche can be very irregular. According to [118], these inhomogeneities lie in structurally-induced, turbulence-induced and topographically-induced pressure differences. Even turbulence-induced pressure differences can cause turbulent peak pressures, which can be up to a factor of 2 over the averages. If there is a flow over or around buildings or obstacles, four effects are of importance:

- During the overflow, the powder snow avalanche has a hydrostatic overpressure. It is highest in the middle of the head and depends directly on the avalanche mass, but not on the velocity. Typical values are between 0.5 and 2 kPa.
- Building sections orientated against the flow direction are exposed to the dynamic pressure. The effective pressure above the resistance coefficients depends on the shape and orientation of the building towards the direction of flow and generally reduces from the middle of the front of the obstacle to the edges.
- Dynamic negative pressure (suction) on side walls and roofs, as follows:

$$p_{suc} = -{}^{1}\!/_{2} \cdot \rho_p \cdot v_p^2 \tag{3.26}$$

where:

ρ_p Density of the powder snow avalanche, in [kg/m^3]
v_p Flow velocity of the powder snow avalanche, in [m/s]

It is generally only compensated to a small extent by the hydrostatic pressure.
- On flat obstacles or avalanche gallery roofs, shear stresses are to be expected in the range of 0.5–5 kPa, where the saltation layer is still pronounced, that is in the avalanche path or an early stage of the runout.

Table 3.6 Height development of powder snow avalanches dependent on the slope inclination

Slope inclination	*dh/dx*
10°	0.05
30°	0.10
50°	0.15

- There are only approximate estimates for the height development of a powder snow avalanche.
- The following height development arise dependent on the slope inclination (Table 3.6). For the calculation of the powder snow avalanche pressure, a linear decrease in the limit of the saltation layer upwards is assumed.

According to [127], the height of the powder cloud h_p is calculated as follows:

$$h_p = 10^{-4} \cdot l_{track} \cdot v^2 \quad [\mathrm{m}] \tag{3.27}$$

where:

l_{path} Length of the avalanche path, in [m]
v Velocity of the avalanche, in [m/s]

3.3.1.5 Impact of massive components (rocks, trunks)

Often solid components such as stones, ice clumps or trees are carried in avalanches and cause an *impulse pressure* (momentum) when they hit an obstacle (Figure 3.19). If this impact is to be assumed, it is simpler to consider a static substitute load (Figure 3.20). This is assumed to be a concentrated load on the avalanche-side wall of an obstacle with a circular impact surface with a diameter of 25 cm [74].

3.3.1.6 Action by vertical avalanche deflection

For the *vertical deflection of flow avalanches* with galleries or buildings, in addition to the snow or avalanche load, the pressure is to be determined by vertical deflection q_u (Equation 3.28). According to recent studies in Switzerland [10], the length of the deflection distance depends on the thickness of the flow avalanche and is to be calculated over a length of 6*df*.

$$q_u = (\rho_f \cdot d_f \cdot v_f^2 \cdot \sin\delta)/l_u \quad [\mathrm{N/m^2}] \tag{3.28}$$

where:

ρ_f Density of the flow avalanche, in [kg/m^3]

d_f Thickness of the flow avalanche, in [m]
v_f Velocity of the flow avalanche, in [m/s]
δ Deflection angle [°]
l_u Deflection path, in [m]:
is assumed with $6.d_f$

Fig. 3.19 Avalanche deposit heavily contaminated by avalanche wood (© Sauermoser)

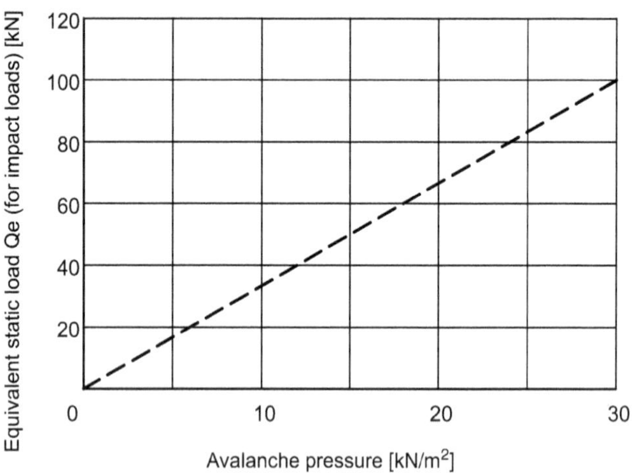

Fig. 3.20 Equivalent static load *Qe* for impact loads depending on the avalanche pressure [74]. This concentrated load is to be assumed as a simultaneous effect with the avalanche load *pf*. This can occur at any point in the effective height of the avalanche [73]

3.3.1.7 Action by wet snow avalanches

The flow behaviour of *wet snow avalanches* described in Section 3.1.2.5 is still largely unexplored. The pressure calculated on an obstacle is therefore approximate according to the Equation 3.18 to Equation 3.21 or Equation 3.23 and Equation 3.24. Wet snow avalanches have a low flow rate, but on the other hand a high density. For the calculation of the avalanche pressure, a density of $\rho = 400\,\text{kg/m}^3$ is recommended [7].

3.3.2 Damage effects of avalanches

3.3.2.1 General remarks

If avalanches hit objects in the avalanche path or in the runout zone, and if these are not sufficiently dimensioned for avalanche pressure, then this will cause damages up to the total destruction of the building or object (Figure 3.21c, d). The nature and extent of the damages are completely different and depend on the type of avalanche, the location of the building in the avalanche path or in the runout zone and the type of building or object.

Voellmy [277] investigated the destructive effects of avalanches in Vorarlberg in 1954 and – after direct observations of damages – he came to the following results:

- The forces that caused the investigated destruction in the avalanche path originated in resistance that the building opposed the flowing snow mass. Observations revealed that, in the cases examined, the increase in force up to the maximum was a relatively slow occurrence. The effect on the building is therefore not to be followed up according to the impulse theory, but one can determine the acting forces based on static principles.
- In most cases, the resulting momentum in the studied avalanches acted from the bottom upwards, and only on very steep slopes was this from the top to the bottom. The vertical forces arising as a result of backlogs in an avalanche caused by an obstacle in the avalanche path can reach about half of the horizontal force (Figure 3.21a).
- Stones, trees and entrained objects carried along caused *battering effects*.
- The destructive power of the air blast preceding the avalanche was low in the evaluated cases and was only strong enough to push in windows and light doors, for which experience has shown a pressure blast of a maximum of 5 kN/m^2 is sufficient. On the other hand, the vortex layers of a powder snow avalanche front could have much the same effects as the pressure and suction effects from a compaction air blast. Destruction attributable to a typical suction effect could not be proven in the destructions in Vorarlberg nor from the data available from Switzerland.

Hagen and Heumader [91] report damage caused by the snow-air mixture of powder snow avalanches, which can only be explained by *gas-dynamic theories* and which are similar to explosion damage. This is demonstrated by simulations of damage caused in Austria by the avalanches in 1984, 1986 and 1988. In particular in the Wolfsgruben avalanche in St. Anton in Tyrol (event 1988), broken canopies and horizontal cracks in

(a)

(b)

Fig. 3.21 Building damage caused by flowing and powder snow avalanches: (a): typical avalanche damage to a building by forces acting vertically upwards; (b): damage to a building caused by a powder snow avalanche: The building is completely filled with snow to under the roof and some windowpanes are broken (© WLV Tyrol). Building damage caused by flow and powder snow avalanches: (c): Parts of buildings scattered in all directions, caused by the turbulence in the powder snow avalanche; (d): Total destruction of buildings by a dry flow avalanche or transition layer (an exact distinction is not possible) (© WLV Tyrol)

the masonry constituted typical damage caused by the pressure of the snow/air mixture deflected vertically and upward.

An approximation of the avalanche forces can be made on the basis of damage, whereby a differentiation in powder or flow components on the basis of Table 3.7 is not possible.

(c)

(d)

Fig. 3.21 *(Continued)*

3.3.2.2 Damage effects by flow avalanches

Typical damage to properties caused by the flowing component of an avalanche are:

- Total destruction of buildings by high avalanche pressures (Figure 3.21d)
- Massive damage to buildings on the lowest level of the building, combined with massive avalanche deposits
- Small-scale delimitation of damage caused by individual flow avalanche arms
- Spatially limited damage effects from individual arms of a wet snow avalanche.

Table 3.7 Damage potential of avalanches in relation to the pressure of the avalanche, according to [7]

Avalanche pressure [kPa]	Damage potential
1	Windows are pressed in and smashed.
5	Doors are pushed in.
30	Wooden constructions and brick buildings are destroyed.
100	Spruce forest is thrown over a large area.
1000	Reinforced concrete structures are damaged or destroyed.

3.3.2.3 Damage effects by powder snow avalanches

The distinction, whether damage to a building was caused by the powder, transition or flow layer of an avalanche is rarely possible because the transitions between the individual layers are blurred. Damage caused solely by the powder component of an avalanche is naturally that already located on the ground in the outermost runout zone of avalanche deposits or in the area of super elevation in avalanche paths.

Typical damage to properties caused the powder component of an avalanche are:

- Broken window panes in the area of the entire building
- Damage to the roof in particular caused by vertical upward forces (Figure 3.21a)
- Damage to chimneys, gutters, wind pipes
- Plastering of snow up to the roof area (Figure 3.21b), deposit densities up to 600 kg/m^3 can be caused by powder snow avalanches [118]
- Traces of inner explosion of buildings caused by penetrating avalanche components
- Widely scattered parts of buildings, especially wooden components for wooden houses (Figure 3.21c)
- Uncovered and removed roofs or roof components.

4 Avalanche hazard assessment and planning of protection measures

Florian Rudolf-Miklau, Christoph Skolaut and Siegfried Sauermoser

4.1 Avalanche hazard (risk) assessment and mapping

4.1.1 Model of hazard assessment and risk concept

Technical defense measures are used to prevent negative consequences of future avalanche events. Therefore, forecasts about the size (magnitude) (Section 2.4.3) and the probability of occurrence of events (Section 2.4.2) and the possible negative (damaging) effects on human welfare, economic activities, the environment and the cultural heritage are required as a basis for action planning.

The *hazard analysis* is the basis for planning avalanche defense measures. This includes the assessment of the location of avalanche catchment areas, the factors for development of avalanches (Chapter 2), the triggering mechanisms, and the avalanche processes (Chapter 3) as well as the potential effects of avalanche events (damages; Section 3.3.2). The graphic presentation of the assessment results in *hazard maps* and *hazard zone plans* (Section 4.2) and is the planning basis for technical avalanche defense measures in accordance with the usual international standards.

A further development of the hazard assessment, including potential avalanche damage, is represented by the *risk concept*, which, according to [27], can be used as a universal model for resolving safety issues in avalanche protection and as a basis for action planning. In accordance with [281], it allows for statements about the practicality, reliability and effectiveness of protection concepts. As the first country in Europe, Switzerland consistently implements the concept of risk in their policies for protection against natural hazards [211].

A schematic diagram of the model for analysis and evaluation of avalanche hazards and risks is shown in Figure 4.1.

4.1.2 Avalanche hazards and risks: definitions and quantification

4.1.2.1 Avalanche hazards and hazard scenarios

For technical avalanche control, only avalanche events are relevant that pose risks to persons or tangible assets [281]. The *avalanche hazard* is a function of the *probability of occurrence* (Section 2.4.2) and the *intensity/magnitude* (Sections 2.4.3 and 2.4.4) of an event and occurs basically independently of concrete risks [26]. For the purposes of hazard and risk assessment, the estimated probability of occurrence (statistical expectation value) of a future avalanche event related to a certain period of time is expressed by the recurrence interval T_N (Section 2.4.2). Figure 4.2 shows the schematic relationship of avalanche activity in an individual avalanche catchment area or for a whole region (as a function of the probability of the avalanche and the avalanche intensity) with the endangered object categories (Section 4.1.2.2).

The Technical Avalanche Protection Handbook, First Edition. Edited by Florian Rudolf-Miklau, Siegfried Sauermoser and Arthur I. Mears.

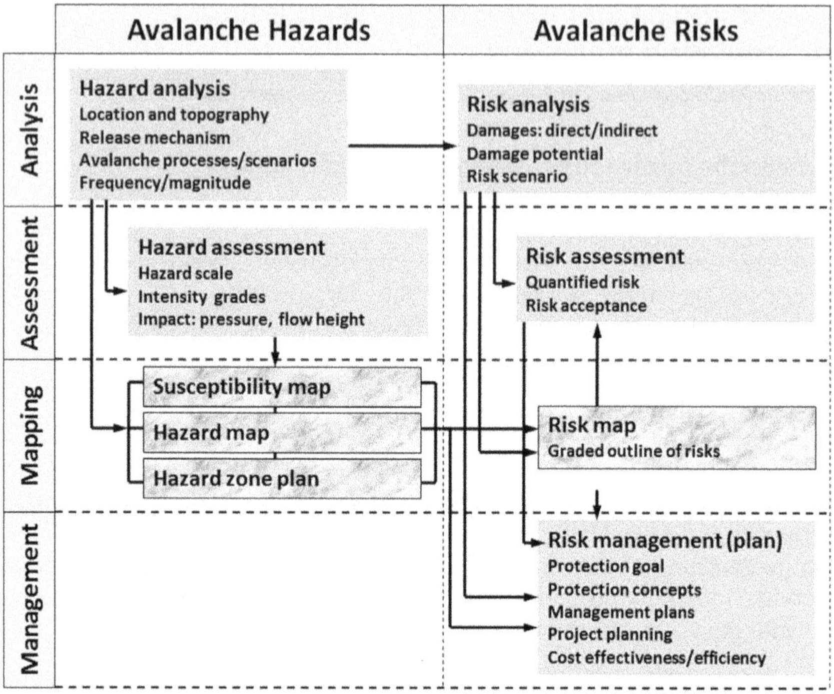

Fig. 4.1 Diagram of the model of analysis, assessment and conceptual presentation of avalanche hazards and risks as a basis for risk management (created according to Heinimann et al. [99], Hübl et al. [114] and Kienholz [138]) (© Rudolf-Miklau)

The analysis of avalanche hazards, which represents a complex set of causes, processes and effects, is not a mono-causal process but must consider a wide range of possible developments, so-called *scenarios*. The scenario analysis has been established as a method to represent the possible processes of future natural events or the associated dangers (damage effects) [114, 138, 235, 253]. Here, not only a specific process is considered, but all possible or essential processes in an avalanche event (within the limits of the recurrent design event of a specific recurrence interval), especially the probable and the extreme case.

The easiest method to quantify the avalanche hazard is represented by the application of hazard and intensity scales on the basis of semi-quantitative criteria. The most commonly used avalanche hazard scale was developed by the avalanche warning service in the Alpine countries (*European avalanche hazard scale* [79]; Table 4.1). Additional scales for the classification of avalanche intensities were published by CAA [41], Rapin [215] and Schaerer and McClung [254] amongst others (Table 4.2).

However, the statistical physical determination of factors of avalanche likelihood and intensity (Section 2.4) is required for engineering questions relating to avalanche hazards.

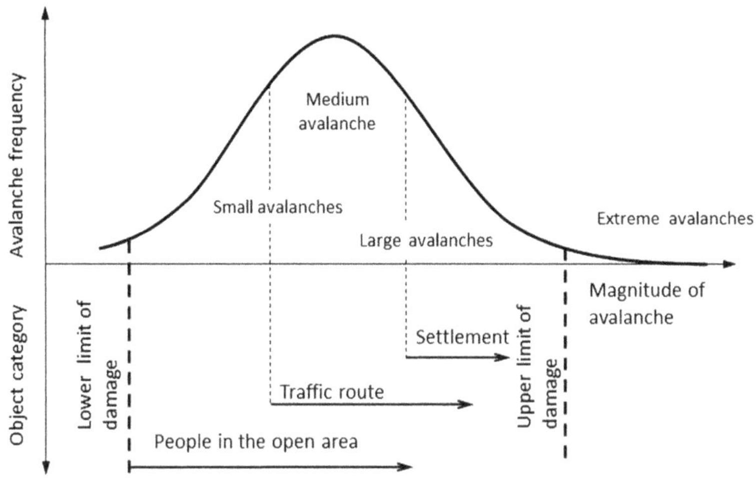

Fig. 4.2 Schematic representation of the relationship of avalanche activity in a region (as function of avalanche probability and avalanche magnitude with the avalanche types and object categories). The illustration shows, amongst other things, that people outdoors are already endangered by frequent avalanches of small dimensions, while settlements are generally threatened only by large avalanches (according to Wilhelm [281])

Table 4.1 European avalanche hazard scale [79]

Danger Level			Snow Pack Stability	Probability of Avalanche Triggering
5	**Very high**		The snowpack is generally weakly solidified and largely unstable.	Many large, often very large avalanches can be expected to be spontaneous even in moderately steep terrain.
4	**High**		The snowpack is poorly solidified on most of the steep slopes[a)]	Avalanche triggering is probable with a low additional load[b)] on numerous steep slopes. In some cases, many medium sized avalanches, often also large avalanches are to be expected.

(continued)

Table 4.1 (*Continued*)

Danger Level			Snow Pack Stability	Probability of Avalanche Triggering
3	**Considerable**		The snowpack is only moderately to poorly solidified on many steep slopes[a)]	Avalanche triggering is probable with a low additional load[b)] especially on the indicated steep slopes. In certain cases, some medium, but occasionally large avalanches are possible spontaneously.
2	**Moderate**		The snowpack is only moderately solidified on certain steep slopes[a)], otherwise generally well solidified.	Avalanche triggering is particularly possible with a large additional load[b)] especially on the indicated steep slopes. Large spontaneous avalanches are not to be expected.
1	**Low**		The snowpack is generally well solidified and stable.	Avalanche triggering is generally only possible with large additional load[b)] at scattered places in the extremely steep terrain[a)]. Spontaneously, only slides and small avalanches are possible.

Explanations:

a) The terrain threatened by avalanches is generally described in the avalanche situation report (altitude, exposure, terrain shape). Moderately steep terrain: slopes flatter than approximately 30 degrees. Steep slopes: slopes steeper than approximately 30 degrees. Extremely steep terrain: slopes which are particularly unfavourable regarding incline (steeper than approximately 40 degrees), terrain shape, crest proximity and ground roughness.

b) Additional load: low: individual skiers/snowboarders, gently moving, not falling; snow-shoe walkers; group with relief distances (>10 m), large: two or more skiers/snow-boarders and so on without relief distances; grooming vehicles; demolition; individual pedestrians/mountain climbers.

Table 4.2 Classification of avalanche intensity (destruction potential) according to the reach, the potential for damage, as well as according to the criteria of avalanche intensity (runout length, avalanche cubic meters, and pressure effect) (compiled to CAA [41], Glade and Stötter [77], Rapin [215], Schaerer and McClung [254])

Order of Magnitude/ Intensity of Avalanche		Classification according to Reach	Classification according to Damage Potential	Classification according to Runout Length/Volume/ Pressure		
				m	m^3	kPa
1	Very low/snow sliding	Irrelevant snow distribution in the slope	Low risk to people	<50	<100	~2
2	Low/ small avalanche	Comes to a stop in the steep slope area (over 30°)	Risk to people: Burials and fatalities possible	<100	<1000	~10
3	Average/ mid-sized avalanche	Reaches the foot of the slope	Risk of destruction for cars, damage risk for trucks and small buildings	<1000	<10 000	~50
4	High/ large avalanche	Covers flatter terrain sections (less than 30°) of more than 50 m, reaches the valley floor	Destruction risk for larger structures and cars, risk to people within buildings	>1000	<80 000	~200
5	Very high/ extreme event	As yet unknown avalanche track and reach	Total destruction of large structures and parts of settlements, high numbers of victims	≫1000	to 400 000	~500

4.1.2.2 Avalanche damage and risk

If an avalanche event has negative effects on people or property, damage occurs. *Avalanche damage* frequently occurs as a *total loss*. According to [222] avalanche damage includes all negatively assessed consequences (effects) of an avalanche event (damage event), whereby the following damage categories are to be taken into account for the technical protection:

- People (deaths, injuries)
- Tangible assets (buildings, traffic systems, infrastructure, vehicles, machines, goods, energy)
- Natural resources
- Economic activities

- Social structures, community
- Cultural goods (cultural heritage)
- Environment

Furthermore, there is a differentiation between direct and indirect damage, whereby the latter is not caused by the event itself, but indirectly by the results of it. These consequential damages include detour costs at roadblocks, lost production by economic enterprises, the non-fulfilment of transport obligations or declines in tourism. Contrary to the concrete damage from an avalanche event, the potential for damage includes all objects (values) that can be damaged by avalanches in a defined area (period), thus corresponding to the maximum possible damage.

According to [222, 261], the *avalanche risk* is defined as the probability of a damage that results from an avalanche hazard meeting people or objects in the hazard zone, therefore depending on the exposure and damage sensitivity of these people (objects). In the further analysis of the risk of avalanche, the person or object (such as buildings, road sections) represents the smallest unit of analysis (*individual risk*, object risk); the *collective risk* as a sum of all object risks or social (community) risks can be derived from this in accordance with [27].

In accordance with [22], the avalanche risk $R_{i,j}$ is a function of the probability of occurrence and the magnitude of damage which consists of the factors of potential for damage and sensitivity to damage:

$$R_{i,j} = p_j \cdot V_i \cdot p_{i,j} \cdot SE_{i,j} \quad [€/\text{Jahr}] \tag{4.1}$$

where:

p_j	Probability of a hazard scenario j [–]
$p_{i,j}$	Probability that an object i is exposed to the hazard scenarios j (exposure) [–]
$SE_{i,j}$	Damage sensitivity of the object i dependent on the risk scenario j (vulnerability) [–]
V_i	Value of the object i, in [€/year]

The exposure $p_{i,j}$ expresses where and how long a person or an object is in the hazard zone (probability of presence). The damage sensitivity $SE_{i,j}$ – expressed as the expectation value of damage to an object as a result of a specific event – includes all direct damage to an object by avalanches. Equivalently, the lethality expresses the probability of the death of a person from the direct effects of an avalanche event.

The damage caused by avalanches can be quantified generally directly in monetary units. The damage value is equal to the financial cost necessary to return the object to the condition it was in before the disaster. The assessment of indirect or consequential damages is more difficult, whereby the biggest problems lie in the clear allocation of costs to the disaster, the availability of more realistic valuations and the demarcation from other costs. Consequential damages are often disregarded in assessing damage or are determined according to standardized methods.

Persons (victims), who are killed or injured by avalanches also represent an economic value, however a direct monetary evaluation of human value is not always possible or desirable. According to [128], human victims are to be seen as the priority avalanche damage, which is why in Iceland the risk assessment is only based on the probability of death, however without assigning a monetary value to human life. Indirectly, however, according to [27] an evaluation in monetary units can be derived indirectly from society's willingness to pay to avoid deaths.

For questions of technical avalanche protection, usually the loss potential is determined on the basis of hazard zone maps. The specific extent of damage in the event case can, according to [27], be determined by overlaying the spatial distribution of the avalanche intensities of different scenarios (intensity maps) with the location of endangered objects, taking into account the damage sensitivity (lethality), $SE_{i,j}$, of the likelihood of occurrence of p_j and the probability of presence, $p_{i,j}$. The quantified damage potential (level of damages) forms the basis for the determination of avalanche risk (Equation 4.1), divided into the components of person risk (individual risk for a person, collective risk) and material risk. In accordance with [27], it is necessary to perform the quantification of the avalanche risk differentiated (risk matrix), taking into account the relevant event scenarios (e.g. scenarios with different return periods: 30, 100, 300 years) and exposure situations.

For the quantitative representation of the exposure to avalanche hazards, according to [27] it is necessary to identify the potentially endangered objects and to describe them in regards to the location, number, type, use and value. In doing so, it is necessary to differentiate between stationary (fixed) objects (such as buildings, transport facilities, infrastructure), temporal or locally variable objects (vehicles, mobile facilities) and people (including place of residence). For variable objects and people, information on the average occupancy and the probability of presence in the hazardous area is also required. If necessary, special situations can also be considered, if within a short period of time, an above-average number of persons or variable objects are located in the hazard zone (e.g. full parking lot for a ski resort in the avalanche hazard area).

The sensitivity to damage (vulnerability) $SE_{i,j}$ can be determined from the tables in the literature [27, 115] as a relative measure for the damage to an object by the effects of an avalanche process (value between 0 and 1) depending on the object type. The same applies to the lethality factor that depends on the exposure of the person – specifically, whether the person is outside in the open air or inside buildings.

4.1.3 Methods of hazard assessment

Effective avalanche defense measures require that the threatening hazard is recognized completely, the position is recorded (localized) and an assessment can be made regarding its causes and effects. The objective of the avalanche hazard analysis is, according to [27], therefore the determination of the relevant scenarios and their probability of occurrence, to enable a forecast about the type, dimension and intensity of the relevant avalanche event.

Table 4.3 Overview of the methods for avalanche hazard analysis (according to [27, 99, 114, 138])

Method	Description and Area of Application
Historical method	This method is based on the evaluation and statistical analysis of reports, witness statements, and chronicles of past avalanche events. This can be used to form conclusions regarding the frequency of avalanche events, less often this can also provide information about the cause, the procedure, the range and spread of the avalanche, as well as any damage occurring. Historical sources are, however, usually incomplete and often subjective in representation.
Morphological method	This method is based on the identification of the avalanche process and the spatial demarcation on the basis of *silent witnesses* in the terrain or on vegetation [12]. Amongst other things, this includes the dendrogeomorphological dating, which provides good information about the frequency of avalanches under certain conditions.
Statistical method	This method comprises the analysis of measurement data and observation series. The derivation of reliable trends and forecasts requires, however, a large amount of data and a sufficient observation duration.
Numerical-mathematical method	This method is based on numerous avalanche models that generally provide good statements about the intensity of events of a specific probability of occurrence and is therefore preferred for determining the hazard scenarios. However, the models always require calibration with natural data and should be validated on the basis of expert knowledge (Section 3.2.5.5).
Pragmatic method	This method is based on the *expert estimation* of hazard potentials by experts and local people. This is generally used to validate or refute the results of the other methods.

Since neither the scientific basics of the avalanche processes is fully clarified, nor sufficient data available on past avalanche events, according to [138] the most important principle of avalanche hazard analysis is compliance with the highest possible redundancy of established procedures and methods. Regarding the procedure, basically the evaluation of past avalanche events (*backward-facing indications*) and prognosis of (possible) future events (*forward-looking indications*) can be selected. According to this principle, the methods listed in Table 4.3 can be selected for the hazard analysis in accordance with [27, 99, 114, 138]. These are based on historical information, observations of nature, measurements, models and expert assessment and must be applied in combination to ensure the principle of redundancy.

Table 4.4 Overview of the data required in practice and information on performing an avalanche hazard analysis (compiled in accordance with [27, 65])

Category	Data/Information
Avalanche register	Basic information about avalanche catchment areas, general location information
Topographic data	Area demarcations, photomaps, digital terrain models (DTM) or digital elevation models (DEM), location of objects in the avalanche track, location of permanent defense measures
Meteorological data/factors	Frequency of avalanche-triggering weather conditions, 3-day (or longer) new snow totals, total depth of snow cower, maximum snow depth, wind conditions/speed, temperature
Nivological data	New snow amount in a specific unit of time, local wind direction and wind speed, air and snow temperature, snowpack structure, stability of the snow, loads
Morphological data	Terrain analysis in the starting zone, in the avalanche path and in the deposition zone: Incline, exposure, terrain shape, roughness, vegetation, obstacles
Event data	Event register, historical sources and witness statements, expert knowledge, *silent witnesses*
Current avalanche situation	Avalanche report (bulletin), avalanche hazard level, surveys by the local avalanche commission, observed avalanche events

The avalanche hazard analysis (see Table 4.4) is limited for various reasons (Table 4.5) with regard to the quality of the results and uncertainty in the accuracy of the statements. The knowledge of these factors is absolutely necessary for the correct interpretation of analysis results and for the formation of the scenario.

4.1.4 Avalanche risk assessment

The *risk analysis* is organized according to [27, 133] into the identification and evaluation of the current and future avalanche hazards and into the analysis of the damage potential. It serves as the most objectively possible determination of the size of the risk for a concrete disaster, for a specific object (person) or a certain area and includes by definition

- the hazard analysis
- the exposure analysis
- the analysis of effects (damages) and
- the risk identification

Table 4.5 Factors for limitations and uncertainties in avalanche hazard analysis (according to [134])

Uncertainty Factors	Causes
Limited data situation	One-off and limited availability of meteorological measurement data and data about the avalanche release (causes, break-off power, stability of the snowpack), observations that are too short, broad lack of direct measurements of the avalanche process (velocity, pressure effects, movement process)
Incomplete and incorrect documentation of events	Missing records and event databases, varying quality of information and documentation, different density and intensity of recordings, different recording standards and data interpretation
Uncertainties in the scenario selection	Incompleteness in the scenario selection, judgement error in the selection of the scenario, incomprehensible processes
Observation of defense measures	Erroneous evaluation of the effectiveness and the condition of the defense measures, erroneous evaluation of the effects on the probability of occurrence and intensity of the avalanche processes, erroneous evaluation of the residual risk

Different loss scenarios are used to determine which factors and conditions contribute to the overall risk. Based on this, the risk assessment serves as proof as to whether the identified risks are above or below the determined protection goals, and therefore whether there is a lack of protection. This can be used to determine the optimum combination of protective measures intended to achieve the protection objective or also to prove the efficiency and effectiveness of measures taken.

Avalanche risk analyses in practice are – depending on the problem – performed for different spatial analysis units, such as based on a municipality (Davos/Switzerland: Fuchs et al. [65, 66];) Galtür/Austria: Keiler and Stötter [131]), for a valley (Pitztal/Austria: Huber et al. [113];) Suldenta/Italy: Keiler et al. [133]) or for larger regions, as well as based on different periods such as the current time, a previous period of time (development over time) or performed as a forecast for future development. The intensity of data collection has to be adapted to the spatial and temporal analysis unit (Table 4.6).

4.2 Mapping of avalanche hazards and risks

4.2.1 Overview

Hazard mapping according to [138] includes the elaboration of scientific basics and the cartographic representation thereof in *hazard maps* (hazard indication maps). This

Table 4.6 Overview of the data required in practice and information on performing an avalanche hazard analysis (compiled in accordance with [27, 113, 114])

Category	Data/Information
Hazard representation	Results of the avalanche hazard analysis Risk map, hazard zone map, map of intensity of modelling
Usage information	Buildings: address databases, spatial register (building stock), building statistics Traffic routes: street and rail register People: demographic statistics, population development, average household size Economic enterprises: company statistics, sales, means of production Work: statistics of jobs by industry Land use: use share (settlements, traffic areas, agricultural land, forest, etc.) Agriculture: operating structure, size, income, promotion database Tourism: number of beds, accommodation statistics, occupancy, accommodation turnover, secondary sales
Damage assessment	Building value: insurance value, real estate valuation, gross volume/plan area of the building; building with special function (churches, hospitals, etc.). Cultural value (intangible)
Protective measures	Location, investments (adjusted for inflation), maintenance costs, service life
Spatial distribution of objects	Land register, land-use planning (space usage plan)
Development over time	Use information from various periods; time series

is used to develop the *hazard zone plans* relevant to spatial planning and/or the transmission of hazard zones in the (more or less) legally binding planning instruments for general or local area planning.

The following categories generally apply for the conceptual representation of hazards and risks:

- Hazard (indication) maps (Section 4.2.2)
- Process maps (intensity maps) (Section 3.2)
- Hazard zone plans (Section 4.2.3)
- Risk maps (Section 4.2.4)

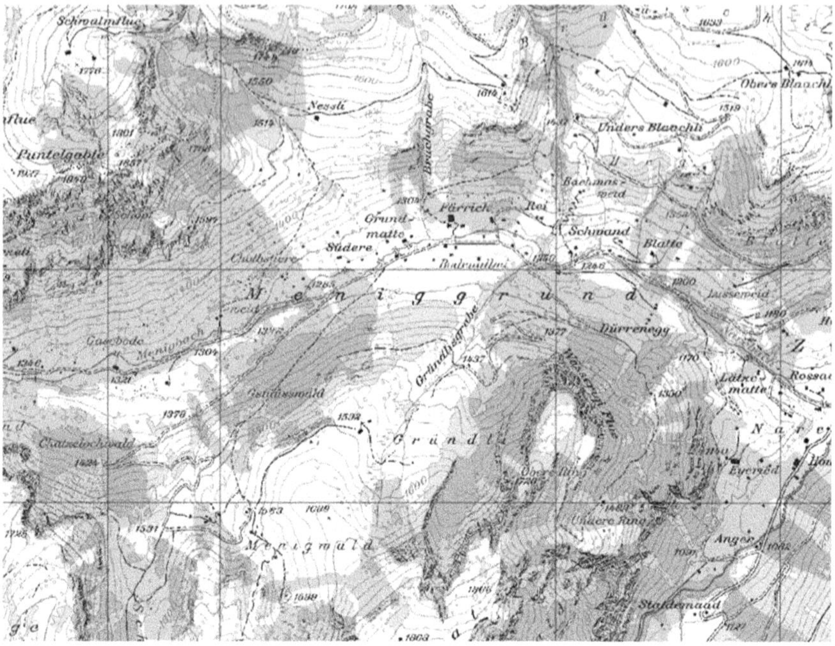

Fig. 4.3 Hazard indication maps: excerpt from a hazard indication map according to the Swiss standard (© Canton of Berne)

4.2.2 Hazard (indication) maps

Hazard (indication) maps provide a rough overview of which areas can expect avalanche danger. These have a purely indicative function, but say nothing about the specific extent of the hazard. If there are no detailed hazard zone plans, then hazard indication maps are usually the only source of information about the location and spatial extent of an avalanche hazard.

Hazard indication maps are created in Switzerland for large territorial units (cantons) with less accuracy (scale 1: 10 000–1: 50 000) (Figure 4.3). In Austria, no hazard indication maps are created, however, the overview maps referred to as *hazard maps* (scale 1: 10 000–1: 50 000) serve a similar function in the hazard zone plans (Figure 4.4).

In Italy, there is an avalanche hazard indication map in the scale 1: 25 000 (CLPV: La Carta localizzazione delle valanghe) for those regions of Italy with avalanche problems. In addition, an avalanche register in the scale 1: 10 000 is kept in some regions of Italy. In South Tyrol, there have been hazard indication maps for avalanches since the 1980s. These have been used in combination with the damage potential categories (intensity of the area use) as a basis for the creation of the hazard zone plans. In Germany (Bavaria), the conceptual presentation of the avalanche hazard is performed solely on the basis of hazard indication maps, detailed hazard zone plans are not created. In France, the

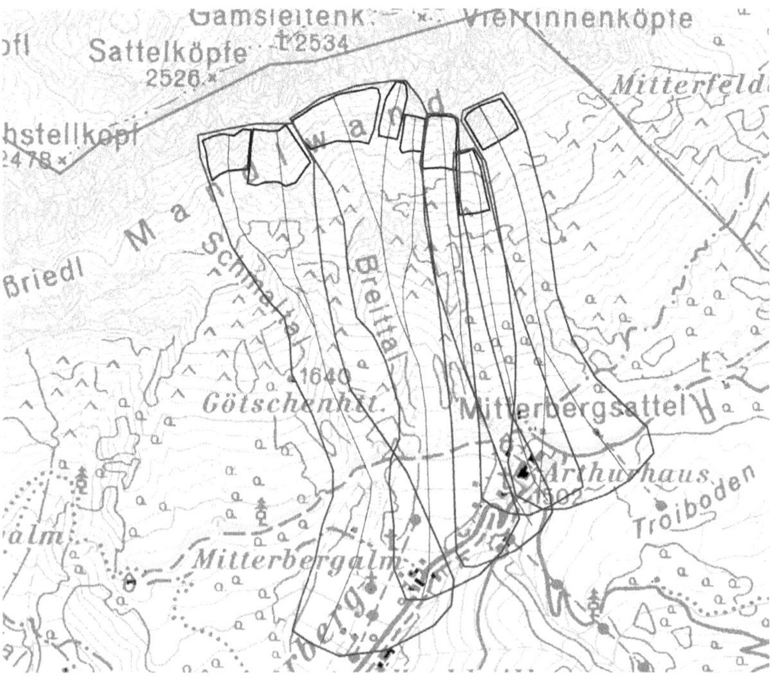

Fig. 4.4 Hazard indication map as part of the hazard zone plan in Austria (municipality of Mühlbach am Hochkönig) (© WLV Salzburg)

assignment of avalanche areas is based on databases of damage events – such as, for example, the EPA (Enquete Permanente sur les Avalanches), which documents the ongoing survey of avalanche activity of approximately 650 major avalanches – but also on the basis of annually updated *Carte des Localization Probable des Avalanches* (CLPA), which covers a surface of approx. 6000 km^2 at a scale of 1: 25 000 [114]. In Spain, susceptibility maps (hazard maps) are available at a scale of 1: 25 000 for parts of the Pyrenees, based on the standards of the French CLPA. In Slovenia, those areas which are vulnerable to avalanches are represented on the basis of the Water Act 2002, and others. The plan is to create hazard indication maps on the principles of water economics on a rough scale (for the entire national territory) and fine scale (for the area of a municipality) [112]. Since 1979, hazard indication maps that give a general indication of avalanche dangers have been created in Norway at a scale of 1: 50 000. (Figure 4.5a) These maps are already available for many areas of Norway. They consist of event maps, which are based primarily on historical data and chronicles, and geomorphological hazard maps, which contain information from terrain and remote sensing. Regional hazard indication maps are also available in Canada and the United States (Figure 4.5b). In Japan, the hazard maps (scale 1: 25 000) indicate the avalanche areas from the starting zone up to the runout zone and are created by the municipality.

(a)

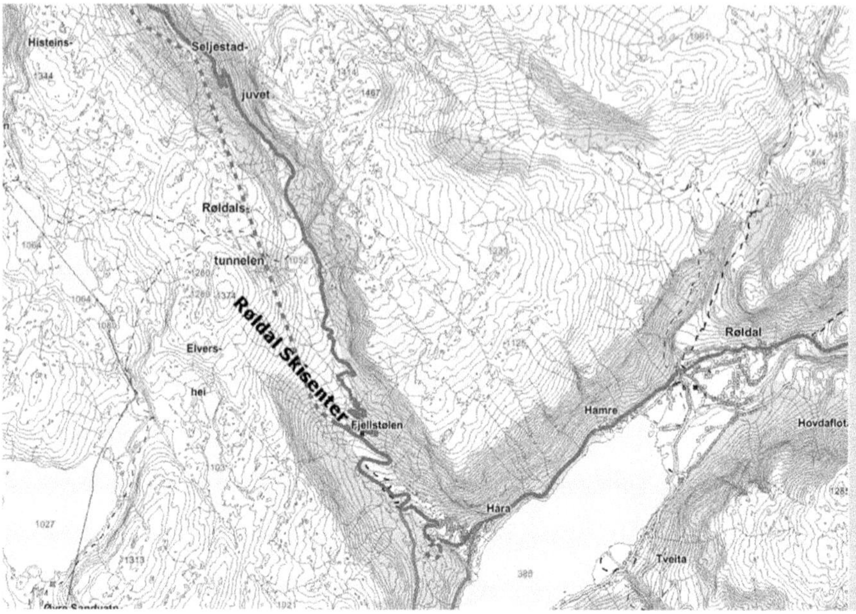

(b)

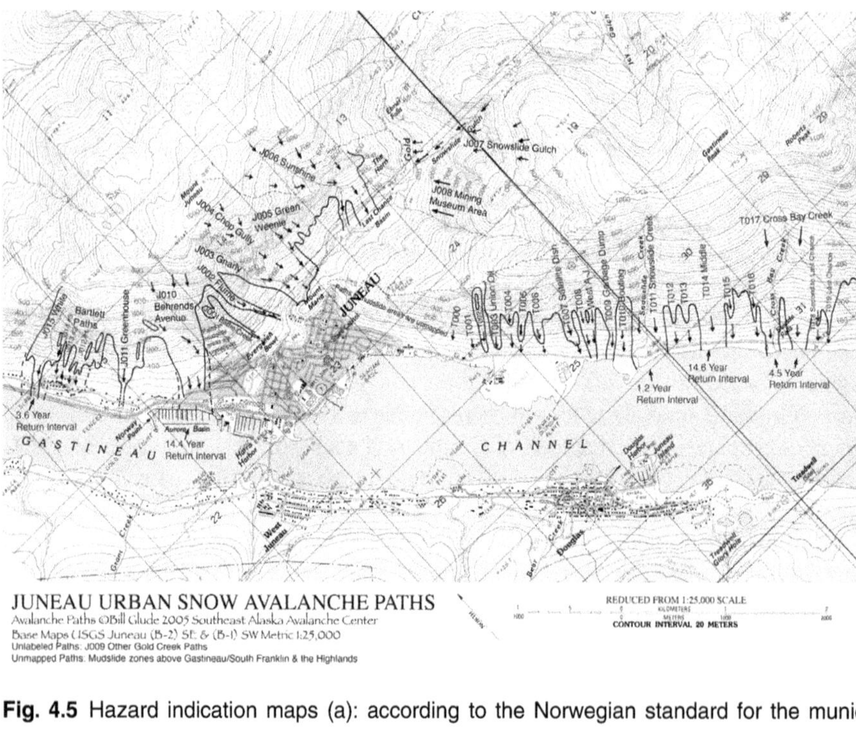

Fig. 4.5 Hazard indication maps (a): according to the Norwegian standard for the municipality Røldal (© NGI) and (b): for the city of Juneau (Alaska/United States) (© South East Alaska Avalanche Centre)

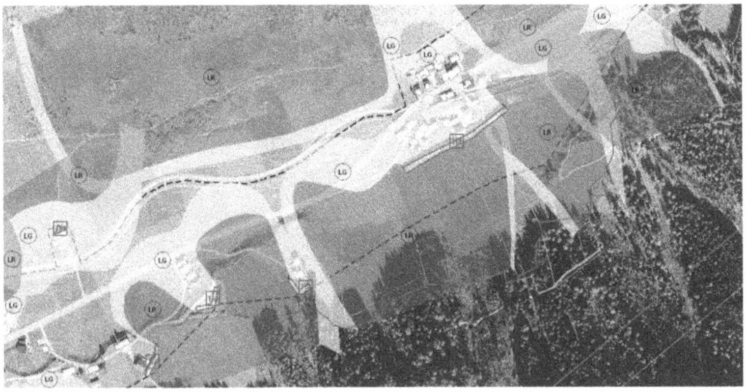

Fig. 4.6 Hazard zone planning in Austria: avalanche hazard zones (© WLV Tyrol)

4.2.3 Hazard zone plans

The *hazard zone plan* represents the frequency and intensity of avalanches on the basis of a land register. In addition to avalanche hazards, these maps also show hazards from floods, rock slides or mass movements (falling rock, rock fall, landslide) in a scale of 1: 2000 to 1: 10 000.

Hazard zone maps in the Alpine countries have the character of professional assessments (expert opinion) with a forecast character (Switzerland, Austria, Italy/South Tyrol, Slovenia), only in France are they directly part of the general spatial planning. They are based on a *recurrent design event* set by law or policy, which sets the intensity of the avalanche effect in relation to a certain return period (see Table 4.7 for details).

In *Austria*, the hazard zone map for avalanches (Figure 4.6) is based on the provisions of the Forest Act of 1975 [55] and was primarily developed as a basis for the planning of defense measures, ranking these measures according to their urgency and for the expert activity of the Austrian Service for Torrent and Avalanche Control. This is however – as in other countries – to be created so that it can also be used in spatial planning as well as construction and safety. Red and yellow hazard zones are assigned on the basis of physical demarcation criteria as total lines of all possible risk scenarios of a recurrent design event.

In Switzerland, according to [154], the results of the avalanche hazard assessment are translated into the red, blue, yellow and yellow-white shaded areas of the hazard map (Figure 4.7) with the help of the intensity/probability matrix (Figure 4.8). In France, the cartographic representation of avalanche hazards has been in the Plan de Prévention of Risques Naturels (PPR) since 1995 on the basis of a relevant regulation from the year 1995 [155]. Hazards are divided into three zones: red (high risk, building prohibition zone), blue (medium risk, bid zone) and green (zone with no identifiable danger) (for a comparison of the hazard-zone map signatures in Austria/Switzerland/France, see Table 4.8). In Norway, the areas at risk of avalanches are shown in detailed hazard zone maps (scale 1: 1000–1: 5000) on the basis of the hazard indication maps and avalanche models according to the safety requirements of the planning and construction

Table 4.7 Overview of the hazard zoning in the countries most strongly affected by avalanches (selection)

Country	Designation	Scale	Zones	Recurrence interval Criteria Avalanche hazard zone	Legal basis Guidelines	Creation Enactment
Austria	Hazard zone plan	1: 2000 – 1: 5000	Red – yellow– white	150 10/1 kN	Forestry Act 1975 [55]. Hazard-zone map ordinance 1976 [75] guidelines for hazard zone planning [36]	Federal Minister of agriculture and forestry/ torrent and avalanche control (Federal)
Switzerland	Hazard zone plan	1: 5000	Red – blue – yellow – yellow/ white – White	300/100/30 30/3 kPa	Federal Forest Act (WaG, 04 Oct. 1991, status: 1 Jan. 2008) [32] Guidelines for the consideration of avalanche hazards in activities with spatial impact [29]	Engineering offices under contract with the cantons Cantonal administration Federal Office for the environment (FOEN)
Italy (South Tyrol)	Hazard zone plan	1: 5000 – 1: 10 000	Red – blue – yellow – yellow/ white – White	300 Additional events occurring frequently (< 30 years)	Decree 42/2008 Execution regulation regarding the hazard zone maps [46] Policy for establishing the hazard zone maps 2008 [219]	Civil engineers by order of the communities Provincial administration

Table 4.7 (*Continued*)

Country	Designation	Scale	Zones	Recurrence interval Criteria Avalanche hazard zone	Legal basis Guidelines	Creation Enactment
France	Plan de Prévention des Risques Naturels (PPR)	1: 10 000	Red – blue – white	100 (300) 30/1 kPa	PPR directive 1995 (décret no. 95–1089 de l'octobre 1995) [155] Law on the prevention of technical and natural risks (loi n 2003-699 of 30 July 2003) [156]	Prefecture Communities
Norway	Avalanche hazard maps	1: 2000 – 1: 5000	Safety levels: Red – blue – green	5000/1000/ 300–	Norwegian planning and construction law 1985, 2005 [157]	Civil engineers by order of the communities Communities
Canada	Snow Avalanche Risk Map	1: 5000	Red – blue – white	300/30 30/1 kPa	Guidelines for Snow Avalanche Risk Determination and Mapping in Canada from the CAA (2002) [40]	Certified civil engineers–Canadian Avalanche Association (CAA)

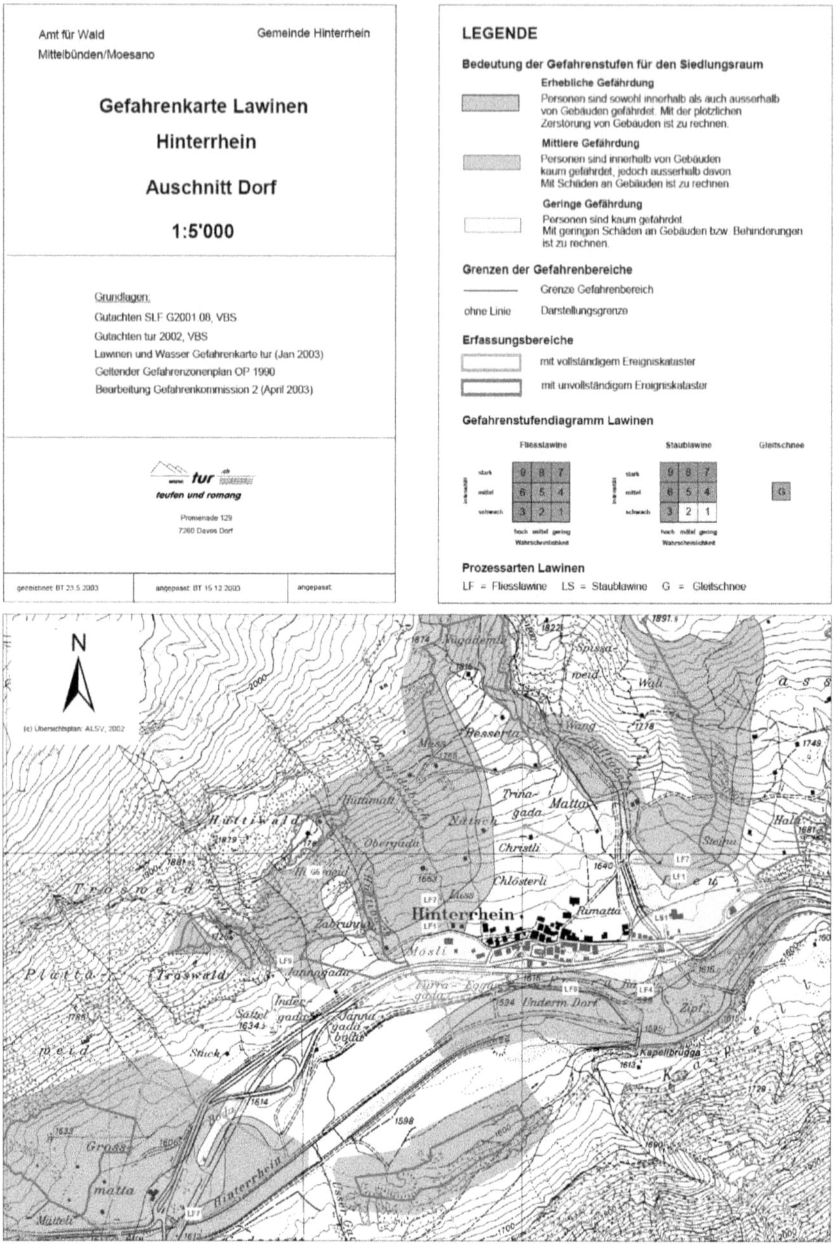

Fig. 4.7 Swiss hazard zoning: (a): hazard zone plan avalanches from the municipality Hinterrhein (village) (© Canton Grisons)

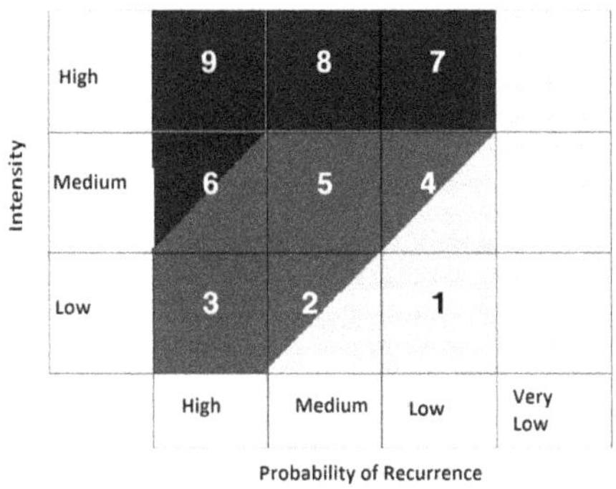

Fig. 4.8 Swiss hazard zoning: (a): intensity/probability matrix (© FOEN)

laws from 1985 [157] for the purposes of local planning and construction. The hazard zones (red, green, blue) correspond to the three safety classes and depend only on the probability of occurrence, however not on intensity parameters (Table 4.9) [236]. Table 4.7 provides an overview of the hazard zoning in the countries most affected by avalanches.

4.2.4 Risk maps

High-value *risk maps* for avalanches were created only in individual studies and so far have not been considered the standard for avalanche protection planning in Europe. However, in some countries there are maps containing simplified approaches for avalanche risk.

In Italy, the cartographic representation of hazards and risks is done on the basis of the implementation of the *national legislative decree for the demarcation of risk zones* in local laws. According to this law, natural hazards, such as avalanches, are considered in these risk areas, which arise from the overlapping of four hazard and vulnerability classes [138] in each instance.

A system was also developed in Iceland for the cartographic representation of avalanche hazards (Figure 4.9), which places the delimitation of hazard zones exclusively on the level of risk. The most important factors of this model are the estimate of the avalanche frequency, the runout length, and the damage sensitivity, which is determined on the basis of individual risk or the annual probability of avalanche death (lethality) [9].

4.3 Planning of avalanche defense structures

4.3.1 Principles of planning

Planning, by definition, is the mental anticipation of measures and activities that appear to be necessary to achieve a goal – in the concrete case the protection against avalanches – (future relation of planning). It should be noted with what effort the objective can be reached and how the sustainability of the protective effect can be checked.

Table 4.8 Definition of signatures for avalanche hazard zones in Austria, Switzerland and France

Avalanche Intensity	Austria	Switzerland	France
High	**Red hazard zone:** Due to the probable damage effects of the rated event for avalanches or due to the frequency of the risk, the constant use of the area for settlement and traffic purposes is not possible or only possible with disproportionally high costs.	**Forbidden area (red):** People are in danger, both inside and outside of buildings. The sudden destruction of buildings is to be expected.	Construction prohibition zone (red)
Medium	**Yellow hazard zone:** All other areas at risk of avalanches and for which the constant use for settlement and traffic purposes is impaired as a result of this hazard.	**Bid area (blue)**: People are rarely at risk inside buildings, but are when outside. Damage to buildings is to be expected, destruction is not to be expected under consideration of certain building requirements.	Restricted zone (blue)
Weak		**Information area (yellow)**: People are hardly at risk. Low levels of damage to buildings or hindrances are to be expected. Significant material damages can occur within the building.	
Residual hazard		**Information area (yellow-white shaded):** Information area that indicates a residual hazard or a residual risk with a very low probability of occurrence.	
Without Hazard	White	White	White

Table 4.9 Safety classes for avalanche protection in Norway based on the probability or recurrence and the type of construction (according to [157, 236])

Safety Class	Maximum Normal Annual Avalanche Probability	Recurring Period of the Avalanche	Building Type
1	10^{-2}	100	Garages, annexes, boat houses
2	10^{-3}	1000	Residences, business buildings
3	$<10^{-3}$	5000	Hospitals, schools, public halls

Table 4.10 Criteria for determining risk in Iceland (see Figure 4.9)

Risk Zone	Lower Limit of Local Risk	Restrictions for the Construction Industry
Zone C	$3 \cdot 10^{-4}$ per year	No new buildings, exception: seasonal use in the summer.
Zone B	$1 \cdot 10^{-4}$ per year	Premises without protective measures permissible, residential buildings only with object protection measures, public buildings: extension only with object protection. Continued development of settlement areas not permitted.
Zone A	$0.3\ 10^{-4}$ per year	Object protection for buildings where large gatherings of people are expected (schools, hospitals).

This section covers the principles and tasks of the planning phase of avalanche defense until the approval (acceptance) of a plan before the start of the implementation of the measures. Planning processes take place in several phases (Figure 4.12) and refer not only to the direct catchment and avalanche effect area but to the region as a whole that the measures directly or indirectly influence. Planning is therefore not only confined to the objective, but has a far-reaching design character. An important feature of planning is the information function for the decision makers, those affected by the planning, the executors and the public in general. Plans for avalanche protection that meet these complex requirements allow for an early analysis of the problems and offer solutions for distribution and conflicts of interest. The negative consequence of failure to comply with these requirements for planning can be the lack of implementation of protection objectives, above-average implementation costs (construction costs) or the failure of concrete protection projects in the implementation phase.

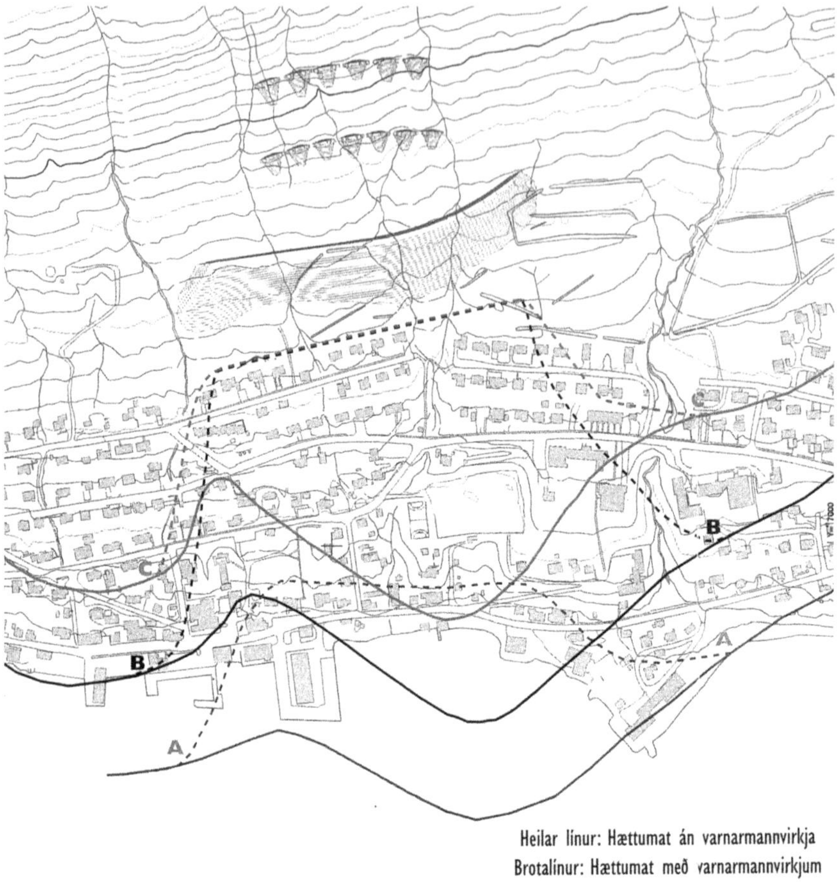

Fig. 4.9 Hazard zoning in Iceland on basis of the avalanche risk as well as criteria for zone delimitation in accordance with Arnalds et al. [9] (see Table 4.10)

The planning procedure identifies all legally standardized planning and decision-making processes in the context of a planning process. In most countries, avalanche protection is not subject to the general construction laws so that specifically legally regulated procedures for establishing plan works apply, depending on which sector (protection of the settlement area, protections of traffic corridors, protection of ski areas; see Section 4.3.3) the planning takes place. For publicly subsidized avalanche control projects, the planning procedure is usually set forth in guidelines (Austria: [34], Switzerland: [27]) or standards. Generally, formalized planning procedures contain conclusions of political bodies for the establishment, determination and financing of the corresponding plans and the measures these contain as well as the participation of the public and the technically relevant authorities or other bodies of public interest. In addition to the legally standardized planning procedures, there are numerous informal planning processes in avalanche protection – especially in the private sector. Since a primarily regional approach is followed in the established planning culture for avalanche protection in Alpine countries, so far no pan-regional guideline for planning has been established, such as for example by the implementation of the European floods directive [218] for the protection against inundations. Most sections in this handbook therefore refer to the parts of the planning processes that cover the hazard and risk analysis, as well as the design, construction and the sizing of avalanche defense structures.

The implementation of technical measures for avalanche defense is carried out on the basis of projects. The *project* is the underlying execution design that represents the planned measures by description, plans, calculations, operating regulations and other necessary documents [34]. In consistent implementation of previous implementations, avalanche protection projects therefore – apart from technical designs – are also to include all documents that are necessary for the successful realization of the project in a legal, social and ecological as well as a political and economic sense.

4.3.2 Objectives of avalanche defense

4.3.2.1 Principles of protection objectives

An integral component of the planning process in avalanche protection is focusing action on a *protection objective*. According to [211], protection objectives are also key elements of risk assessment, which decide which aims risk management shall pursue. According to [222], these define the desired level of safety and enable a monitoring of success after implementing protection measures. Protection objectives depend significantly on the protection needs of those affected, on the social environment, on the legal framework, on the responsibility of the beneficiaries of the protection, on the economic efficiency as well as the social and environmental friendliness of the measures [100].

When distributed over the risk concept, this means that protection objectives are based on the distinction between (conditionally) accepted and non-accepted risks. Protection objectives can be quantitative or qualitative in nature and relate to the need for security by an individual, a group of people or society (the common good). Furthermore, when determining protection objectives, note that certain risks are taken voluntarily, others involuntarily (such as avalanche risk in an open ski area or in permanent settlement areas).

From a technical point of view, protection objectives according to [210] correspond to the determination of limit values of safety requirements for defense measures. Thus, a publicly accepted risk level is anchored at the same time and – at least in theory – risk scenarios can be compared for different places and different hazard types.

In avalanche defense, protection objectives serve to prioritize projects, the evaluation of the economical, ecological and social effects and the selection of the most efficient and most economical protection option. The main protection objectives of avalanche control are the protection of life and health of people, the protection of settlements and economic areas, protection of transportation routes, infrastructure and supply systems, protection of the economy as well as the protection of the environment (Table 4.11).

Table 4.11 Overview of the hierarchy of avalanche protection objectives modified in accordance with [11]

Avalanche protection	Direct protection effect	Protection of immaterial values	Protection of people	Life, health
			Protection of cultural values	Cultural heritage, monuments, land improvement
			Safety	Sense of security, general level of safety, quality of life
		Protection of material values	Damage prevention	Settlements, public institutions, economic and production facilities, traffic facilities, utilities, defense structures
			Prevention of follow-up costs	Costs for the disaster relief, costs for restoration
	Side effects	Structural improvements	Parallel usages	Other water uses, recuperation and leisure
			Use conversion	Accessibility, higher quality use, infrastructure improvement
		Change of living space	Cost-efficient intervention	Redistribution, change of purchase
			Social intervention	Changing the settlement structure, changing the social position
			Ecological intervention	Changing the ecosystem, the environmental burden
		Sustainable development	Maintenance and operation	Management of the catchment area, monitoring and maintenance of protection measures

4.3.2.2 Quantitative and risk-based protection objectives

As quantified or technical protection objectives, normally the maximum permissible event intensities (such as avalanche pressure, range) are defined for each return period or object type, in some countries only the recurrence probability of events is defined (*design event*).

Another way of quantifying consists of the representation of protection objectives as well as maximum permissible risk. In Switzerland, according to [210], a general protection objective for the individual mortality risk (for people with low self-responsibility, individuals) a maximum value of 1×10^{-5} is recommended. Higher risks are the responsibility of the individual. As marginal costs for the collective risk of death, which correspond to the same safety measures to avoid the last possible death, 5 million Swiss francs were set. For measures to prevent property damage, including consequential costs, no more than 1 Swiss franc is to be spent for defense measures to prevent a damage of 1 Swiss franc (cost-benefit ratio <1). This value is also to be applied to the overall risk (personal and material risk), however this requires that risks to people are monetized. The value applied corresponds to the amount of money that society is ready to pay to safeguard the life of one person (Section 4.1.2.2).

Risk-based protection objectives were also set and made binding in other countries (such as Iceland, Norway), in Austria and Germany, however, so far no such provisions have been made for avalanche protection. There are limits in the application of risk-based protection objectives, particularly with regard to the lack of quantification of certain risk sectors and in the limited, political and public acceptance of decisions about defense measures, which are based solely on economic considerations. In certain countries, there are also ethical and political reservations against the financial assessment of risks to people.

The (general) objective system shown in Table 4.12 is used by planners and authorities for specific conservation projects. For state-funded avalanche protection in Austria, the protection objectives set forth in the hazard zone maps and the relevant technical guideline [34] are to be used.

Table 4.12 Objective system for defense measures against natural hazards (according to [220], modified in [222])

Overall Objective	Protection of People and Objects	Optimal Resource Implementation	Promotion of Environmental Quality
Criteria for evaluation	Remaining residual risk	Efficiency	Good ecological condition
Indicator	Amount of damages and probability Vulnerability and protection deficit Risk assessment	Cost-benefit analysis Cost-effectiveness analysis Utility analysis	Ecological risk analysis Adaptation to the landscape image Space requirement

A differentiation of protection objectives for concrete measures is necessary when the determination of the desired safety is devalued according to the use category or the value of the objects to be protected (protection goods). As an instrument to illustrate the devalued protection objectives, the *protection objectives matrix* is used in Switzerland (Table 4.13).

Table 4.13 Example of a protection objectives matrix as a benchmark for public investment in avalanche protection in Switzerland, (modified according to [29])

Object category			Acceptable risk		
Tangible assets	Infrastructure systems	Commercial values	Return period		
			1–30 often	30–100 rarely	100–300 very rarely
Location-bound structures (except for special structures)	Ski routes Mountain trails	Alpine wasteland natural landscapes	3	3	3
	Hiking trails and cross-country trails Agricultural and forestry paths Municipal supply lines	Alpine meadows	2	3	3
Uninhabited buildings (economic buildings)	Municipal traffic paths, trans-regional supply lines	Forest with protective function, agriculturally used areas	2	2	3
Occasionally inhabited single buildings and hamlets, stables	Trans-regional traffic routes, main supply lines, cableways, ski slopes		1	1	2

(continued)

Table 4.13 *(Continued)*

Object category			Acceptable risk		
Tangible assets	Infrastructure systems	Commercial values	Return period		
			1–30 often	30–100 rarely	100–300 very rarely
	National traffic connections, ski and chairlifts		0	1	2
Closed settlements, enterprises, construction zones, campsites, leisure and sports facilities, other public places	Stations for traffic and transport systems		0	1	1
Special risks, special sensitivity to damage	Special risks, special sensitivity to damage	Spring protection areas	Determined in individual cases		
Intensities:	0: no intensity 2: average intensity	1: weak intensity 3: strong intensity	Complete protection	Protection up to weak intensity	

In terms of the level of safety to be applied uniformly throughout the country, this serves as a benchmark for the description of the protection objectives depending on the object type (type of protected good) in a probability-intensity diagram [30].

4.3.3 Sectorial protection concepts

4.3.3.1 Principles

Protection concepts are the strategies underlying the planning of avalanche protection and serve the concrete implementation of the protection objectives to achieve the desired level of safety. According to [34], the protection concepts include the most efficient and economic combination of active and passive measures with a permanent or temporary effect (Section 1.2.1) and cover the entire avalanche catchment (the feed and starting zone, the avalanche path or the deposit zone), therefore this considers all relevant environmental and economic factors. These are therefore also called *integral* or *comprehensive protection concepts*.

An essential element of planning specific protection measures is therefore the *variant planning*, in other words the comparison of possible combinations of measures regarding the technological suitability, efficiency, economy as well as the ecological and social compatibility. The variant planning is to be performed specifically for each protection project, however, depending on the object category not only are different safely levels to be applied (Table 4.13) the choice of measures is also to be differentiated with regard to the protection effect. For example, different safety concepts are decisive for settlements, transport routes or ski resorts.

The following section provides a brief overview of the characteristics of protection concepts for these important sectors of avalanche protection.

4.3.3.2 Protection concept for settlement areas

According to the protection objective matrix (Table 4.13), a *closed settlement area* (Figure 4.10) is the area within which only the slightest risk is acceptable and the highest safety level is to be applied. This is the area where the citizens invest their private wealth and spend the most time, and where therefore the length of stay of people is highest. The subjective feeling of security should also be the strongest where people live. The actual risk must be therefore considerably lower than, for example, on traffic routes or in other areas of daily life.

The focus of the protection concept is thus not only the safety of structures like homes and associated infrastructure facilities, but first and foremost the protection of the life and the health of the inhabitants. It may not be assumed that the inhabitants will always be in their homes when there is a hazard or that they will remain indoors despite warnings. Therefore, a protection concept for settlement areas must consider the areas between and around the houses and also the accessibility of the buildings.

A hazard-zone plan (Section 4.2.3) or a risk map (Iceland) (Section 4.2.4) is the basis for the protection concept for settlement areas in the Alpine countries and Scandinavia (Section 4.2). Basically, it is the responsibility of the spatial planning authorities to keep hazard areas free of settlements – if this has not already been done before the hazard mapping.

Protection concepts for settlements whose protective effects justify a decrease in size of hazard zones may only include permanent defense measures in Switzerland and Austria [161, 265]. Temporary defense measures are only appropriate in exceptional cases for the protection of settlements. Permanent technical defense measures are considered by the population to be the most effective measures for the protection of settlements and meet this requirement with proper planning and execution. However, for the construction and maintenance of technical avalanche defense measures, the responsible parties must also consider that these only ensure protection to a certain level, which is why protection concepts must also take precautions against increased risk or residual risk [211]. Permanent technical defense measures are subject to ageing as well as wear and tear, part of the safety concept is therefore also the ongoing maintenance and monitoring according to the applicable directives (Austria: [204];) Switzerland: [163]; Sections 7.2.2 and 7.2.4).

Fig. 4.10 Avalanche protection for an enclosed settlement area in Galtür (Tyrol) (© WLV Tyrol)

4.3.3.3 Protection concepts for traffic routes and supply lines

The focus of the protection concepts for traffic routes is not only the safety of the structures themselves, but also that of the road users. The avalanche risk for traffic routes is therefore determined to a central degree by the frequency and number of avalanches, which threaten the traffic routes or the duration of presence of vehicles (cars, trains) and road users (people) in the hazard zone. The range or intensity of the avalanche effect is often less decisive than its frequency. Avalanches with a more frequent rate have a greater impact on the safety of road users and are to be given priority in the framework of a safety concept. Avalanches that rarely occur basically represent a lower risk for road users, are however more difficult to assess due to their being less experience with regard to their risk [282].

Fig. 4.11 Avalanche defense measures for safeguarding of traffic routes: Avalanche Gallery St. Anton/Arlberg (Tyrol) (© Sauermoser)

In addition to the protection of human life, vehicles and the transport system itself, the most important protection objective is the maintenance of mobility and accessibility in protection concepts. Since the presence of road users in the hazard zone is time limited, the focus of avalanche defense measures for traffic routes (Figure 4.11) is on temporary defense measures. In critical avalanche situations, the authorities or the maintenance crew can block the traffic route and therefore cause a temporary restriction of mobility. Blocking of main traffic routes is generally to be assessed differently than blocking of secondary roads (or rail lines), the possibility of a bypass (emergency route) leads to a different assessment than in the case of a total block (only access to a valley). The nature and duration of the block depends highly on the containment of the hazard time based on the weather conditions. Furthermore, the ability to reduce the risk time through active measures is decisive. Depending on the importance of the traffic corridor and the number of avalanche paths, a permanent defense measure can also be economical.

Avalanche protection concepts for traffic routes rely first and foremost on the activity of the avalanche warning service and avalanche commissions. Avalanche warning services create their prognoses based on meteorological and nivological data and supply the basis for the concrete decision of the avalanche commission on blocking or opening traffic routes, mainly based on their knowledge of the area. Measures for the artificial release of avalanches are an option in principle to reduce the hazard time and thus the duration of the block. Temporary avalanche defense measures however have an error rate or subjectivity in the decision (precautionary block) so that the permanent protection of the traffic route by means of an avalanche gallery, an avalanche tunnel or snow supporting structure is the most effective, but also the most expensive protection concept. This judgement will change in near future with increasing reliability of artificial release and monitoring technology.

4.3.3.4 Protection concepts for ski areas

Ski areas are usually vulnerable to avalanche hazards because they are located in Alpine areas and steep slopes are a prerequisite for skiing. In ski areas, there are a lot of people in potential avalanche terrain at the same time during the winter season, which is why special protection concepts for avalanche protection are required [239].

In 1975 in Austria, after severe avalanche events in ski areas with numerous fatalities, the *avalanche decree* from the Federal Ministry for Traffic [37, 38] was enacted to regulate avalanche safety in ski areas. This determined that both the area itself – the stations and supports including the associated parking – must be permanently avalanche-safe or, if this is not possible for natural reasons, these must be secured with permanent defense measures. Furthermore, at least one exit pertaining to the area, which corresponds to the capacity of the area, must also be permanently avalanche-safe or secured. Since 2003, the cable car legislation in Austria (Federal Law I no. 103/2003 amended), which covers the establishment and operation of funicular railways, cable railways (cable car, chairlifts/lifts) and T-bar lifts in the framework of the concession procedure, requires a safety analysis to determine all operating risks (including avalanches) as well as the necessary protection measures. In the prescribed security report, temporary avalanche protection measures or the combination of permanent and temporary measures can be determined [37]. Temporary avalanche protection is, however, only allowed for additional tracks and replacement equipment in developed ski resorts under the condition that permanent avalanche protection is not possible. Moving away from the principle of the exclusive permissibility of permanent avalanche defense measures is justified by the fact that the quality of the technology of temporary avalanche protection (Chapter 9) has increased greatly in recent years so that this can have even better protective effects in subareas than permanent defense measures.

In Switzerland, avalanche safety in ski resorts is regulated by *directives of the Federal Office for Transport* [31] and the *Swiss cableways* (SBS) [252]. Contrary to the Austrian guidelines, different levels of security are defined for terminal buildings, support sites and runs (Table 4.14). SBS policy determines that slopes are basically to be designed so that they are avalanche-safe under normal winter conditions. Absolute security from avalanche and snow slab hazards cannot be achieved for the marked slopes, however.

4.3.4 Planning process for technical avalanche defense measures

4.3.4.1 General planning procedures in avalanche defense

The planning process for avalanche defense projects should use the latest technology and the planning steps should be set forth in the standards and guidelines. The objective is to follow a standardized procedure to meet all requirements for the planning of avalanche defense measures. The standard planning procedure is generally opposed to a planning culture established in engineering practice, which corresponds to the technological, organizational, legal and political framework of the country.

Table 4.14 Flow and powder avalanche return durations to be considered for areas and slopes according to [31]

Object Category		Return Period					
		For Areas				For People	
		10 yr	30 yr	100 yr	300 yr	1–2 yr	100 yr
Accessibility tracks	Stations		•		•		
Transport tracks	Stations	•		•			
Accessibility and transport tracks	Support routes artificial structures	•		•			
Accessibility and transport track	Slope routes (without rescue track)					•	(Information)

An *accessibility track* is a system with a shuttle function into the snow sport area. A *transport track* is a system without a shuttle function.

The normal, practical approach (Figure 4.12) for the planning of publicly subsidized avalanche defense projects in the Alpine countries, especially in Austria, will be described briefly as follows:

Protection requirements generally arise either due to recurring avalanche events with damage or by a perceived lack of safety, for example, as the result of the creation of a hazard-zone plan. The concrete requirement is manifested by the request of a municipality or a corporation of public law (regulation association) as a client and applicant (prospective) for state subsidies. In a first step, the priority of the avalanche defense project is classified (on the basis of the hazard zone map or the risk map), in the case of a large number of projects on the basis of regional planning (studies) or general planning (preliminary planning phase). The actual process of designing (planning phase) begins with the definition of the objective of the protection objective and the development of a protection concept including the alternative variants (variant study). The determination of the planning basis follows, then the draft (conception) of the measures (variants), the decision making, the calculation of costs as well as proof of the efficiency (cost-benefit assessment). The results of the planning are summarized in a project, which forms the basis for the investment decision, the technical approval and the statutory authority procedure. The environmental impact assessment and public involvement in avalanche defense projects is integrated into the authority procedure according to the prevailing practice. Completion of the planning phase is the acceptance of the project by the client and the granting of legal, technical and financial approvals.

This planning procedure corresponding to practical customs requires a high public acceptance for avalanche defense and well-established procedures between the client,

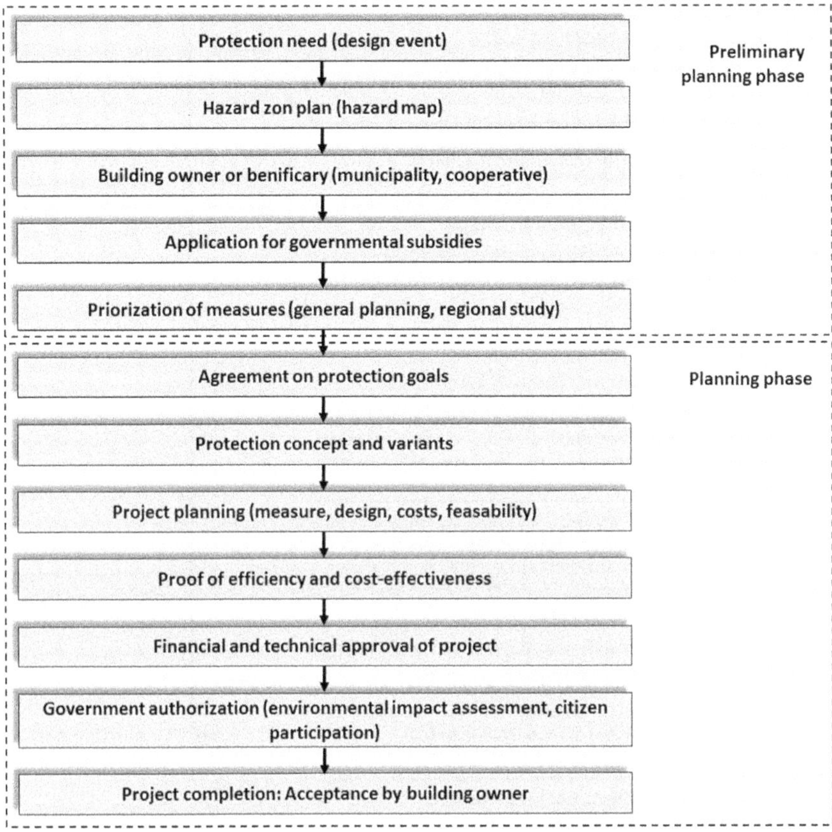

Fig. 4.12 Planning process in practice for publicly subsidized avalanche defense projects in Austria (simplified) (© Rudolf-Miklau)

the state institutions and the authorities. The procedure can cause high efficiency in planning, is however relatively inflexible in the event of anomalies (such as conflicts of interest, financing problems, lack of environmental compatibility).

This takes into account a newly developed [211], highly formalized procedure in Switzerland for action planning based on risk concept (Figure 4.13). The trigger for the planning processes is a suspected security deficit, which represents the basis for performing a risk analysis and assessment (Section 4.1). Based on the results of the risk assessment, the objective is set and specific combinations of measures and measure variants are developed. Proof of the measures' effectiveness comes from the determination of the risk reduction, whereby the risk evaluation process again passes through the accepted exposure situation after implementation of the measures. The economic evaluation of the proposed measures occurs based on a *limit cost criterion*, whereby the chosen variant must be less than this criterion. Other factors for the selection of the protective variants are the aspects of the proportionality, acceptance, and ecological compatibility. The early involvement of the population, NGOs and policy makers is

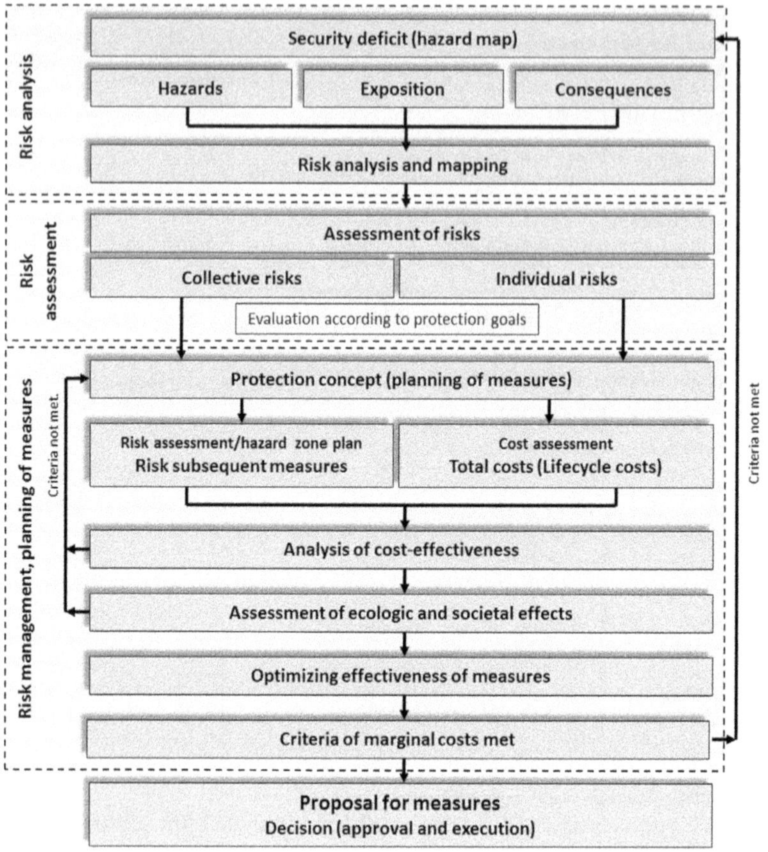

Fig. 4.13 Formalized planning process in accordance with the risk concept established in Switzerland, according to [211] (simplified)

sought in order to secure this 'soft' decision criteria. This also includes considerations for the future use of space under consideration of the residual risk. The choice of measures and decision about projection and implementation is based on this comprehensive pre-planning and evaluation of the effects.

4.3.4.2 Design of avalanche defense structures

This handbook presents the basics of projection for specific technical avalanche defense measures. *Projection* is the process of designing and preparing the necessary basis, the technical documentation and the planning documents for the execution of avalanche defense measures (establishment of defense structures). Table 4.15 summarizes the principal approach for projection and individual projection steps at a glance. The procedure is a recommendation; the actual design process is to be determined on a case-by-case basis.

Table 4.15 Overview of the project steps set out in this handbook for the design and construction of avalanche defense structures (the planning steps each refer to the corresponding chapters/sections)

Step	Task	Chapter Reference
0	Plan initiation (safety deficit)	Chapter 4
1	Collection of basic data (meteorological, nivological, morphological; frequency and magnitude of avalanche events; protective effect of the forest)	Chapter 2
2	Hazard analysis	Section 4.1.3
2.1	Historical method: avalanche chronicle	—
2.2	Statistical method: extreme value statistics	—
2.3	Numerical-mathematical method: avalanche modelling	Section 3.2
2.4	Morphological method	—
3	Risk analysis (exposure analysis, analysis of effects, risk assessment)	Section 4.1.4
4	Decision-making depiction of hazards and risks	Section 4.2
4.1	– Creation of hazard zone maps (hazard maps)	Section 4.2.3
4.2	– Creation of risk maps	Section 4.2.4
5	Definition of the protection objective (recurrent design event)	Section 4.3.2.1
6	Definition of the protection concept, including the comparison of potential variants	Section 4.3.3
7	Proof of the efficiency and cost effectiveness of the protection concept	—
8	Definition of the function and category of planned defense structures	Section 5.2.2.1
9	Survey of the basic design criteria	Section 6.1
10	Design of the protection structures (location, determination of construction types, choice of the geometric dimensions of the constructions, determination of important components)	Sections 6.2.1–6.2.4
11	Determination of relevant influences and effective combinations	Section 6.2.5

(continued)

Table 4.15 (*Continued*)

Step	Task	Chapter Reference
12	Determination of the relevant limit states and design situations	Chapter 6
13	Proof of stability and serviceability	
14	Constructive detailed planning	
15	Preparation of all necessary planning documents	—
16	Check the fulfilment of the safety objective after completion of the planning process	—
17	Determination of the rules for the operation of the defense measures	Section 7.2.2
18	Project planning completion: Acceptance and approval of the planning	—

5 Structural avalanche protection: defense systems and construction types

Wolfgang Schilcher, Stefan Margreth, Siegfried Sauermoser, Christoph Skolaut, Michael Mölk and Florian Rudolf-Miklau

5.1 Principles of structural avalanche defense

The main protection strategies for structural avalanche defense – on the one hand the prevention of avalanche release with snow supporting structures or the control of the snowdrift and snow accumulation in the starting zone, on the other, the deceleration or deflection of avalanches in motion form – have already been discussed in Section 1.2.2. Within the framework of integrated avalanche protection concepts, permanent technical defense measures are, to the extent possible, combined with temporary avalanche defense measures (artificial release of avalanches, avalanche warning, Sections 1.2.3, 9.1, and 9.2), in order to attain an optimal protection for all hazard scenarios.

The suitable defense systems and construction types are chosen relative to the protection goal, and must be adjusted according to the concrete hazard situation. This depends on the principles of planning (Chapter 4) and design (Chapter 6), and in particular the action on the structure (Section 3.3). Furthermore, it is essential to know which safety concept should be used as a foundation for the measures taken, and the amount of residual risk that should be regarded as acceptable. This is determined according to the type of objects to be protected (damage potential), the assessment of the potential hazards and risks (Sections 4.1.2.1 and 4.1.2.2), but also the consequences that may occur if the protective structure fails. When choosing the measures (construction types), the criteria of durability and lifecycle costs are becoming increasingly more important. The monitoring and maintenance efforts for the protection systems and the potential effect of extreme events are also relevant. The environmental and landscape compatibility with the measures accounts for further important selection criteria.

Because of the large number of criteria, it is practically impossible to give a simple solution for the correct selection of protection measures. However, a depiction of the protection systems and construction types for the avalanche defense should make it possible for the reader to choose the optimal measures (combination) when taking the individual criteria into account.

5.2 Structural avalanche defense in the starting zone

5.2.1 Overview and classification

Defense measures (Table 5.1) taken in avalanche starting zones usually counter the formation of avalanches. This includes snow supporting structures (Figure 5.1a), snowdrift control structures (Figure 5.1b), and snow glide protection structures [241].

The most important and frequenty applied structural measures at the avalanche starting zone are *snow supporting structures*. Using different construction types, the snowpack is supported, thereby significantly reducing the probability of avalanche release. The construction elements are designed in terms of the static action of the snowpack on the

The Technical Avalanche Protection Handbook, First Edition. Edited by Florian Rudolf-Miklau, Siegfried Sauermoser and Arthur I. Mears.

Table 5.1 Overview of defense structures in the avalanche starting zone: classification

Measure	Position	Effect	Protection Goal	Projected Action
Snow supporting structures	In the starting zone	Support of the snowpack against avalanche release, minimization of avalanche formation	Protection of settlement areas and infrastructure against avalanche damage	Slope-parallel snow pressure in cases of extreme snow depth
Snowdrift control structures	Outside the starting zone/ accumulation zone, respectively at the edge of the starting zone	Deposition of drifting snow outside the avalanche starting zones, prevention of extreme snow cornices	Easing the snow load of snow supporting structures in the starting zones through the reduction of the snow volume	Wind action, extreme snow depth, settling snow pressure
Snow glide protection structures	In the starting zone	Damming or limiting snow gliding and creeping, limited support of the snowpack, reduction of the glide factor	Protection of forest areas against snow-mechanical damage, reduction of the slope-parallel snow pressure for the snow supporting structures	Slope-parallel snow pressure, required effective height

slope. They are set up and constructed in such a manner that they also can catch small avalanche movements within the technical construction. This means that they cannot develop into destructive avalanches and not cause damage to the technical defense structures. The snow supporting structures are used particularly frequently in Switzerland and Austria.

The *snowdrift control* structures are usually constructed in the snowdrift zone. The structures change the wind field and thereby influence the snow deposition in the affected area. This dumps drifting snow outside the avalanche starting zone and reduces the potential fracture height. Snowdrift control structures usually support the effects of the snow supporting structures and are often constructed at the same time.

(a)

(b)

Fig. 5.1 (a) snowpack-stabilizing structures with steel snow bridges: Avalanche Heiligenblut (Carinthia) (© Sauermoser) and (b) snowdrift fence in steel (© WLV Salzburg)

The *snow glide protection structures* prevent or slow the continuous or sudden glide of the snowpack on the ground or reduce the creeping of the snowpack. The technical snow glide protection is erected to protect afforestation and to reduce the glide factor, this can also be in combination with the snow supporting structures. These structures also provide limited protection against the formation of ground avalanches.

5.2.2 Snow supporting structures: construction types

5.2.2.1 Protection effect of snow supporting structures

According to the Swiss Guideline [163], the effect of snow-supporting structures rests on the fact that the creeping and (possibly) gliding snowpack is set against a supporting surface that reaches the snow surface (Figure 5.2). This creates a back-pressure effect, that is the creep and glide velocities are continuously reduced into the direction of the obstacle. Within the back-pressure zone, which practically covers a slope-parallel distance of at least three times the vertical snow depth, additional slope-parallel compressive stresses are created. These are absorbed by the support surface, diminishing the snow slab forming shear and tensile stresses that are in the back-pressure zone. In case of avalanche release, the snow supporting structures prevent the old snow cover from being carried along, and limit the area size in which shear fractures could spread. The braking effect of the structures keeps the velocity of released snow volumes within bounds. Finally, the reservoir capacity of the snow supporting structures is impacted positively.

Snow supporting structures limit the avalanche hazard by reducing the avalanche probability and the avalanche size. During the avalanche winter of 1999, the measures had a very positive effect; according to one assessment, snow supporting structures in Switzerland were able to prevent 300 destructive avalanches [246].

5.2.2.2 Classification

The snow supporting structures (Figure 5.3) are classified according to the criteria of service life and reaction of the supporting surface (rigid, flexible). The choice of construction material is related to the service life. There is a differentiation between

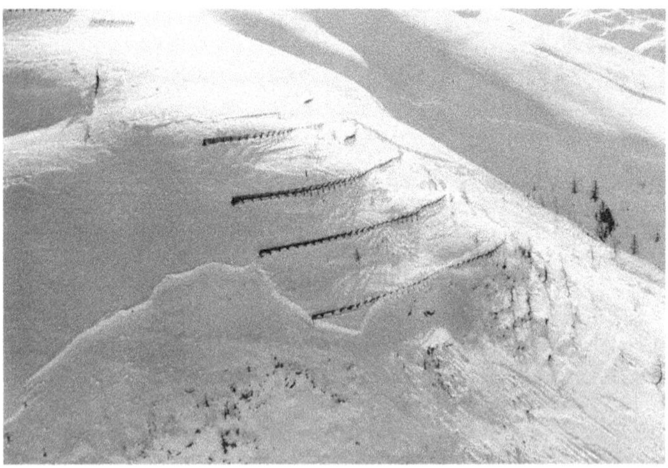

Fig. 5.2 Release of a slab avalanche next to the avalanche defense measures at Nolla in Goms (Upper Valais canton, Switzerland) in February 1999. The fracture height is approx. 2 meters. The snow mass was stabilized in such a manner that no greater slab avalanche was released (© Margreth)

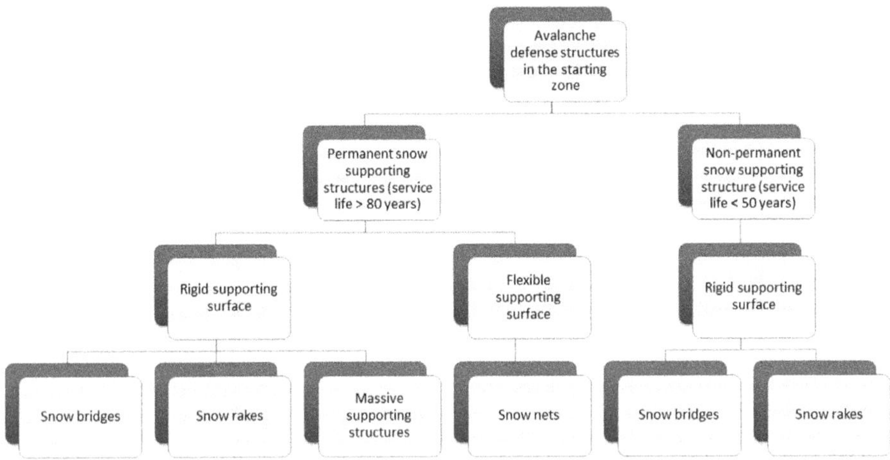

Fig. 5.3 Avalanche defense structure systematics: classification according to service life and reaction of the supporting surface (© Margreth)

snow supporting structures with a service life of more than 80 years, which are usually constructed from steel, and snow supporting structures with a service life of less than 50 years, which usually have smaller structure height and are constructed from wood. (The designations common in the past, *permanent* or *temporary*, are no longer used in the Austrian technical standards [203] for this classification. In Switzerland, *temporary avalanche defense structures* is an established technical term.) Furthermore, one differentiates between structures with rigid supporting surfaces (snow bridges, snow rakes, massive supporting structures), and construction types with a flexible supporting surface (snow nets see Table 5.2). Sometimes combined snow supporting structures with a bearing construction in steel and a grate made of wood are erected.

The expected *service life* for snow supporting structures made of steel is at least 80 years. Given the condition of the oldest structures of this type, we may conclude that when projecting and constructing according to the state-of-the-art and drawing on local experience, such a lower service life can be upheld if maintenance is carried out as planned. Snow supporting structures with a shorter service life are predominantly used in the restoration of protection forests and as a protective measure at lower altitudes, where one may expect a swift development of afforestation. For these purposes, predominantly wooden snow bridges are used. Similarly, combined steel/wood snow supporting structures are only utilized where protective growth is to be expected within approx. 50 years.

The *supporting structures with rigid supporting* surfaces include, in particular, *snow bridges* (Figure 5.4) and *snow rakes*. Snow rakes were used, for instance, at the beginning of the systematic avalanche defense structure on Arlberg. Nowadays, snow rakes made from steel are rarely used because the assembly of intermediate structures is more cumbersome compared to steel snow bridges. In Switzerland, wooden snow rakes

Table 5.2 Selection of construction type: advantages of (•) steel snow bridges and snow nets

Criteria for Construction Type Selection	Steel Snow Bridge	Snow Net
Material costs	•	
Transportation weight		•
Assembly efforts	•	
Foundation efforts in loose material	•	
Foundation efforts in bedrock	•	•
Limitations to the landscape		•
Maintenance efforts	•	
Adaptation to structured topography	•	
Rock fall hazard		•
Snow retention in case of loose snow	•	
Suitability on creeping slopes		•
Increased requirements for corrosion protection	•	

are the standard snow supporting structure for non-permanent avalanche defense structures.

Massive snow supporting structures in the shape of wall or earth terraces, as well as freestanding walls, are amongst the oldest construction types when it comes to avalanche defense structures. Until about 1940, they represented the standard defense

(a) (b)

Fig. 5.4 (a) Snow bridges in steel in continuous rows of structures: Bachertal Avalanche (Neustift in Stubaital, Tyrol) (© Rudolf-Miklau); (b) Wooden snow bridges (© Schilcher)

(a) (b)

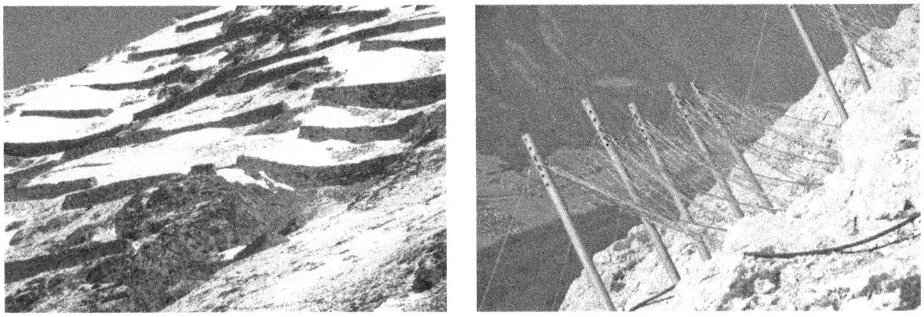

Fig. 5.5 (a) Massive snow supporting structures (wall terraces) on Schafberg (Switzerland) (© Margreth); (b) Snow nets on Hohe Munde (Telfs, Tyrol) (© Rudolf-Miklau)

method. Particularly because of the predominantly lower upslope structure height, the avalanche prevention effect became limited (Figure 5.5a).

Snow nets (Figure 5.5b) are also categorized as *snow supporting structures with a flexible supporting surface*. Instead of the rigid crossbeams, a flexible net of steel cables is used. Compared to steel snow bridges, they are less sensitive with regard to creep movements in the substratum and falling rocks, but are difficult to anchor on soil with low load carrying capacity.

5.2.2.3 Construction types: snow bridge of steel

Snow bridges are snow supporting structures with a grate consisting of horizontal crossbeams. The classic snow bridge consists of two supports, two girders, and depending on the structure height, a varying number of crossbeams (Figure 5.6).

The girder connection on the upslope foundation, the support connection on the girder, and the support connection on the foundation plate are jointed with bolts. The grate consists of pressed or rolled hat-shaped steel beams 4 meters long. Two girders and the crossbeams are screwed together forming the supporting grate. The grate is upheld by two supports, and is erected on the plane normal of the slope at an angle of 10° to 15° facing downslope. The angle between girders and supports is 60° to 70° (Figure 5.7c).

Usually, the snow bridges are erected with a lateral distance of 2 meters (Figure 5.7a). The intermediate distance between structures is closed using profile beams (connecting crossbeams), which are placed on the cross beams, and fastened with clamping bars. These connecting crossbeams (together with the clamping bars) are designated connecting structures (Figure 5.7b and d). The somewhat larger shape of the connecting crossbeams compared to the main crossbeams makes it possible to have some clearance when erecting the snow bridges, and thereby allow for adjustment of the structures according to the structure of the terrain. Steel snow bridges are predominantly erected for a snow thickness between 2.5 and 4.5 meters, in special cases up to 5 meters.

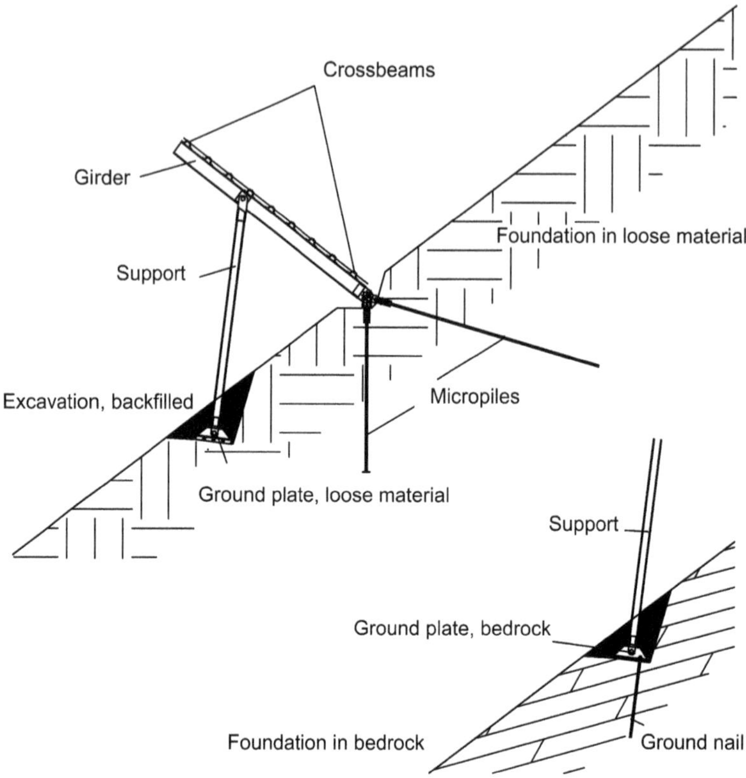

Fig. 5.6 Classic steel snow bridge (three hinge frame) D = 4.0 meters (Sketch of construction type) (© WLV Vorarlberg, sketched Herbert)

The construction design of the steel snow bridges was, during construction development, subjected to significant changes and improvements. The original construction design of the steel snow bridges as a three hinge frame was developed in series for the first time in the mid-1950s by Austrian and Swiss experts together with the company Voest-Alpine (Oesterreichische Alpine Montangesellschaft). The top construction (crossbeams, girders, supports) has only changed insignificantly until today. The design was carried out according to the *Swiss Guidelines for defense structures in avalanche starting zones* [50].

Until the mid-1970s, steel snow bridges were predominantly anchored with concrete foundations for girders and supports. In rare cases, ground plate foundations or pile drive foundations were used to anchor the girders. Next, combined concrete/explosive anchor foundations, as well as pure explosive anchor foundations were used to ensure reduced excavation and concrete cubature when anchoring the girders. Since the beginning of the 90s, the micropile foundation with a high degree of mechanization and minimal excavation work for girders has become standard (Section 5.2.3).

Fig. 5.7 (a) Steel snow bridge D 4.0: Three hinge frame, single structure (Madlein Avalanche, Ischgl/Tyrol); (b) Single structures combined with a connecting structure (Madlein Avalanche, Ischgl/Tyrol) (© Rudolf-Miklau); (c) Steel snow bridge D = 4.0 with micropile and ground plate foundation (© WLV Vorarlberg); (d) Steel snow bridges in closed structure rows (Großtallawine, Galtür/Tyrol) (© Rudolf-Miklau)

In order to reduce the total weight and the foundation force, a *pressure bar structure* was developed in Switzerland in the seventies (Figure 5.8). The V-shaped supports transfer the load at terrain level through a foot construction directly to the foundation. The downslope foundation consists of a micropile, which has an inclination of 15° to the slope face, and a tension anchor, which is connected perpendicularly. Depending on local topography, the length of the supports and pressure bars can be adjusted using telescopes or by cutting to desired length. Compared to the classic three hinge frame, the pressure bar structure reacts more sensitively with regards to settlement, however, because of the stiff frame it does not collapse, even if the downslope foundation fails. In Switzerland, the pressure bar structure is still one of the most frequently used construction types.

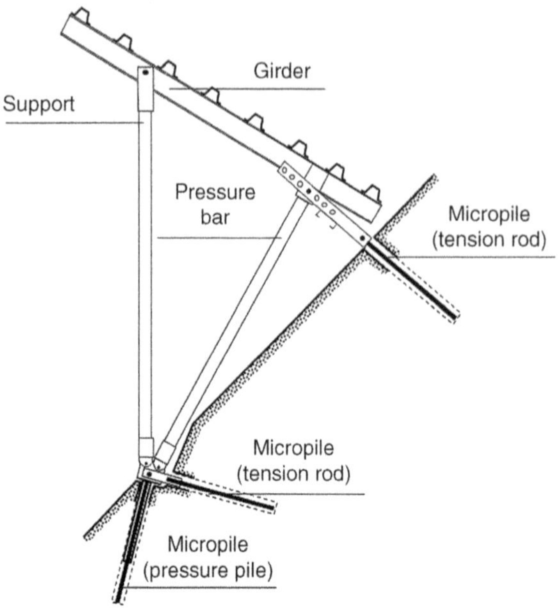

Fig. 5.8 System sketch of pressure bar structure: V-shaped supports carry the load at terrain height off the foundation directly through a foot construction (© Margreth)

5.2.2.4 Construction type: snow net

Snow nets consist of flexible-mounted swivel supports kept downslope with guywires and upslope with triangular or rectangular, flexible steel cable nets, which are connected to the upslope anchors by connecting cables (Figure 5.9a). Because of these cable connections, which can be adjusted lengthwise using wire clamps, there is a certain amount of flexibility when choosing the exact position of the upslope and downslope anchor points, which is advantageous under difficult topographical conditions, or when working on creeping soil.

The steel cables in the top construction are galvanized stranded cables. The nets and guywires are attached to the head of the support using adjustable bolts or shackles. The supports are manufactured from pipe or HEA steel profiles. If the supports have a ball joint at the foot, they are flexible in all directions. The supports point upslope opposite vertically at 10° to 15°. The nets sag by at least 15%. Since the geometry of a snow net can change during the course of the winter, depending on snow thickness, flexible foundation methods such as wire rope anchors and ground plates have been established, since they are relatively desensitized to changes in the force direction (Figure 5.9c and d). The distance between supports usually varies between 3.5 and 4 meters. The structure heights are between 2 and 4.5 meters. Because of the modular construction of the nets, it is possible to erect rows of structures of any length. In order to attain sufficient side stability, also when using rectangular nets, triangular nets are used in the end structures. The flexible support surface somewhat reduces snow pressure against rigid works, and the mounting weight is smaller. Compared to steel snow bridges, it is

(a)

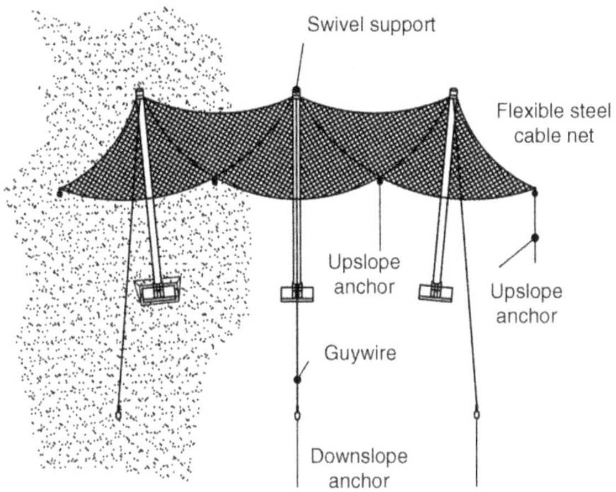

(b)

Fig. 5.9 (a) System sketch of snow net with ground plate (© Schilcher); (b) Triangular steel cable net with covering net strips placed in a crosswise sequence (© Margreth); (c) System plan for the snow net GL 1004d (Dk = 3.5, N = 2.5, f 0 = 1.1) (© Geobrugg); (d) System plan for the snow net TS-LV (© Trumer Schutzbauten)

particularly the upslope anchoring forces that are very large, which means that good foundation conditions (e.g. existing bedrock) are a prerequisite for the use of snow nets.

The classic nets in triangular or rectangular shape consist of peripheral net cables, and 6 to 8 mm thick mesh cables that are fastened with clamps. The mesh size is usually 200 to 250 mm. With this mesh size, the retention capacity of the nets may be insufficient in case of an avalanche formation. The problem was solved by covering the nets with a tight-meshed wire mesh (mesh size 5 × 5 cm) completely or by strips (mesh size

(c)

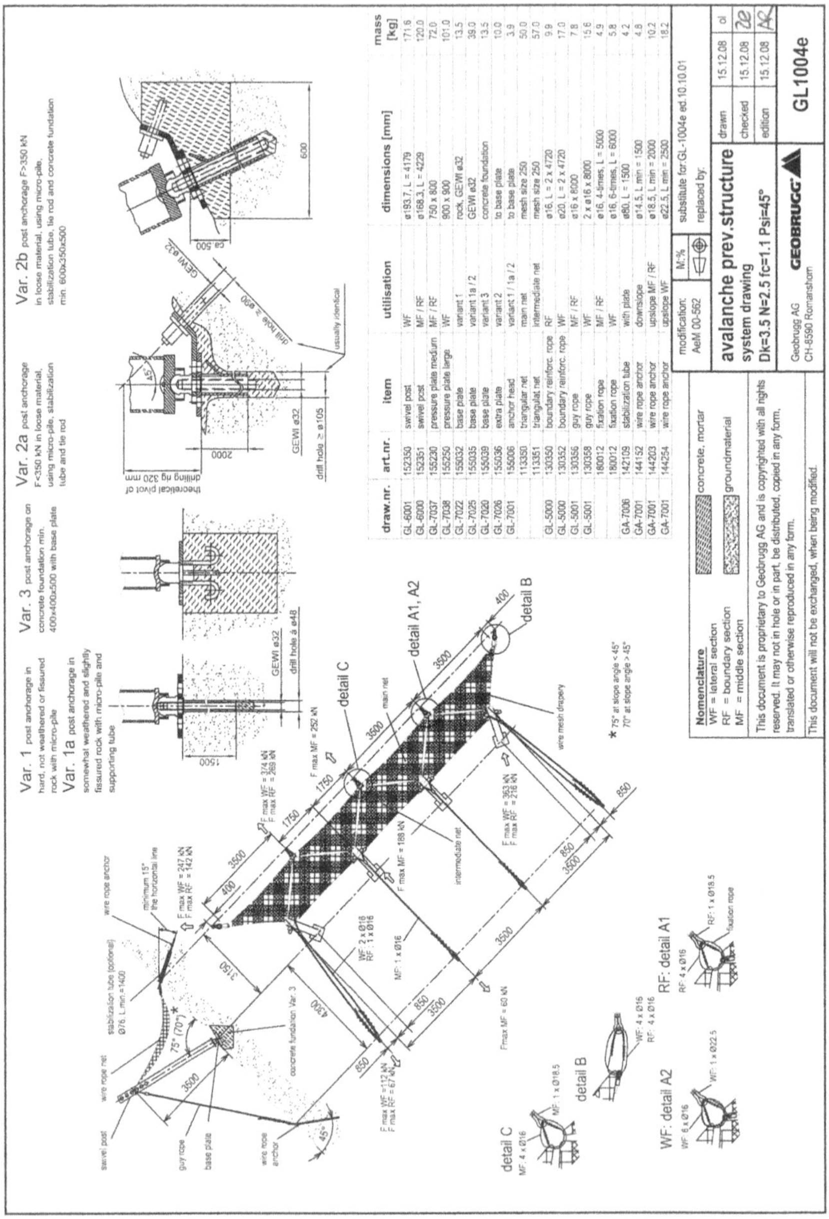

Fig. 5.9 (*Continued*)

(d)

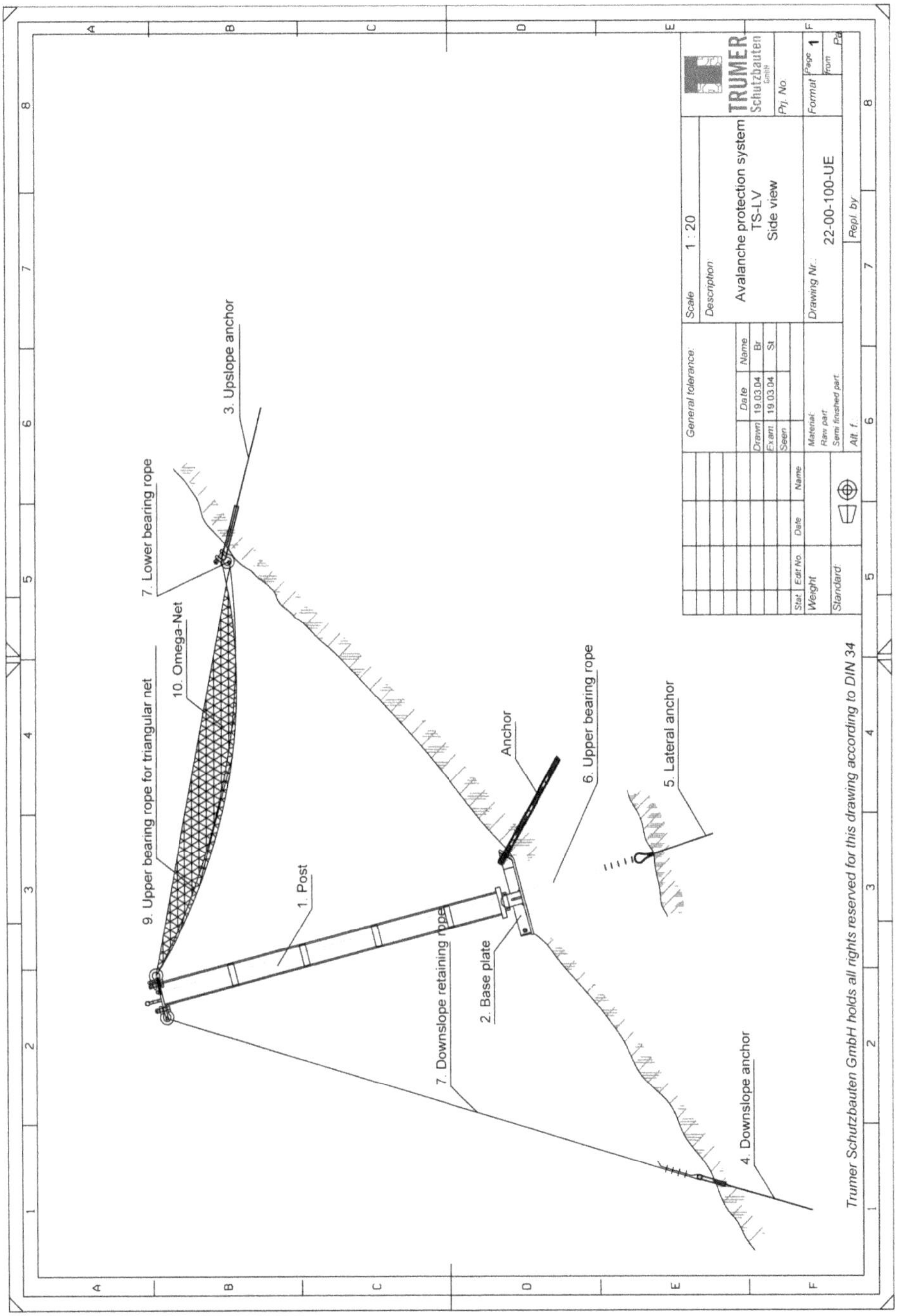

Fig. 5.9 *(Continued)*

2.5 × 2.5 cm). (Figure 5.9b) During the last few years, the manufacturers of special nets for avalanche defense have brought products with greater performance on the market (e.g. Omega-net®, TECCO-mesh®).

The first snow nets were built in Switzerland in 1951. To begin with, the supports were manufactured from wood; however, a move to steel was soon made. Since the first snow nets were produced completely empirically (without theoretical foundation and structural design), great damage occurred during the starting phase. Based on these experiences, the system was improved significantly in the course of time. Today, the supports are approx. 4 times as strong, and the damage potential is therefore comparable with that of steel snow bridges (Section 7.2.3.4). Snow nets are used today all over the world.

5.2.2.5 Construction type: combined snow bridge

The *combined snow bridge* consists of a bearing construction in steal with round log crossbeams and originally had the same geometry as steel snow bridges (three hinge frame, Figure 5.10a). The wooden crossbeams were chosen for economic reasons. Today, the financial argument for choosing wooden grates is barely relevant. The combined snow bridge was used primarily in Austria in the last 40 years as a light construction type for maximum snow thickness of up to 3 meters. It is used solely in such areas where one expects development of protective forest stands within 50 years.

For this reason, the original construction of the combined snow bridge was changed. The girders are dug approx. 1 metre into the ground. The girder-support pairs are – depending on load – placed at a sideways distance of 2.5 to 3.5 meters. Natural robinia round wood is used, since when using wood without impregnation, it is unproblematic to dispose of the wood at the end of the service life. By digging down the supports and

(a)

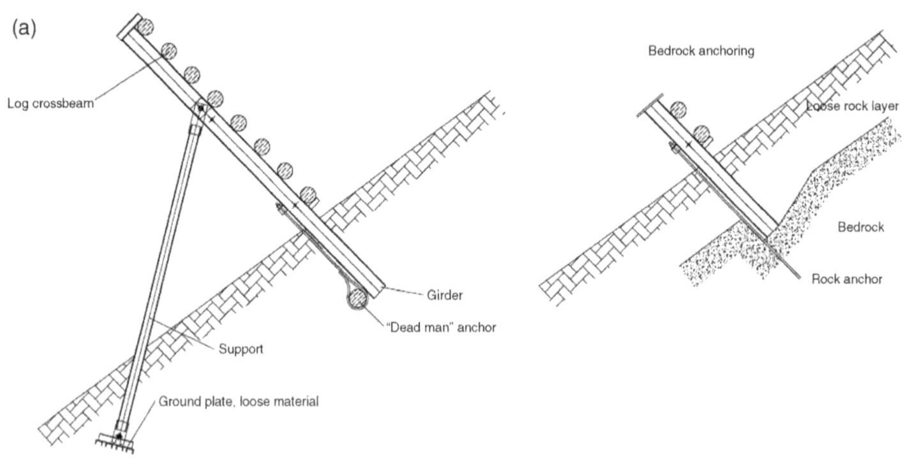

Fig. 5.10 Combined snow bridges D 2.5 with robinia round wood crossbeams: Austrian System (a) System sketch (© WLV Vorarlberg); Austrian System (b) construction type (© Schilcher); Swiss construction type Rempar Grischun: (c) system sketch (© CrestaGeo); (d) construction type (© CrestaGeo)

(b)

(c)

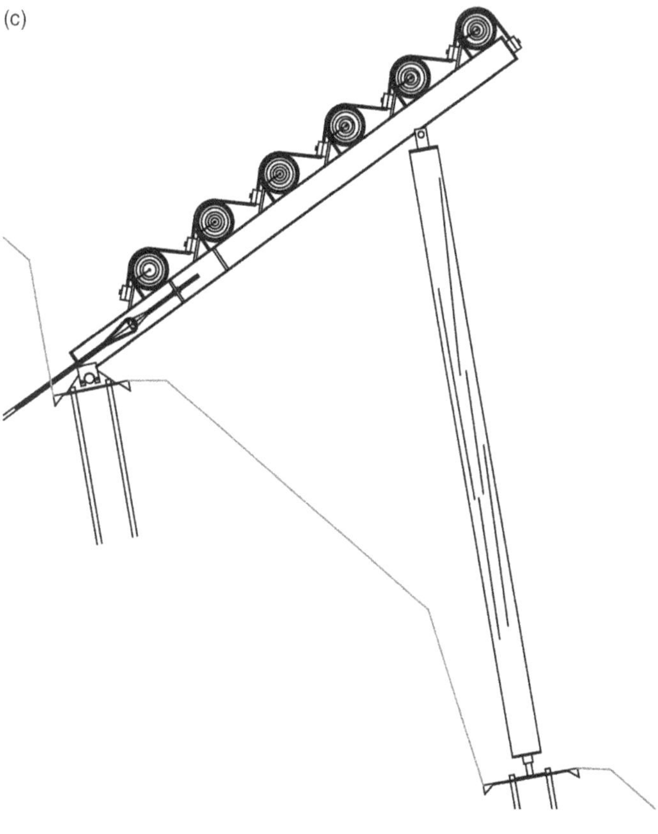

Fig. 5.10 *(Continued)*

(d)

Fig. 5.10 (*Continued*)

girders, one attains a lateral stability, and no further stiffening measures are required on the sides. When using a *dead man anchor* (Section 5.2.3.3) for the foundation of the girder, the slope-parallel force and the traction are transferred via the dead man to the ground. When using rock anchors or explosive anchors to absorb the traction, ground plates are used to transfer the slope-parallel force. The pressure force from the supports is transferred through ground plates in the ground (Figure 5.10b and d).

In Switzerland, a similar combined snow bridge has been used with success in the latest years, the so-called Rempar Grischun®. It is a combined snow supporting structures with steel girders and round wood crossbeams (sweet chestnut, robinia or oak). The traction anchoring uses wire rope anchors or explosive anchors, the support foundation uses a ground plate from steel [45] (Figure 5.10c). This construction type is generally used to protect afforestation in combination with technical protective measures against snow glide, which additionally has an effect on the reduction of the glide factor as basis for the design of combined snow bridges.

5.2.2.6 Construction type: snow bridge and snow rakes of wood

Snow bridges and snow rakes of wood are used primarily to protect afforestation below the forest limit. Service life is generally less than 50 years.

In Austria, snow bridges are usually used, and there are two different construction types widely available (Figure 5.11a and b). The type 1 wooden snow bridge is composed of two snow glide supports, which are connected with round wood to the snow bridge. A snow glide support consists of two round wood girders, which are screwed together at the upper end with a round wood support. Construction types with a length of 5 to 6 meters are common for snow thickness of 1.5 to 2.8 meters. The type 2 wooden snow

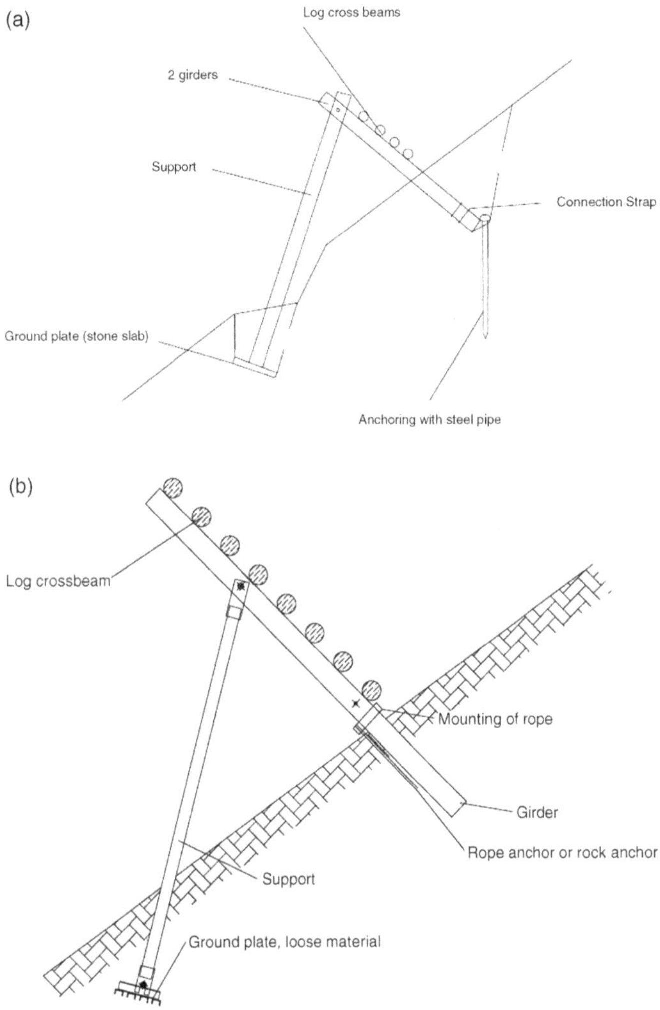

Fig. 5.11 Wooden snow bridges: (a) type 1, D 1.5; (b) type 2, D 3.0 (© WLV Vorarlberg)

bridge has the geometry of a classic snow bridge and consists of two round wood girders, two round wood supports, and round wood crossbeams. The connection between supports and crossbeams is carried out using screwed in steel profiles. The grate length is 4 to 5 meters at snow thickness of 2 to 3.5 meters. When necessary, wood or steel strips are screwed onto the grates and supports to serve as a lateral bracing.

In Switzerland, almost exclusively round wood snow rakes of type SLF are used for non-permanent avalanche control structures (Figure 5.12). Snow rakes are preferred because the upslope anchor of the so-called sleeper can be carried out easily on the entire length of the structure, because the angled woods beams of grate are less likely to rot, and because the young wood suffers less damage as it grows through the snow rakes.

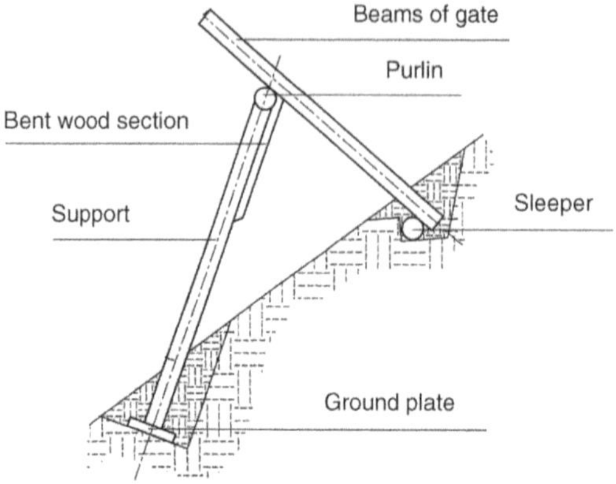

Fig. 5.12 System sketch of wooden snow rakes type SLF (© Margreth)

The grate transfers the snow pressure force upslope to the sleeper and downslope via a purlin to the two round wood supports. Both supports and the purlin are stabilized with two bent wood sections in order to increase the transverse stiffness of the structure. The foundation for the supports is for all types constructed of ground plates from steel or concrete. The supports for the snow bridges are extended, and are dug into the bedrock and secured with wire rope anchors, or are solidified in the loose material with injection anchors. For wooden snow rakes, the upslope sleeper is often dug into the soil, with or without pole anchoring.

5.2.2.7 Historical construction types of snow supporting structures

In the history of avalanche defense, the empirical development engineering work brought forth construction methods (Table 5.3) that today are designated historical. These are no longer used and are only relevant in terms of maintenance (Section 7.2.4).

Table 5.3 Historic Construction Types for Avalanche Defense Systems

Construction Type	Description
Earth terraces	*Earth terraces* were erected with a width of 0.5 to 5 meters. The downslope side slope was erected with sod slabs with a batter of approx. 3 : 1 and a maximum height of 2 meters. Cohesive soils and a good turf were a prerequisite for such construction types. The protective effect against avalanche formation is limited to small snow height.
Avalanche dry walls	The *avalanche walls* consist of dry-laid masonry and have a downslope batter of 5:1. The upslope effectual height is usually limited to 2 meters. Larger concrete walls were also erected and backfilled upslope, creating an effectual height of 4 to 5 meters.

(*continued*)

Table 5.3 (*Continued*)

Construction Type	Description
	Since 1951, no new avalanche walls have been erected (Figure 5.13a).
Arlberg snow rakes	The *Arlberg snow rakes* were developed simultaneously with the construction of the Arlberg rail line and consist of 4 m long railway tracks, which are encased in concrete at a distance of 3.5 meters and stick out 2.5 to 2.8 meters from the ground. The tracks are fastened upslope and sideways with steel cables that were drilled on site. The individual tracks are also connected to steel cables. 5 cm strong, round or fissured wooden banisters are secured to the steel cables. The anchoring of the upslope and downslope steel cable is carried out in bedrock or on stones in the loose material (Figure 5.13b).
Snow rakes/bridges of light metal	*Snow bridges* or *snow rakes of aluminium* were used in Switzerland for the first time in 1951. The first construction types consisted of voluminous aluminium bended profiles with a thickness of 3 to 4 mm or of light metal press profiles. By using aluminium, no corrosion protection was required. Compared to steel, the weight of a snow bridge made in aluminium was approx. one-half (heaviest single parts approx. 70 kg.). The thin-walled profiles were, however, not very resistant to falling rocks, and there were often stability problems (e.g. local bulging). Because of the high material costs, aluminium is no longer used as a construction material.
VOBAG snow bridges	*VOBAG snow bridges* consist of pre-fabricated concrete elements. The beam weighs up to 1000 kg. The VOBAG snow bridges are sensitive to falling rocks, and are, because of the substantial weight, difficult to erect in the terrain. This construction type was used in Switzerland until the mid-seventies. The pre-stressed concrete elements have turned out no longer to be sustainable. Today, many VOBAG snow bridges are in a bad condition, and must be replaced systematically.
BEDO structures	In the eighties, a construction type was used in Switzerland, where the grate consists of vertical and horizontal crossbeams. In BEDO construction types, the main structures consist of snow rakes, and the connection structures of snow bridge.

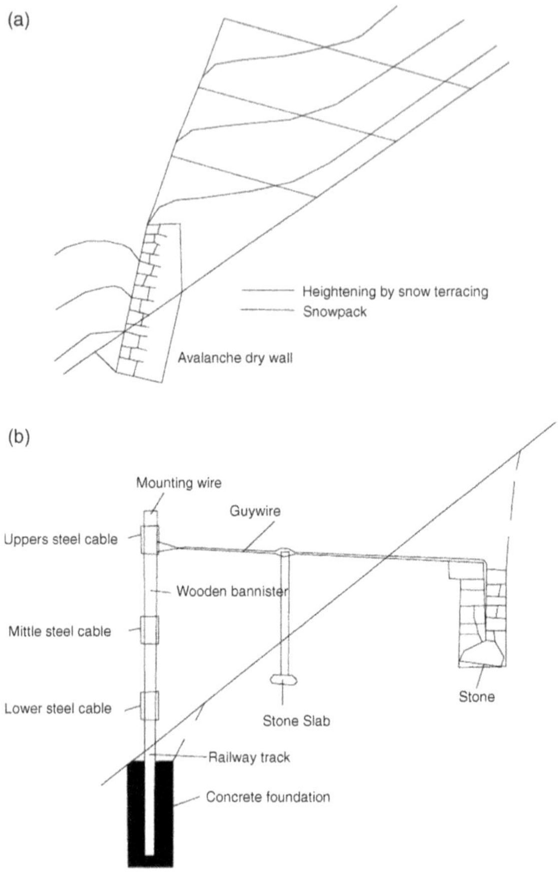

Fig. 5.13 Historic construction types for avalanche defense systems: (a) avalanche dry walls; (b) Arlberg snow rakes (© WLV Vorarlberg)

Furthermore, in the last decades, new construction types have been developed by certain companies, or by state-run avalanche control services. These systems have, until now, been used only to a limited degree and under specific conditions (special construction types) (Table 5.4). This means that they as a rule do not conform to the current state-of-the-art for avalanche defense systems [163, 203].

5.2.2.8 Type approval test

Since 1977, the Federal Office for the Environment (FOEN) in Switzerland has carried out *type approval tests* of standard construction types and the corresponding foundation based on the Swiss Guidelines for defense structures in avalanche starting zones [163]. If state subsidies are applied for avalanche defense structures, only officially tested and approved construction types, or anchor grout may be used for these works.

Table 5.4 Special construction types for avalanche defense systems: overview (examples)

Construction Type	Description
Hanging snow supporting structures	During the development of the *hanging snow supporting structures*, it was attempted to make the mounting process as simple as possible. For this construction type, a support is no longer required. The snow supporting structure is mounted upslope on one or more anchor points. Thereby, mounting in even very difficult terrain becomes possible, and for soil with poor load carrying capacity, the anchoring point can be shifted to soil with a higher load carrying capacity using a cable extender. The problematic part of the mounted structures is that the connection between the support surface and anchoring is subjected to the complete snow pressure. This means that this construction type can overturn upslope.
Hanging snow bridges	A *hanging snow bridge* is a type developed in Austria by the Forest-technical Service for Torrent and Avalanche Control from the Arlberg snow rakes. Following the avalanche disaster in 1954, rail-tracks were also made available as emergency assistance. Using these rail tracks, a snow bridge was constructed with an upslope tensile anchor (optional) using round steel or ropes. The angle between grate and upslope terrain was approx. 70° (Figure 5.14).
Glarner hanging snow supporting structures	In the so-called Glarner *hanging snow supporting structures*, the grate is connected with the anchor point via a framework construction. The construction lies freely on the soil. The construction type was only used on singular occasions in very steep, rocky terrain (Figure 5.15a).
Ombrello snow supporting structures®	In the *Ombrello snow supporting structures*, which is offered by different producers, the support grate is made up of square cable nets, which are secured onto steel profiles placed crosswise. The support grate is connected to the anchoring via a steel rod and four steel cables. This type is particularly common in Italy (Figure 5.15b).
SIFA A® and SIFA C®	By using a second support, parallel to the first support, the weight of the girder could be reduced. Both supports were founded with a micropile and the girder with a tension anchor, which made a very efficient drilling of the anchor points possible. In the 1980s, these types – SIFA A and SIFA B – were built extensively in the Brenner Pass. In Switzerland, type 1986 was approved for a short time period

Fig. 5.14 Snow hanging bridges: D = 3.0 with wooden crossbeams (© WLV Vorarlberg)

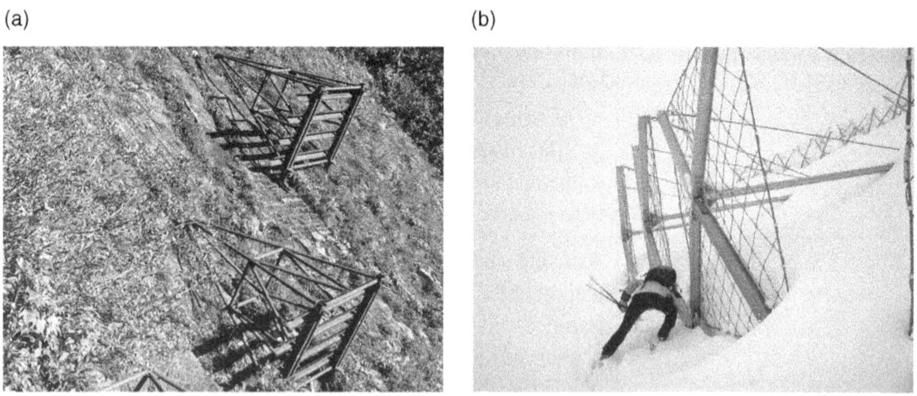

Fig. 5.15 (a) Glarner hanging snow supporting structure and (b) Ombrello snow supporting structure (© Margreth)

During a first phase, the structural design and construction plans, as well as the suitability of the constructions type, are tested with regards to snow and avalanche mechanics. The actions must be determined according to the Swiss Guideline [159]; the proof of structural strength must be carried out according to the relevant technical standards. In the second phase, the practical usability is tested. For new snow supporting structure types, test structures are erected, and must be observed for at least two winters. In case of constructional changes to existing types, usually only a mounting test is required. If both phases pass successfully, a construction type may be approved. On the home page of the Federal Office for the Environment (www.bafu.admin.ch), you can find type lists containing all approved construction types. This type approval is partially accepted in other European countries. During the last years, the usability of different snow net systems has been tested in field tests in Austria [214].

5.2.3 Foundation and anchoring of snow supporting structures

5.2.3.1 General remarks

Depending on different parameters (snow height, glide factor, construction type etc.), supporting the snowpack can generate a significant force that must be transferred to the ground or bedrock using suitable foundations. The following section provides an overview of the common and historic methods of foundation and anchoring of snow supporting structures.

5.2.3.2 Methods of foundation (anchorage)

The choice of anchor and foundation methods in avalanche defense systems usually depends on the construction type used and the substratum conditions. A further factor is the possibility of using modern drilling and injection techniques, as well as manipulating loads in the area of the avalanche starting zone.

For *tension anchors*, stiff rod anchors, pairs of micropiles (two micropiles, which are connected with bolts and transfer the pressure and tension components separately), self-drilling anchors or rope anchors are used. Stiff rod anchors (micropiles) are predominantly used on cohesive soil or bedrock when the drilling hole stands without casing. For unstable substratum, self-drilling anchors are a reasonable alternative. Sometimes *explosive anchors* are used for non-cohesive loose stones. Rock anchors can absorb traction and transverse force, which is the reason why it is advantageous to use flexible wire rope anchors. Transverse force, which occurs on upslope rope anchor of snow nets through the deflection of the anchor force, must be taken into account when dimensioning the anchors. At suitable sites (dense, stable soil), it is sufficient to install an additional stability pipe. At non-accommodating sites (humus-like, loose soil; large snow slipping), the erection of a concrete foundation is recommended. For *anchoring (foundations) under pressure*, ground plates without anchors (micropile), ground plates with micropiles or micropile systems are used, which transfer the pressure and tension components separately into the foundation soil (pair of micropiles).

Micropiles are foundation piles stressed by pressure or tensile strength (Figure 5.16a) [67]. These structural components can transfer a built-up pressure or tensile force to a

(a)

(b)

Fig. 5.16 Foundation of snow supporting structures: (a) upslope using micropiles stressed by tensile force and pressure and (b) downslope using ground plate made of steel (© Rudolf-Miklau)

layer with load carrying capacity or reduce it via the skin friction. The diameter is 90–110 mm (drilling piles). Micropiles consist of a tension or pressure rod made of steel and placed in a drilling hole (diameter: 32–40 mm) and a cover of anchor grout. When required, the micropile can be covered with a textile sock (net anchor), in order to avoid great loss of grout.

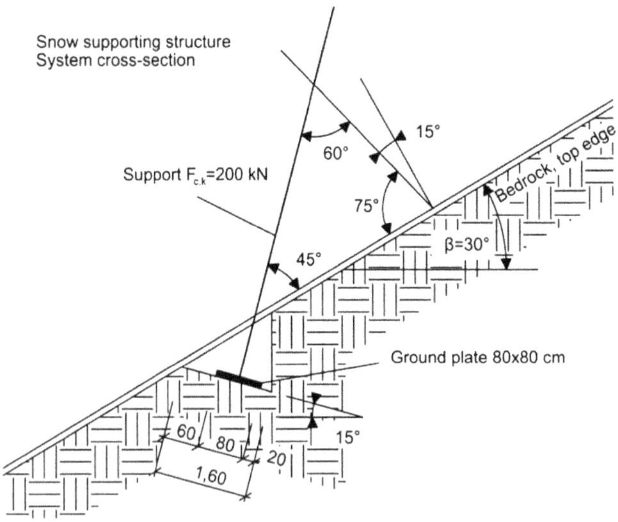

Fig. 5.17 Geometry of the ground plate foundation, according to [106] (© Hofmann)

In order to transfer the pressure force to the ground, *ground plates* can also be used if the foundation soil permits (Figures 5.16b, 5.17 and 5.18b). In this connection, one must ensure that there is sufficient distance from the downslope edge of the ground plate to the terrain surface in order to enable for a geotechnical proof of the safety for ground shear failure. Ground plates are used exclusively for the foundation of the supports.

Wire rope anchors, with steel cables as reinforcing elements, are coated along the total anchor length by means of injection of anchor grout (Figure 5.18a).

In order to manufacture *exploding anchors*, an explosive charge is placed in the borehole deep via a metal pipe and lit. This creates a cavity and leads to the formation of a pear-shaped body of grouting material. Following this, a reinforcing element (e.g. threaded rod) is placed into the exploded metal pipe, or borehole, and then injected with grouting material.

5.2.3.3 Historical foundation methods

In the development of foundations for avalanche defense systems, the procedures described in Section 5.2.3.2 have proven to be the most effective and have established as methods up to the state-of-the-art. Over time, a number of different foundation methods were used, which today are deemed historical and are no longer used in practice (Figures 5.13–5.15). The knowledge of these methods, however, is important with regard to assessment of damages to existing defense structures (Section 7.2.3.4), and for assessing the service life with regard to the maintenance. Table 5.5 provides an overview of historical foundation methods in Austria and Switzerland.

(a)

(b)

Fig. 5.18 Foundation of Avalanche Nets: (a) wire rope anchor for the upslope foundation of an avalanche net – anchor head with guywires (© Rudolf-Miklau); (b) support with pivot on ground plate for an avalanche net. (The support is connected with the ground plate via a splint-safe bolt connection. The ground plate sits on the soil without anchoring and is secured upslope with ropes.) (© Rudolf- Miklau)

Table 5.5 Historical foundation methods for technical avalanche defense in Austria and Switzerland: Overview

Method	Description
Dead man anchor	During the early phase of the avalanche control, the technical possibilities on steep slopes were strongly limited, and hence the foundation, to begin with, was produced using manually dug sleepers (*Dead Man Anchor*) and ground plates.
In-situ foundations	A significant advancement in the foundation of avalanche defense structures was the possibility to manufacture and manipulate in-situ concrete on the site of the avalanche formations. *In-situ foundations* were used for the first time in the fifties pursuant to the Swiss Guidelines [50] which were valid at the time in Austria and Switzerland. In that connection, both the upslope girder foundations and the downslope support foundations were manufactured in concrete. The girder foundation, which is under tensile stress, was constructed in solid form in order to compensate for the traction that ensues from the shear stress between concrete head and soil. When the soil conditions were bad, this foundation method required excavation depths of 1.5 to 2.0 meters.
Explosive foundation	For the method of *explosive foundation*, which is still partially used today, a hole is drilled into the loose material, and protected against caving in with a steel pipe. There are differences between the Austrian and Swiss methods of explosive foundations. In Austria, a sharpened, closed pipe was rammed in without drilling. Drilling only took place if you hit a stone or bedrock with the pipe. In Switzerland, the hole was drilled, just as for a micropile. The pipe had to be perforated or slit so that the grout also could escape on the length of the pipe, thereby ensuring grouting along the entire length of the pipe. Thereby, the load deflection was not limited to the grout body in the excavated cavity. Following this, an explosive load (approx. 30–100 g Donarit gelatine) was placed in the drill hole, and a cavity was blown out, which then was filled with anchor grout. Finally, the anchor was placed in the cavity through the pipe. After the hardening, the grout body took on the form of an underground pear, which was able to carry the anchor tension [124].

(continued)

Table 5.5 *(Continued)*

Method	Description
In-situ foundations with explosive foundation	Further development led to a combining of in-situ foundations and explosive foundation. Using this method, the traction is transferred by parallel explosive anchors, so that the cubature of the upslope foundations could be reduced significantly. This only required the dimensioning of one in-situ foundation to the ground pressure force. The support girder for the steel snow bridge was extended by 1 metre and concreted into the in-situ foundation. The downslope foundations were executed cylindrically, and the framing was done using half frames that could be shut [140]. The downslope foundation was at an early stage carried out with ground plates made of steel [98].
Tyrolean explosive anchor (parallel rod anchor)	In Tyrol, pre-fabricated foundations (*parallel rod anchor*, support plates in steel) were already used in the 1970s. Starting in 1976, the parallel rod anchor method developed by *Eliskases* was put into use. With this method, two parallel rod anchors (St 360 B) were welded onto a steel plate (*Tyrolean Parallel Rod Anchor*). The anchors were placed in two drill holes drilled on top of each other in the loose rock and filled with grout. The snow bridge beam was connected directly via a joint (bolts) with the anchor plate. The same method was used for anchoring in solid bedrock; however, the drill holes were not blasted [49]. This foundation system was not statically sound, and did not live up to the Swiss Guidelines [52]. Damage was predominantly caused by the welding seams tearing apart when the steel plate was not placed directly on the soil. In spite of the technical shortcomings, the Tyrolean Explosive Anchor was for years considered to be the most effective and economic foundation method for avalanche defense systems [140], but it is no longer in use today (Figure 5.19).
Total explosive foundation	Since the beginning of the 1980s, both anchors and support foundations with explosive anchors have been used in Switzerland (total explosive foundation) [124]. By using specific structure types (locking mechanism), the beam foundations are now also loaded with traction in a constant direction. In order to ensure a rational operation of the drilling caddy, the supports downslope were founded on a so-called anchor tripod, which resulted in double safety – also against tilting caused by uneven settlement (Figure 5.19). The method turned out to be time and money consuming, and required particularly precise work [140].

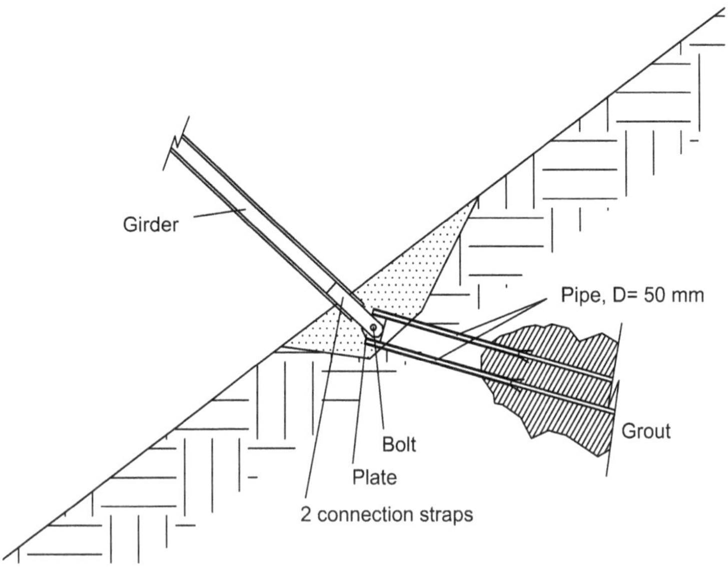

Fig. 5.19 System sketch of the parallel rod anchor with girder connection, according to [49] (© WLV Tirol, Herbert)

5.2.4 Snowdrift control structures

5.2.4.1 Effects and classification of snowdrift control structures

Snowdrift control structures include construction types that influence the snow deposition. By using the wind force, the snow depositions can be influenced by the control systems in a way that prevents the formation of large cornices and positively influencing the snow mechanics of the snowpack. This happens as the wind trajectory meets with an obstacle, which leads to a whirling of the air, and thereby deposition of snow (*snowdrift fence*), or which increased the wind speed incrementally and thereby cause the snow particles to be transported further on (*wind baffle, wind roof*).

Snowdrift control systems are not only used for avalanche defense, but also alongside of roads to prevent snowdrifts. For avalanche control, they are predominantly used in combination with snow supporting structures in order to prevent snow accumulation in the starting zone or the formation of cornices along a ridge (Figure 5.20). Table 5.6 contains a classification of the snowdrift control systems according to function and effect.

The effect of snowdrift control structures is strongly connected to the correct placement relative to the main wind direction; awareness of this fact, as well as the related morphological factors is a prerequisite for an optimal effect. In order for a snowdrift control system to be effective, observations must have been carried out throughout many winters. In special cases, wind measurement at the location where

Fig. 5.20 Combination of an avalanche defense structures (snow supporting structures) with wind baffles on a ridge (© WLV Vorarlberg)

defense structures are to be erected will provide valuable additional information. In that connection, particular attention should be paid to the small-scale relief-conditional wind situation, which can deviate significantly from the wind measurements from the surrounding weather stations (ridge winds). Below you will find an overview of the most important construction types for snowdrift control systems.

Table 5.6 Classifications of snowdrift control systems according to function and effect

Effect	Function	Construction Type
Structures that prevent the transportation of drifting snow	Braking the wind speed, deposition of the drifting snow	Snowdrift fence with vertical beams
		Snowdrift fence with horizontal beams
Structures that further the transportation of drifting snow	The wind is deflected in such a manner that the drifting snow in particular avalanche starting zones is continuously brushed away, or the snowpack is prevented from carrying out its normal build-up because of turbulence effect.	Wind baffle
		Wind roof

(a) (b)

Fig. 5.21 Snowdrift fences: (a) with wood panels and irregular tear off edge and (b) with horizontal steel boarding (© Sauermoser)

5.2.4.2 Construction type: snowdrift fence

Ideally, *snowdrift fences* are constructed on high plateaus or ridges above snow supporting structures, in order to minimize snow drifting into the avalanche starting zone (Figure 5.22). The structures that are constructed facing the main wind directions have the best effect; deviations relative to the inflow angle lead to a significant drop in effect [267]. Deviations up to 25° are tolerable [266]. Snowdrift fences can be erected with both horizontal beams (grate) (Figure 5.21b) and vertical beams (grate) (Figure 5.21a). Usually steel girders with wooden boarding are preferred; height offset vertical boarding is known to cause irregular tear off edges on the fence, and thereby an increase in turbulence. Snowdrift fences can be erected with or without ground level gap. Depending on the quality of the substratum, the foundation is carried out with anchors in combination with ground plates or concrete pillars.

5.2.4.3 Construction type: wind baffle

Wind baffles are placed in areas of potential cornice formations (Figure 5.23a and b). The boards affect the airflow and therefore influence the currents, high turbulences in

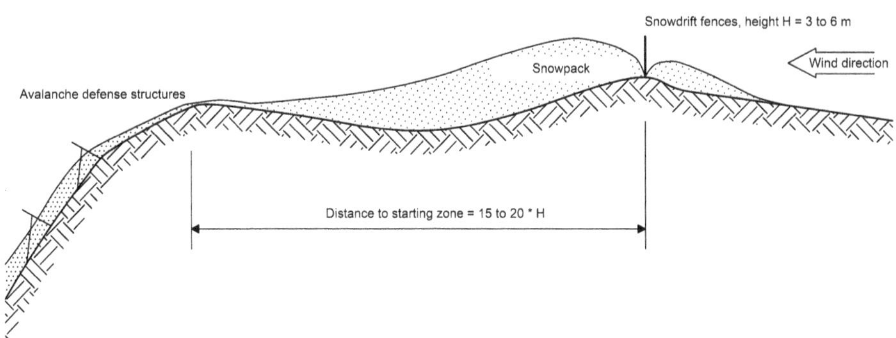

Fig. 5.22 Placement of a snowdrift fence (© Margreth, illustrated by Herbert)

(a)

(b)

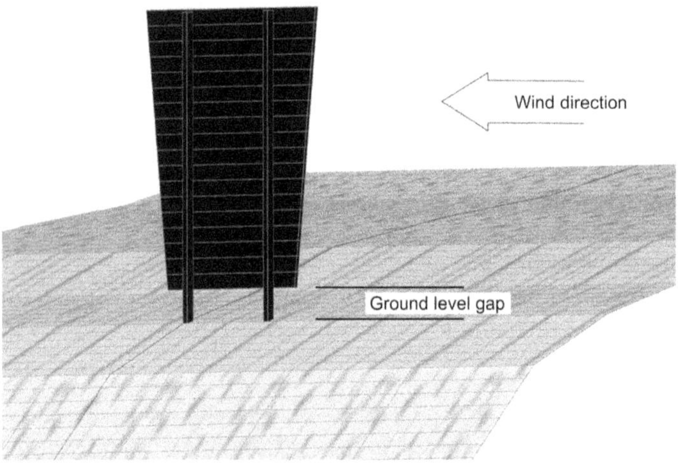

Fig. 5.23 Wind baffles: (a) wooden wind baffles along a ridge crest on the Jaufen pass (South Tyrol) (© Sauermoser); (b) system sketch of a wind baffle according to [203] (© WLV Tyrol, Herbert)

the immediate vicinity of the boards create snow erosion; a scour is formed. The snow is deposited around the scour. The thickness and density of the snow deposition are inhomogeneous. Thereby the formation of a continuous cornice is prevented. Just like wind fences, wind baffles are constructed in combination with snow supporting structures. If a cornice falls onto the upper defense rows, damage may occur to the rows, or they may be overloaded by drifts from the accumulation zone. By erecting a row of wind baffles along the ridge crest, this can be prevented (Figure 5.20).

Experiments on the effect of different types of wind baffles were carried out by Wopfner and Hopf [284]. Since local hardening zones form around scours because of denser stratification, wind baffles effect stabilization of the snowpack and, in the end, disintegration of connected tensions fields. Wind baffles are built *with* and *without ground level gap*. A hardening of the snowpack results in both types, a disintegration of the tension field, however, only for wind baffles with ground level gap. The best shape turned out to be a board with a ground level gap of one metre [7]. The trapeze shape turned out to be better than the rectangular shape. Experiments have shown that the trapeze shaped wind baffles cause scours that are approx. 30% larger. In Switzerland, the cross-shaped wind baffles without ground level gap have turned out to be successful.

5.2.4.4 Construction type: wind roof (jet roof)

Just like wind baffles, *wind roofs* (jet roofs) are used to prevent the formation of cornices along ridges (Figure 5.24a and b). Through the positioning in the beginning lee, the wind speed in the nozzle increases, and the snow drifts far onto the slope. Wind roofs do not reduce the total amount of snow that is blown in, however the roof has the effect that the snow is transported on and deposited on the slope. A disadvantage of this structure is that the drifting snow is deposited on the avalanche slope. Wind roofs are constructed together with snow supporting structures in order to prevent excessive snow or damage to the top rows. According to [7], wind roofs should have a leeward side opening with a height of 1 metre, the windward side opening should have a height of approx. 2 meters. The inclination of the closed roof should be slightly larger than the terrain inclination in order to attain a nozzle effect. Wind roofs can be constructed in wood or steel.

5.2.5 Snow glide protection structures

5.2.5.1 Protection effects

Snow glide protection structures have the task of keeping the slow movements of the snowpack, the gliding, and the creeping (Section 2.3.4.2) to a minimum, and thereby reduce the forces acting on obstacles. Snow glide protection structures are in particular use to protect afforestation.

Slopes pointing from SW to SE are in particular risk of snow glide, if the slope is above 30°, and have a low level of roughness. Within snow supporting structures on slopes with a low degree of roughness, the snow-mechanical loads on plants are so great that these plants can only grow if snow glide control measures are in place [151]. When planning snow glide protection measures, one must define whether the area concerned is exclusively subjected to snow glide or is also avalanche prone.

5.2.5.2 Snow glide protection methods: overview

The most extensive and systematic studies on the suitability of different methods of snow glide protection were carried out by In der Gand [116]. However, he only studied manual methods of manufacture because, at the time of the studies (1955/56 to 1961/

(a)

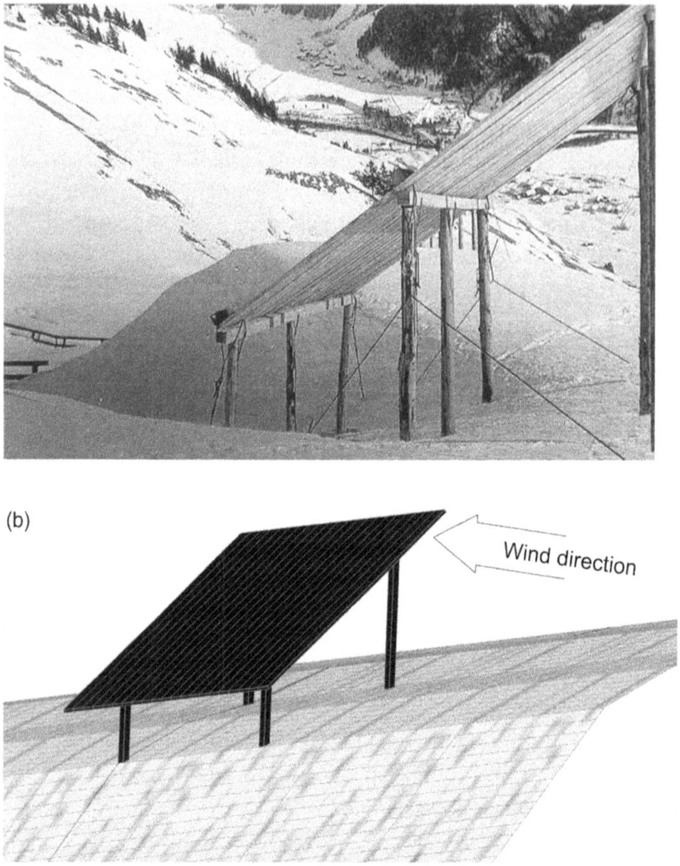

Fig. 5.24 Wind Roofs: (a) wooden wind roofs (© WLV Tyrol); (b) system diagram of a wind roof according to [203] (© WLV Tirol, Herbert)

1962), no suitable terrain-compatible mini-excavators were available. Table 5.7 contains a complete overview of the snow glide protective measures.

In general, measures with a high density (placement on the area) are better suitable to stabilize the snowpack than single elements, which only provide punctual protection [116, 151]. The loss of small trees in the areas examined by In der Gand from 1955 to 1963 in areas with berm steps, berms, and pales was 0–28% less than in areas without protective measures and with snow supporting structures against snow glide (8–57%).

The dimensioning of snow glide structures on slope-parallel and slope-vertical snow pressure is not possible due to the difficulty of defining the correlation between the single elements of the snow glide protection and the snowpack. The effect of the measures summarized in Table 5.7 is based on decades of practical experience.

Table 5.7 Overview of snow glide protection structures: Production and effect (advantages and disadvantages)

Type of Measure	Description	Advantage	Disadvantage
Array of pales	Single wooden pales in different spatial arrangement rammed manually or by machine into the ground	Narrow spatial arrangement possible	Only possible in deep soil
Array of pales guyed	Single wooden pales in different spatial arrangements rammed manually or by machine into the ground and guyed with ropes or wires	Also possible in flat soil	Extensive maintenance in connection with controls and reconstruction
Berms/Berm steps by machine	Slope terraces of different widths made with mini-excavator or plough	Cheap machine-made production on deep soil	Susceptible to erosion, edge loading
Berms/Berm steps manually	Manually produced slope terraces	No machines needed	Work-intensive, susceptible to erosion, water balance on slope destroyed
Plate berms Pit-planting	Manually produced mini-berms 40 × 40/60 × 60 cm	No machines needed	Cost-intensive, susceptible to erosion, water balance on slope destroyed
Pressure mound	Manually produced earth mound (250 pieces/ha)	Good effect on deep soil	Only possible on deep soil, susceptible to erosion, water balance on slope destroyed
Snow glide tripods	Tripods from robinia/sweet chestnut wood (250 pieces/ha)	No erosion, different densities possible, cheap	Transport of material difficult

(*continued*)

Table 5.7 *(Continued)*

Type of Measure	Description	Advantage	Disadvantage
Snow supporting structures against snow glide	Steel snow bridge for small snow thickness, wooden snow bridges, combined snow bridges	No erosion, different densities possible, cheap	Only isolated effect
Round wood sleepers	Flat/-lying round wood from robinia wood: D = 30 cm Anchor with two wire rope anchors or threaded rods	Difficult foundation work	Low effect depth

5.2.5.3 Construction type: array of posts

According to [116], the array of posts is only effective if the pale spatial formation is denser than 2 meters spacing. The following recommendations are in effect for the spatial arrangement according to Table 5.8.

For the array of posts to have an effect, along with sufficient density in the poles it is imperative that there is a ratio of post lengths above the soil to the post in the soil of 2 : 1, and at least 1 : 1, which In der Gand [116] assesses to be ideal. Any deviation from this basic rule leads to a great malfunction in the array of posts, which has been confirmed by studies in Switzerland [151]. The required post diameter for round wood is 10 cm, and 16 cm for half-round wood.

Table 5.8 Guideline values for the post spacing (horizontal distance at level line and distance in the fall line) in dependency of the extreme snow depth H_{ext} and the slope gradient when the posts are placed offset according to [116]

Slope Gradient	Post Distance in Meters			
	H = 1.5 m	H = 2.0 m	H = 2.5 m	H = 3.0 m
<30°	2.0	1.7	1.5	1.3
>30°–35°	1.9	1.6	1.4	1.2
>35°–40°	1.8	1.5	1.3	1.1
>40°–45°	1.7	1.4	1.2	1.0

The choice of wood depends on the required usage time of the array of posts. Fungus-resistant wood species such as sweet chestnut, robinia, oak, or pressure-impregnated wood have a service life of up to 30 years, but when using larch, spruce or pine, one should count on maximum 10 to 15 years.

Since an array of pales, according to [116], is usually not possible due to the flat soil in the montane and subalpine regions, a system is occasionally used in Austria where the pales are guyed with wires [104]. The steel or wood posts with a length of 2 meters are rammed 0.5 meters into the soil at a gradient of approx. 10°. The wooden posts are subject to the same longevity conditions as regular array of posts. In order to increase longevity, sharpened steel pipes with a diameter of 48 mm are also used. The wires used are galvanized 4 mm wires. The wires must be placed in such a manner, that they hang loosely when not under tensile strength. Each post head is connected to the next with the next offset pales above close to the ground, so that an elastic total system is formed.

5.2.5.4 Construction type: snow glide tripod

The construction of *snow glide tripods* from robinia wood or pressure-impregnated wood is the most commonly used snow glide protective measure in the Alpine Region. The effectual height of the tripod is between 0.8 and 1.5 meters. According to the different effectual heights and related load, different foundation methods are used. For tripods with an effectual height of 0.9 meters, it is sufficient to bury the girders to a depth of 0.5 meters, and secure them against being pulled out using one-metre long steel nails. For greater effectual heights, it is necessary to use explosive foundations, which on the other hand leads to a sharp increase in costs [233].

The tripods can be placed in a unitary grid (Figure 5.25), which however may not exceed five meters spacing, or in groups. It has been shown to be prudent to construct groups of approx. 10 blocks, where the distance between the individual tripods may not exceed maximum 1.5 meters. This corresponds to the clump structure of the forest in subalpine areas, and a group of tripods can be planted with approx. 30–50 seedlings (Figure 5.26). Experience has shown that by placing upslope crossbeams close to the ground, one can additionally reduce snow glide.

5.2.5.5 Construction type: berms

Berms are earth terraces of varying width, which are placed continuously or non-continuously (plate berms) on slopes in order to reduce the scope of snow glide movements. The construction can be carried out using a climbing excavator. The manual construction of berms is no longer carried out due to the high costs of constructing by hand. Broad terraces [151] reach a width of 0.8–2.0 meters.

What is problematic is the impairment of the water level on the slope caused by the berms, and the furthering of erosion. The studies carried out by *Leuenberger* [151] showed that pressing the berm edges is particularly problematic when the berms were constructed by deposition and excavation, and the slope gradient was above 30°. The effectiveness of berms as snow glide protection decreases as the berm distance increases. In unstable slope areas, which have a tendency for water logging or increased

Fig. 5.25 Combination of avalanche defense structures with technical snow glide protection measures (snow glide tripods) in the Lenzerwald Avalanche (Tyrol) (© Sauermoser)

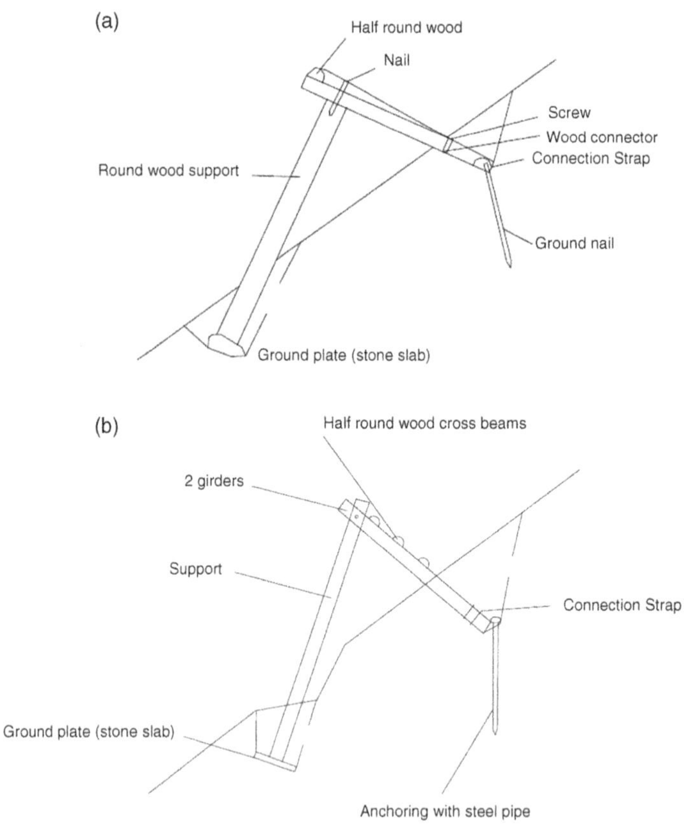

Fig. 5.26 Snow glide tripod (System sketch): (a) tripod H ≤ 1 meter; (b) tripod H = 1–1.5 meters (© WLV Vorarlberg)

erosion, one cannot construct berms. In all cases, a geotechnical assessment of the slope stability must be carried out.

Special types of berms are *plough berms*, which are built by pulling a berm plough horizontally to the slope using a ground skidding winch [103]. Plough berms only have a low effectual height, and can only be constructed on soil with a thick humus cover. *Pressure mounds* are a special type of berm, which only can be constructed in deep loamy soil. Sod slabs are cut out of the soil by hand and constructed with the sod end facing down toward hills. Compared to berms, they have the advantage that they do not influence the slope water drainage as strongly, as the water can flow out between the mounds.

5.3 Structural avalanche defense in the avalanche path and deposition zone

5.3.1 Classification

If the release of avalanches in the starting zone cannot be prevented, or only at unreasonably high costs, measures in the avalanche path or deposition zone constitute an alternative to diminish the energy from the avalanches and shorten the run out distance. Often measures are also used in the avalanche path or deposit zone in combination with smaller measures in the starting zone (partial defense of the starting area).

There is a basic differentiation between

- Avalanche deflecting and guiding structures
- Avalanche catching and retarding structures, as well as
- Bypassing (bridging) structures and special defense structures

Table 5.9 gives an overview of the defense measures used in practice in the avalanche path and deposit zone, and their effects on flow and powder avalanches.

5.3.2 Longitudinal defense structures (construction types to guide and deflect avalanches)

5.3.2.1 Protection effects of avalanche deflecting structures

The *avalanche guiding and deflecting structures* are constructed in the form of dams or walls, which contain the avalanche in its path directionally stable, or for deflecting it in another direction, or to deflect it away from objects that need to be protected. The effect is predominantly restricted to an avalanche's flowing component; however, for a completely developed powder component from mixed-snow avalanches, or a pure powder avalanche, it is very small.

Splitting wedges and *roof terraces* are systematically considered defenses to deflect avalanches. However, they lie directly in front of an object in need of protection and are therefore coupled to the object protection measures (Sections 8.1.4.2 and 8.1.4.3).

5.3.2.2 Construction type: guiding wall

Guiding walls have the objective of limiting the avalanche lateral expansion and directional deviation in terms of width, where for example gully shaped paths on

Table 5.9 Overview of the technical avalanche defense structures in the avalanche path and deposit zone

Designation		Placement	Effect on	
Category	Construction Type		Flowing Avalanches	Powder Avalanches
Guiding and deflecting structures	Dam/wall	Avalanche path	++	–
Catching and retarding structures	Dam/Wall	Deposition zone	++	–
	Fall ground	Avalanche path	+	– –
	Avalanche-retarding mounds, cones, wedges	Avalanche path Deposition zone	+	– –
	Avalanche breaker	Avalanche path	++	+
Special construction works	Gallery, tunnel	Avalanche path Deposition zone	++	++
	Pipe Bridge	Avalanche path	++	++
Object protection	Roof Terrace	Avalanche path Deposition zone	+	+

++ Large effect.
+ Effect in place.
– Lesser effect.
—Negligible effect.

fan shaped talus cones end up in the flatter valley bed. For slope avalanches, guiding walls also effectively prevent sideways expansion (Figure 5.27).

Avalanche guiding walls constructed as homogenous earth walls or walls constructed in steel concrete, and often are built in connection with galleries and tunnels near roads or railways, in order to reducing their length to an absolute minimum. The direction of the guiding walls is parallel to the avalanche path. The avalanche is not deflected off its main course.

5.3.2.3 Construction type: deflecting dam (wall)

Deflecting dams – as opposed to guiding dams – cause a flowing avalanche to deflect to a desired direction, or counter the sideways break-out following a directional change in the avalanche path. Deflecting dams have practically no effect on a completely developed powder component from mixed-snow avalanches or pure powder avalanches.

Fig. 5.27 Guiding walls prevent the sideways expansion of flowing avalanches in the direction of the settlement area (Ischgl/Tyrol) (© Rudolf-Miklau)

The effectiveness of the dam depends on the position of the dam in the avalanche path (and thereby the velocity of the avalanche at the dam), on the dam height, on the terrain gradient, on the deflection angle, and on the geometrical shape of the dam. The smaller the deflection angle, the easier the directional change is 'accepted' by the avalanche. If the angle is too large, the flowing avalanche might overleap the dam, or avalanche parts might be deposited, followed by an unwanted pre-filling and reduced effectual height of the dam.

The maximum possible deflecting angle can be calculated according to [127]. For practical purposes, it is recommended to reduce the maximum calculated deflection angle by 10°. On the side facing the avalanche, the dam should have a gradient of at least 60°, which requires using construction methods to stabilize the slope (e.g. reinforced earth, stone pitching). A dam inclination of less than 30° facing the avalanche was at least deemed less efficient in small-scale tests, while slope inclinations of 60° nearly corresponded to the maximum effect of vertical walls [127]. Deflecting dams can be constructed as earth dams, slab stone walls or steel concrete walls (Figure 5.28a and b).

When positioning the guiding or deflecting dams, it is necessary to carry out a precise analysis of the width of the avalanche path using hazard indicators, avalanche chronicles or simulations in order to ensure the planned functionality. If the dam is situated at the edge of the path or outside the path only parts of the avalanche will be affected directionally.

5.3.3 Transverse defense structures (construction types to catch or retard avalanches)

5.3.3.1 Protections effects of avalanche catching or retarding structures

Avalanche-retarding and *avalanche catching structures* are placed transverse to the flow direction, directly facing the avalanche. The function of this defense is to catch

(a)

(b)

Fig. 5.28 Avalanche deflecting structures: (a) avalanche deflecting wall in steel concrete, Hanggerbach Avalanche, Tux/Tyrol (© Sauermoser) and (b) deflecting dam, Steinbichele Avalanche, Neustift/Stubaital/Tyrol (© WLV Tyrol)

the entire avalanche volume or reduce the avalanche velocity, whereby the avalanche run out distance is reduced.

One may differentiate between the following construction types:

- catching dam
- fall ground
- retarding cone/hump, retarding wedge
- avalanche breaker

When placing the avalanche-retarding systems, one must take into account that the effect of these measures is so much the greater, the more the natural terrain aids that facilitates avalanche deposit. By increasing the roughness, the velocity of extreme avalanches and thereby the run out distance is reduced. Already in the middle of the nineteenth century, avalanche-retarding systems were used to secure higher-lying settlements or alpine huts.

5.3.3.2 Construction type: catching dam (wall)

The main idea of the *catching dam* (catching wall) is that, with the help of an artificial obstacle, an avalanche can be made to stop completely and then deposited (Figure 5.29a and b). In order for a catching dam to be effective, the effectual height must be correct, as must the placement in the avalanche path, the geometrical formation, and the storage volume of the deposition zone. For this reason, avalanche catching dams should only be constructed in places where the avalanche velocity already is low, and if not, a safety distance to the objects to be protected must be kept. Avalanches can be caught and deposited in their entirety, but a partial deposition of an avalanche, and thereby a partial overflow of a dam, can also be part of a safety concept. This must be taken into account when placing a dam. Furthermore, the dam gradient on the avalanche side is paramount, and gradient should be as steep as possible. The geotechnical requirements, such as sufficient compaction of the cover material and drainage of the storage capacity, must be taken into account when constructing a dam. When designing catching dams, the effect of the avalanche plays a lesser role, as the volume of the dam usually plays a much more defining role. When designing the dam height, the pre-filling of the deposition zone facing the avalanche must be taken into account using previous avalanche results. Thereby, the effective height can be reduced significantly for the recurrent design event.

Catching dams are evaluated as having a negligible effect on fully developed powder components from mixed-snow avalanches and powder avalanches. Laboratory experiments, however, suggest that the powder component can also be significantly influenced if the dam height accounts for approx. 40% of the avalanche flow height. Then the velocity reduction of the powder component behind the dam can account for 20–40% [127].

Catching and retarding systems (Section 5.3.3.3) can be combined in an effective manner. The avalanche flow velocity can be reduced by avalanche-retarding defense systems, and thereby the effective dam height of the catching system can be reduced.

5.3.3.3 Construction type: avalanche-retarding cone

As a precursor to the avalanche-retarding systems, at the beginning of the last century simple *avalanche-retarding cones* were constructed from earth material and erected offset in large numbers. The idea originally came from the splitting wedges that were found by old houses in the avalanche paths. The offset set-up was supposed to cause a branching of the avalanche in the runout zone and a reduction of the run out distance. Retarding cones were often heaped using simple earth construction methods, or paved using quarry stone, in order to increase the effect (Figure 5.29a).

(a)

(b)

Fig. 5.29 Massive avalanche wall, constructed in two parts to protect the village centre of Galtür (Tyrol): (a) in overview; (b) seen from the inflow side with the Alpinarium in the background (© Rudolf-Miklau)

(a) (b)

Fig. 5.30 Avalanche-retarding cones: (a) retarding cones in the Heiligenblut Avalanche (Carinthia/Austria); (b) avalanche-retarding cones in Neskaupstadur (Iceland), designed according to [127] (© Sauermoser)

New experiments by Hákonardóttir and Johannesson [93] show that it is possible to cause a velocity reduction of 20% using avalanche-retarding cones with one row of cones and of 30% using two cone rows (Figure 5.30b), if the following criteria are met:

- The height of the avalanche-retarding cone must be at least 2 to 3 times higher than the avalanche flow height.
- The avalanche-retarding cone on the avalanche side must have a gradient of at least 60°, and the relation of width to height of the mounds must be 1.
- The cones must be placed systematically across the total flow width of the avalanche; the intermediary space between the trapeze shaped mounds may not exceed the crown width of the cones.
- The avalanche-retarding cones must be placed in the avalanche run out area; in the avalanche, the mounds have almost no velocity-reducing effect.

5.3.3.4 Construction type: avalanche breaker

Avalanche breakers are solid structures of high-quality steel concrete. These directly affect the process of flowing and power avalanches because of, on the one hand, the turbulence of the snow-air mixture, and on the other hand, the back-pressure of the flowing components from the mixed-snow avalanche. These thereby significantly reduce the runout distances.

For avalanches with confined avalanche paths, avalanche breakers can be efficient alternatives to constructing large avalanche defense systems or avalanche-retarding and catching systems, if erecting those is not possible for topological or technical (economical) reasons. The avalanche breakers placed in the avalanche path are sometimes combined with catching dams and/or retarding cones at the beginning of the deposit zone. In case of specific morphological circumstances (e.g. rocky scarp underneath the formation zone), avalanche breakers were also used on wide avalanche paths [256].

(a) (b)

Fig. 5.31 Avalanche breaker: (a) structures in the Mühlauerklamm avalanche (Innsbruck/Tyrol) (© Skolaut) and (b) in the Gamsleiten avalanche (Obertauern/Salzburg) (© WLV Salzburg)

Until now, five avalanche breakers in concrete steel have been constructed in Austria. The avalanche breakers in the Mühlauerklamm avalanche (Innsbruck/Tyrol) (Figure 5.31a), in the Hörndlgraben and Hasental avalanches (Salzburg) and in Rastetzenbach avalanche (Bad Hofgastein/Salzburg) are placed in confined avalanche paths. In the Gamsleiten avalanche (Obertauern/Salzburg) (Figure 5.31b) nine avalanche breakers were constructed on a wide avalanche path.

The initial observation of the effect of avalanche breakers within the framework of a protection project in Rastetzenbach in Salzburg showed that the construction could disturb the snow-air mixture in the boundary layer between flowing and powder components. This effect is magnified if the construction is situated on a natural grade changing in the avalanche path. The effect of the avalanche breaker in the Mühlauerklamm avalanche (Tyrol) was examined in a small-scale test [132] under simplified conditions. This small-scale test showed a velocity reduction at the scale of 40–60% compared to the velocity of the reference avalanche, that is before coming into contact with the avalanche breaker.

5.3.4 Avalanche galleries and tunnels

5.3.4.1 Construction type: avalanche gallery (tunnel)

Avalanche galleries and *tunnels* are direct defense measures for transport roads in the avalanche path or deposit zone. They protect roads or railroad tracks against the effect of avalanches by bridging or bypassing them (Figure 5.32a and b). Avalanches galleries differ from avalanche tunnels with openings they have on the side facing the avalanche. The construction of avalanche galleries or tunnels is particular effective when dealing with large starting zones with narrow, clearly defined avalanche paths [10, 59, 257].

When planning avalanche galleries, it is necessary to know the type and size of the effect, and the width of the avalanche path. The sizes can be calculated using hazard

(a)

(b)

Fig. 5.32 Bridging structures: (a) avalanche gallery in Kühtai (Tyrol); (b) gallery in the Mahdermähder avalanche (Schmirntal/Tyrol) under construction (© Sauermoser)

indicators, avalanche chronicles, and numerical avalanche simulations. In order to design a gallery, the depositions of an avalanche or deflection forces are by far the most important. For this reason, the gallery for the flowing avalanche should constitute as small an obstacle as possible. Slope angle changing should be placed as far above the gallery roof as possible. Since galleries also often protect against falling rocks, the effects that occur during these processes should also be calculated by a geologist and be incorporated in the dimensioning.

The path width, or the width of possible avalanche deposits, is the criterion for determining the length of a gallery or tunnel. For financial reasons, guiding walls are frequently constructed upslope on the gallery up to the clear delimitation of the avalanche path in this section. This is to protect the portals on both sides of the avalanche deposit. In avalanche galleries, constructions from half frames (i.e. single stem frames with upslope supporting wall or deposition consoles anchored in bedrock) have been used, as have arched constructions made from steel concrete in previous years. Today, monolithic concrete constructions without dilation joints are preferred. The size of the downslope opening is dependent on whether the backflow from an avalanche can occur through the openings (Figure 5.33a and b). This is the case when the gallery is in the avalanche runout zone, and the terrain downslope to the gallery is so flat that avalanche deposition occurs. If there is a gallery immediately in the avalanche path, and no downslope avalanche penetration is to be expected into the gallery, the downslope openings may be made larger.

5.3.4.2 Construction type: avalanche-secure pipe bridge

If a transport road crosses an avalanche path at a sufficient height, an avalanche-secure *pipe bridge* may provide the solution. A pipe bridge is a bridge deck with a closed overhang (Figures 5.34 and 5.35). These types of constructions are implemented when crossing avalanche formation zones with a high probability of powder avalanche incidents. Because of the high cost and the complex construction, these types of

(a) (b)

Fig. 5.33 Different variations of the downslope cladding of the openings in avalanche galleries: (a) cladding with concrete beams; (b) wood cladding (© Sauermoser)

Fig. 5.34 Pipe bridge in Großer Gröben (Bschlaberstrasse/Tyrol) (© WLV Tyrol)

Fig. 5.35 Pipe bridge in Rorbach avalanche (St. Gotthard Pass, Swiss Federal Railways) (© Margreth)

protective constructions are only carried out in narrow, deeply incised avalanche paths with sufficient foundation possibilities on both sides (bedrock if possible). Pipe bridges leading over avalanche lines exist in Austria, for example in Großer Gröben (Bschlabs/Tyrol) [59] or in Switzerland in Rohrbachgraben at St. Gotthard pass (Wassen/Kanton Uri) [217].

6 Structural avalanche defense: design and construction

Stefan Margreth, Jürgen Suda, Michael Mölk, Robert Hofmann, Peter Gauer, Wolfgang Schilcher, Christoph Skolaut, Art Mears and Siegfried Sauermoser

6.1 Normative bases of design

6.1.1 EUROCODE and national standards in Austria, Germany and Switzerland

The technical assessment of defense structures in permanent technical avalanche control systems in Europe is based on the Eurocode construction engineering standards. The entire Eurocode (EC) standard system is made up of ten areas (standard series). The series EC 0 (EN 1990) regulates the framework for the structural planning; the series EC 1 (EN 1991) sets a standard for the action on the structures. The individual construction type standards are derived from these. For these structures, the following are relevant: EC 2 (EN 1992) – Concrete and Steel Concrete, EC 3 (EN 1993) – Steel Construction, EC 5 (EN1995) – Wooden Construction and EC 7 (EN 1997) – Geotechnical Assessments, EC 8 (EN 1998) – Assessment in the event of earthquakes. In general, a Eurocode standard consists of a European standard general document, which is designated with EN (e.g.: EN 1992-1-1), and a national application document (NAD), in which each country is allowed to enter national definitions under specific items in the basic document. In Austria, these national documents are designated with B (e.g. ÖNORM B 1992-1-1). In Germany, the designation NA is placed at the end (e.g. DIN EN 1992-1-2/NA). In Switzerland, construction is carried out in accordance with the SIA structural standards (Swiss Engineering and Architecture Association). The SIA structural standards are, to a large extent, based on the Eurocode standards. The standards relevant for assessing avalanche control systems are separated according to usage area and are shown in Table 6.1.

Besides the standards shown in Table 6.1, there are also special guidelines available for assessing snow supporting structures. In most cases, the defense structures are assessed according to the Swiss Guideline 'Defense structures in avalanche starting zones' [163]. In Austria, the ONR 24806 [203] is used for the assessment and construction design of the avalanche snow supporting structures. This ONR is harmonized with the EUROCODE.

6.1.2 American national standard (ANSI) and Canadian standard (CSA)

Because avalanches are not national issues in the United States or Canada, ANSI standards pertaining to avalanche defense structures do not exist. Construction must conform to general structure standards. In the US and Canada, engineers rely on the experience and standards developed in Europe.

The Technical Avalanche Protection Handbook, First Edition. Edited by Florian Rudolf-Miklau, Siegfried Sauermoser and Arthur I. Mears.

Table 6.1 Overview of the most important standards for the construction-technical assessment of avalanche snow supporting structures in Austria, Switzerland, and Germany

Topic area	European Basic Document	National appendices and other national transposition standards		
		Austria	Switzerland	Germany
Concrete and steel concrete				
General rules	EN 1992-1-1	ÖNORM B 1992-1-1	SA EN 1992-1-1; SIA 262.001	DIN EN 1992-1-2/NA
Supporting structures	EN 1992-3	ÖNORM B 1992-3	SA EN 1992-3; SIA 262.004	—
Steel supporting structures				
General rules	EN 1993-1-1	ÖNORM B 1993-1-1	SA EN 1993-1-1; SIA 263.001	DIN EN 1993-1-1/NA
Connections	EN 1993-1-8	ÖNORM B 1993-1-8	SA EN 1993-1-8; SIA 263.008	DIN EN 1993-1-8/NA
Steel tendons	EN 1993-1-11	ÖNORM B 1993-1-11	SA EN 1993-1-11; SIA 263.011	DIN EN 1993-1-11/NA
Wooden supporting structures				
General rules	EN 1995-1-1	ÖNORM B 1995-1-1	SA EN 1995-1-1; SIA 265.001	DIN EN 1995-1-1/NA
Geotechnics				
General rules	EN 1997-1	ÖNORM B 1997-1-1	SA EN 1997-1; SIA 267.001	DIN EN 1997-1/NA
Spread foundations		ÖNORM B 1997-1-2	—	DIN 4017
Pile foundations		ÖNORM B 1997-1-3	—	DIN 4014
Earth pressure assessment and supporting structures		ÖNORM B 1997-1-4	—	DIN 4085
Terrain fault		ÖNORM B 1997-1-5	—	DIN 4084
Measurements and structural observations		ÖNORM B 1997-1-6	—	—

(continued)

Table 6.1 (*Continued*)

Topic area	European Basic Document	National appendices and other national transposition standards		
		Austria	Switzerland	Germany
Foundation soil examination	EN 1997-2	ÖNORM B 1997-2	SA EN 1997-2; SIA 267 002	DIN 4020

6.2 Design of avalanche defense structures in the starting zone

6.2.1 General rules for designing avalanche defense structures

The *snow supporting structures* have the task of preventing avalanche release, or limiting snow movements to an extent where they are harmless. Completely developed avalanches develop a pressure that usually cannot be countered by the snow supporting structures [163]. For this reason, the assessment of the snow supporting structures must be carried out within the framework of the total protection system and the individual snow supporting structures. Since the snow supporting structures must be designed in such a manner that the snow supporting structures can penetrate the snowpack, even at extreme snow height (H_{ext}), it becomes essential to define the necessary structure height. The structures are not designed to withstand a load case of a snow height exceeding the structure height; this must be done by choosing a sufficient structure height. However, this is often difficult, as the snow height in terms of relief and exposure can vary greatly over a small area (Figure 6.1). The extent and scope of the required defense structures depend on the location of the objects to be protected (Figure 6.3).

Fig. 6.1 Snow-covered snow supporting structures in the Großtal Avalanche (Paznauntal, Tyrol) (© Sauermoser)

6.2.2 Design snow height

6.2.2.1 Methodology

The calculation of the *design snow height* H_{ext} (extreme snow height at a structures site) can be calculated using the following equation [163]:

$$H_{ext} = H_{\max} \cdot \frac{\overline{H_{ext}}}{\overline{H_{\max}}} \quad [m] \tag{6.1}$$

where:

$\overline{H_{ext}}$ Area average of the extreme snow height H_{ext} over a larger terrain section at the time of extreme snow coverage (on average a maximum of once during the assessment period) according to Figure 6.2a,b (or other reliable sources), in [m]

H_{max} Maximum snow height during a winter at a location of a snow supporting structure, measured using gauges or probes, in [m]

$\overline{H_{\max}}$ Area average of the maximum snow height H_{max} over a larger representative terrain section at the time of H_{max}

Note that the snow height distribution within a construction area changes significantly over a small area, dependent on topography and exposure, but that the distribution pattern remains the same over years. Before planning snow supporting structures, it is therefore recommended that snow stakes be installed in the future construction area. Over several years, these will be able to observe representative points, or to map snow-free patterns, which could give information regarding the snow distribution in the construction area. According to [163] it is recommended to carry out 25–100 gauge measurements or probes/ha in order to calculate the maximum snow height $H_{\max}$. After constructing the snow supporting structures, it is also necessary to observe the snow distribution for several years, as the snow supporting structures change the wind field immediately over the slope, and thereby the distribution of the snow height. Therefore it only afterwards becomes clear whether the choice of structures height was indeed correct.

6.2.2.2 Extreme snow height in Switzerland and Austria

In Switzerland, and partially in Austria, the area averages for extreme snow height ($\overline{H_{ext}}$) can be used to determine the structure height according to Figure 6.2a.

Snow height measurements from the WSL Institute for Snow and Avalanche Research comparative observation stations and from the automatic ENET Stations (SLF/MeteoSchweiz) were used as a basis for the depictions in the Swiss Guideline [163], with measurement series between 10 and 66 years. The map has been calculated using a unitary return period of 100 years, and is valid for the Swiss Alps. The altitude dependencies valid in the four zones can be found in Table 6.2.

Using [163], the Central Bureau for Meteorology and Geodynamics (ZAMG) [149] carried out a zoning of the area averages for extreme snow depths for Tyrol and Vorarlberg. A return period of 150 years is defined for the area average for extreme snow heights. It was possible to use data from a total of 119 measuring stations from

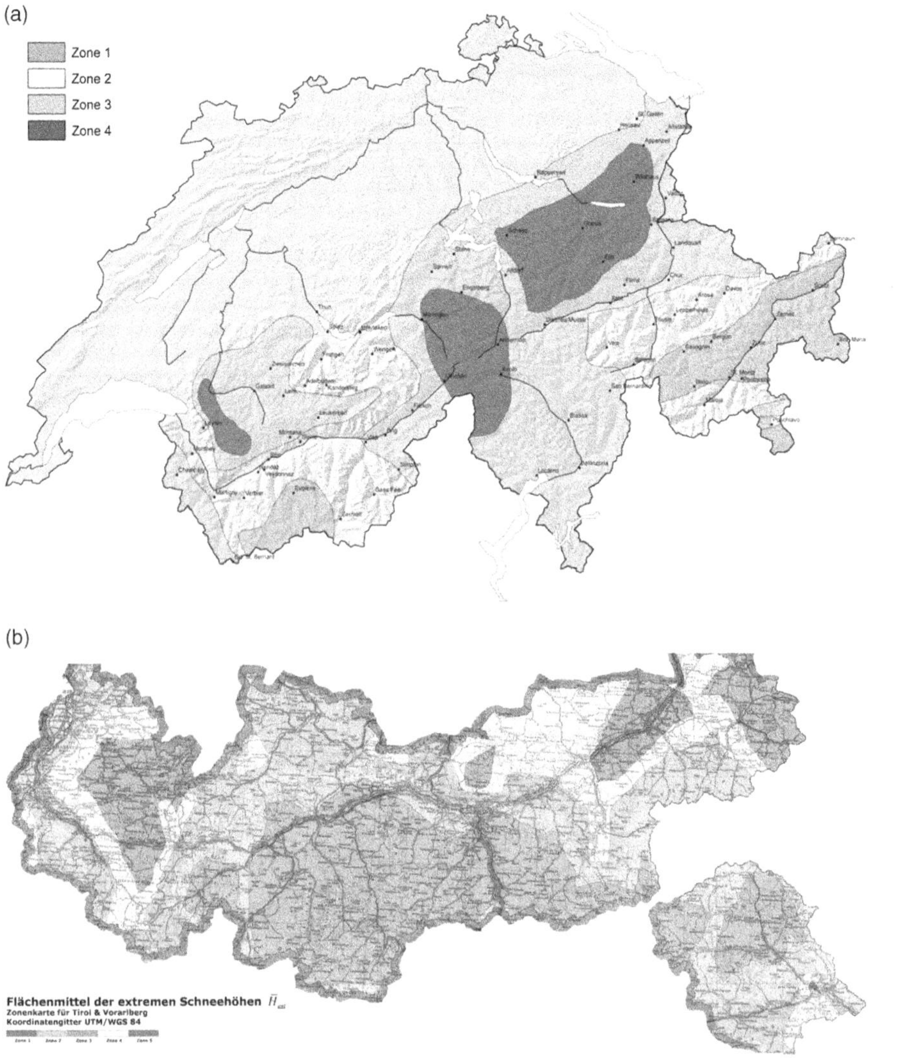

Fig. 6.2 Area average of the extreme snow heights Hext (a): in Switzerland (according to [163]) and (b): for Tyrol and Vorarlberg (according to [149])

the EVA-database (ZAMG) as a basis. The data comes from series spanning many years from the Hydrographical Service (HD), from stations in the Tyrol avalanche warning service area, and from ZAMG. For usage in Tyrol and Vorarlberg, it was necessary to expand the four-step zone scale of the Swiss Guidelines to a five-step scale. The result of the calculation is shown in the zone map for Tyrol and Vorarlberg (Figure 6.2).

Table 6.2 Altitude dependencies from the 100-year events for the total depth of snow cover in Switzerland (Figure 6.2a), or the 150-year events for the total height of snow cover for Tyrol and Vorarlberg (Figure 6.2b) (Z is the height asl in [m]; the colours are taken from the original maps for Switzerland [163] and Austria [149])

Zone	Altitude factor	Formula	Colour according to Figure 8.4		Altitude gradient
			Switzerland	Austria	
1	1	$\overline{H}_{ext} = 1.00 \cdot (0.15 \cdot Z - 20)$	Blue	Blue	0.15 cm/m
2	1.3	$\overline{H}_{ext} = 1.30 \cdot (0.15 \cdot Z - 20)$	White	Green	0.20 cm/m
3	1.65	$\overline{H}_{ext} = 1.65 \cdot (0.15 \cdot Z - 20)$	Pink	Orange	0.25 cm/m
4	2	$\overline{H}_{ext} = 2.00 \cdot (0.15 \cdot Z - 20)$	Red	Yellow	0.30 cm/m
5	2.3	$\overline{H}_{ext} = 2.30 \cdot (0.15 \cdot Z - 20)$		Red	0.35 cm/m

6.2.2.3 Extreme snow height in USA and Canada

Extreme snow height is highly variable with climate region throughout mountainous areas of North America. Extremes probably vary over a factor of 3 or more, possibly ranging from 3 m in continental climates which are 1000 km or more inland to 10 m or more in the deepest snow accumulation zones adjacent to Pacific Ocean moisture sources. Snow height maps are available locally or regionally but these are highly generalized and should not be depended on for the local design of avalanche protection. Engineers must rely on local experience in design.

6.2.3 Position of protected objects

The position of protected objects determines the scope, and in particular the vertical *extent of snow supporting structures* (Figure 6.3). If it is far from the toe of a slope and placed at the end of the potential runout path (A), with a slight reduction of the avalanche volume, the runout distance can be shortened so much that the buildings are no longer in danger. A corresponding slight elongation of the snow supporting structures is required downhill. Only a part of a potential release area needs to be secured. If the objects to be protected are near of the toe of a slope (B), and if a significant reduction of the runout distance is required to secure the buildings, the snow supporting structure must be delimited downward in such a manner that avalanches from remaining starting zones cannot reach the buildings. In a worst-case scenario, the objects to be protected are within the avalanche path (C), which means that the snow supporting structure must reach to just above the object to be protected in order to prevent small avalanche formations underneath the snow supporting structure as well. In this case, bear in mind that also small avalanches can form within the snow supporting structure, which is why the protection of properties within the avalanche path is very limited with a snow supporting structure (in terms of residual risk). Corresponding measures must be taken in particular if a formation area leads into the avalanche path without a gradient change

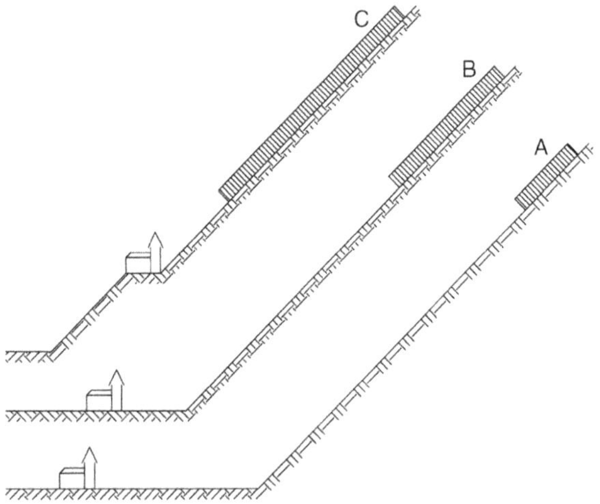

Fig. 6.3 Enlargement of a defense area dependent on the position of the property to be protected: A – property is positioned away from the foot of a slope, B – property is positioned directly at the foot of a slope; C – property lies directly in the avalanche path (© Sauermoser)

or narrowing, and the delineation of the potential starting zone cannot be determined from a morphological point of view.

6.2.4 Static systems for avalanche defense structures

The static systems for snow supporting structures can generally be divided into *massive systems* (earth terraces, retaining walls: Figure 6.4a, b), and *compartmentalized systems* (Figure 6.4c to f). The compartmentalized systems are composed of rigid beams, girders, pressure and tension struts, as well as flexible tension struts (mainly cable). The loads are transferred into the ground at point foundations.

The most frequently used system when talking about compartmentalized snow supporting structures is the one depicted in Figure 6.4c. In this system, the supports hinged at both ends and the girder above ground form a triangle. The girder foundation must in this case transfer the directionally variably tension and pressure forces into the ground, as appropriate for the construction types in question. The support transfers an axial directionally-constant pressure force into the ground. For the system in Figure 6.4d, the girder, the support and the pressure bar a rigid triangle stiffens this layer through a truss. For this system, only a tension force is formed in the anchor axis at the upslope girder foundation. The loads on the support foundation are thereby higher and not constant in the load direction, and are therefore absorbed by a micropile.

The system in Figure 6.4e consists of a girder foundation and two support foundations. Because of the swivel rod between the girder and the foundation, the girder foundation is predominantly subjected to traction. A snow net consists of a swivel support which is secured upslope by a flexible wire rope net and downslope with a guy. (Figure 6.4f) The swivel support transfers pressure force to the ground. The anchor of the guy and wire

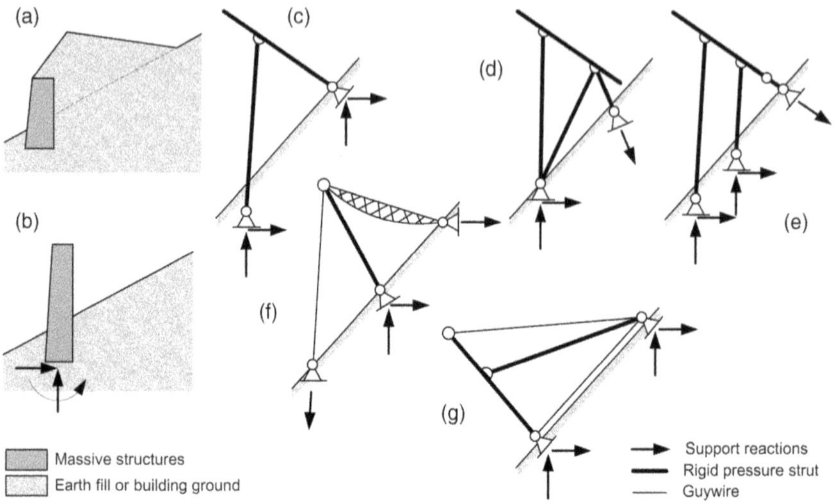

Fig. 6.4 Static systems for avalanche defense structures. Massive systems: (a) earth/masonry terraces, (b) supporting walls; compartmentalized systems: (c) to (e) rigid systems, (f) snow net, (g) Glarner hanging truss (© Suda, Margreth)

rope net is only subjected to traction. For the system in Figure 6.4g, the supporting plane is anchored via traction struts (poles) on the upslope support. The rigid pressure strut between the girder and upslope foundation serves the purpose of absorbing downslope wind forces.

6.2.5 Actions on snow supporting structures

6.2.5.1 Overview and classification

The predominant action on snow supporting structures is the *static snow pressure*, which appears because of a local braking of the creep and glide movement in the snowpack on a steep slope (Figure 6.5) (Section 2.3.4.2). Through a suitable placement of the snow supporting structures (Section 6.2.6), it is possible to ensure that the dynamic loads from avalanche actions do not become greater than the static snow pressure. Because of end-effects, the snow pressure at the end of a row of snow supporting structure or a single series is increased. Because of end-effects, there is a strong increase of snow pressure forces on slim objects or elements such as steel cable. When the snow pressure (e.g. because of irregular topography) does not act normal on the supporting plane, additional lateral force ensues. All effects of the snow pressure are simultaneous and maximum.

Since the dead load of a snow supporting structure is much smaller compared to the snow pressure, it is generally neglected in the design. Because of wind, a snow supporting structure can also be affected by lifting forces. Depending on the location, there can be further factors as a result of earthquakes, avalanches, falling rocks, mass movements, or solifluction. The calculation formulas below, used to determine the

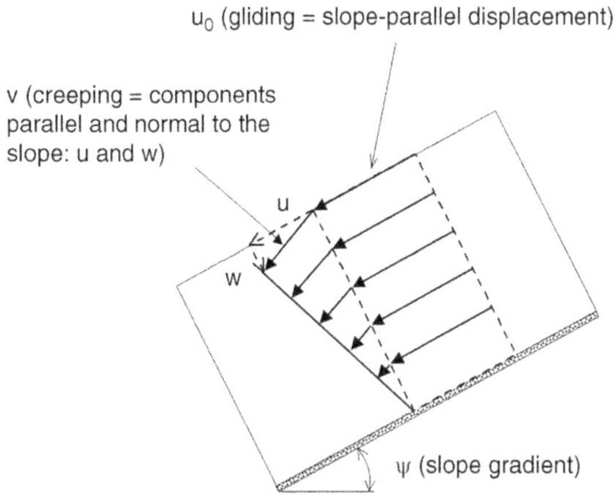

Fig. 6.5 Creeping and gliding in the snowpack: During creeping, the translation movement on the soil is zero and maximum on the snow surface. During gliding, the entire snowpack is displaced uniformly (according to [163])

actions, are based on the Swiss Guideline for Defense structures in avalanche starting zones [163].

6.2.5.2 Snow pressure

The theoretical basis for the snow pressure assessment was largely derived for the assessment of the actions on snow supporting structures. Haefeli [13] recognized that a so-called back-pressure zone forms behind the snow supporting structures, a zone with a limited length, within which additional compressive stress and reduced shear stress forms. The back-pressure length forms the basis for the one-dimensional theory of the snow pressure on a support wall. The snow pressure theory was, in particular, developed further by Salm [225] to such an extent that it could be put into practice.

The WSL Institute for Snow and Avalanche Research (SLF) measures snow pressure forces on snow supporting structures for different structures in Switzerland (Figure 6.6a). During the winter of 1998/99 when there was very heavy snow, maximum snow pressure forces of 46 kN/m were measured, which corresponds to approx. 87% of the assessment snow pressure in the Swiss Guideline [163]. The measurements carried out and experience with avalanche defense systems show that the action on snow supporting structures can be depicted quite well using the following calculation formulas.

McClung [168] developed a continuum-mechanical snow pressure model, which more precisely depicts the snow pressure process, compared to the model developed by Haefeli [13] and Bucher [28]. The calculation formulas from the Swiss Guideline [163] give slightly more conservative values for the snow pressure compared to McClung, whose model has not been applied practically. Snow pressure measurements in Norway and Iceland show that the snowpack there – compared with the Alps – is generally warmer and denser by typically 500 kg/m^3, but that the snow

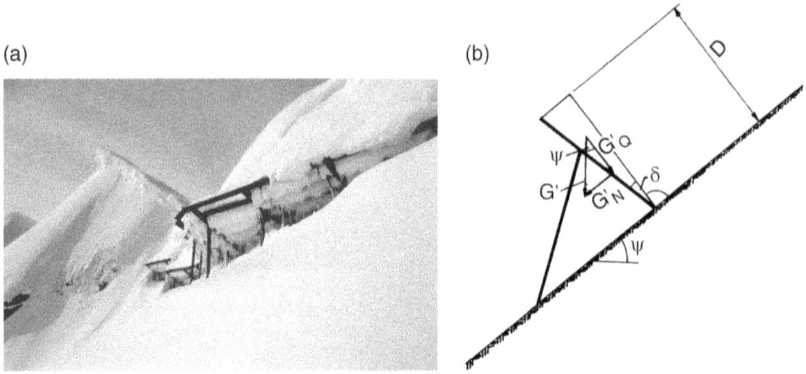

Fig. 6.6 (a): Snow pressure load on a snow supporting structure (© Forest Office of the canton Grisons); (b): Additional factor to snow pressure for non-vertical support surface (according to [163])

creeping is smaller [126]. In order to determine the snow pressure actions in Iceland, it is recommended to use the calculation formulas from the Swiss Guideline [163], and to take the greater snow density into account with an increase of the glide factor by a factor of 2 [124].

By taking a triangularly shaped distributed creep profile and gliding into account, the *slope-parallel components of the snow pressure* S'_N is calculated as follows on a rigid, endlessly long support surface, vertical to the slope and at level:

$$S'_N = \rho \cdot g \cdot \frac{H^2}{2} \cdot K \cdot N \quad [\text{kN/m}] \tag{6.2}$$

where:

S'_N Slope-parallel snow pressure component per length unit of support surface (in the contour line) [kN/m].
S'_N can be calculated in a simplified manner as uniformly distributed across the height.

ρ Average snowpack density, dependent on elevation and slope exposure [t/m^3]. The average snow density ρ can be set to 0.27 t/m^3 for a snowpack structure that appears in the Alps at an extreme snow height at an elevation of 1500 m. On Weissfluhjoch (Davos, Switzerland) at 2650 m, at the time of the snow depth max. of 366 cm, a snow density of 0.29 t/m^3 was measured. The change in this start value of 0.27 t/m^3 with elevation and slope exposure can be corrected with the altitude factor f_c or glide factor N. For load model 2 (see Chapter 6.2.5.4), where the structures is only partially snowed in, an increased snow density of $\rho = 0.40$ t/m^3 is used.

g Gravitational acceleration, constant [= 9.81 m/s^2]

H Vertically measured snow height [m]

K Creep factor, which is dependent on the slope inclination ψ and density ρ.

It can to an extent be calculated as follows:

$$K = (2.50\rho^3 - 1.86\rho^2 + 1.06\rho + 0.54) \cdot \sin 2\psi \tag{6.3}$$

N Glide factor, which indicates the increase of the snow pressure given a glide movement of the snowpack on the soil.
It can be determined as a function of the soil class, which describes the roughness and the slope exposure pursuant to Table 6.3. If the terrain is steeper than 45°, the criteria have to be applied strictly when defining N. For slope inclinations of less than 35° a somewhat milder definition can be made for N. Severe snow gliding can be reduced through snow glide measures (Section 5.2.5).

Assuming an average snow density of 0.27 t/m^3 and a slope inclination of 45°, the calculation is simplified for the slope-parallel snow pressure S'_N as follows:

$$S'_N = H_K^2 \cdot N \cdot f_c \quad [\text{kN/m}] \tag{6.4}$$

where:

H_K Vertically measured structure height [m]
f_c Altitude factor, which takes into account an increase of the average snow density with sea level altitude and the associated increase of the creep factor.
The increase of the altitude factor is assumed to be between 1500 m and 3000 m to 2% per 100 m and can be calculated as follows:

$$f_c = 1 + 0.02 \cdot \left(\frac{Z}{100} - 15\right) \tag{6.5}$$

The *snow pressure component normal to the slope* S'_Q on a snow supporting structure appears when the settlement movement of the snowpack is prevented because of the roughness of the support surface. It is calculated on a rigid, endlessly long support surface, vertical to the slope and at level line, as follows, where S'_Q simplified can be assumed to be distributed evenly across the height:

$$S'_Q = S'_N \cdot \frac{a}{N \cdot \tan\psi} \quad [\text{kN/m}] \tag{6.6}$$

where:

a Coefficient, dependent on the transverse number m of the snow type with uniaxial pressure.
For loose, new snow, which is greatly compressible, $a = 0.5$; for dense old snow or ice, $a = 0$. According *to the Swiss Guideline* [163], in order to calculate snow pressure component normal to the slope, a varies between 0.35 and 0.5, and most unfavourable case should always be taken into account.
Ψ Slope gradient [°]

For *non-normal support surfaces*, an additional factor must be taken into account. The snow pressure theory was developed for a rigid, slope-normal wall. The downslope

Table 6.3 Soil classes and glide factors (according to Swiss Guideline [163])

Soil Classes	Glide Factor N	
	Exposure	Exposure
	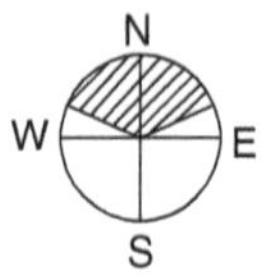	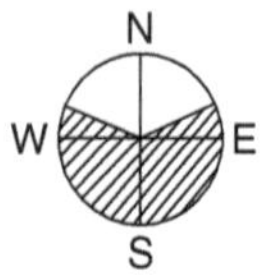
	WNW-N-ENE	ENE-S-WNW
Class 1		
Rough boulder debris (d* ≥ 30 cm).		
Terrain with strong presence of small and large boulders.	1.2	1.3
Class 2		
Areas covered with larger alder bushes or dwarf pine at least 1 metre in height. Humps overgrown with turf and low bushes (height of humps more than 50 cm). Strongly developed cow paths. Large boulders (d* approx. 10–30 cm).	1.6	1.8
Class 3		
Short-stalked turf, interspersed with small bushes (heather, Alpine roses, bilberries, alder bushes, dwarf pine less than 1 m in height). Fine boulders (d* ≤ 10 cm) interspersed with turf and small bushes. Weakly developed humps overgrown with turf and small bushes up to 50 cm height, also interspersed with smooth turf and small bushes. Turf with little developed cow paths.	2.0	2.4
Class 4		
Smooth, long-stalked closed turf. Smooth rock plates with layering parallel to the slope. Smooth slope debris mixed with earth. Swampy depressions.	2.6	3.2

d* is the average block size, which is essential for the roughness of the soil surface.

inclination δ of the support surface on rigid snow supporting structures is normally 10° to 15° and for flexible snow nets about 30°. That is why, in addition to the components S'_N and S'_Q, the weight G' of the snow prism is taken into account. This runs through the support surface and a slope-vertical level, which runs through the intersection line of the soil surface and support surface (Figure 6.6b). If the support surface runs upslope of this level (e.g. vertical snowdrift fence), no additional factor is required. Assuming an increased snow density of 0.3 t/m³, the weight of the snow prism G' is calculated as follows:

$$G' = 1.5 \cdot D_K^2 \cdot \tan\delta \quad [\text{kN/m}] \tag{6.7}$$

where:

D_K Effective grate thickness [m], where $D_K = H_K \cdot \cos\psi$
δ Angle between support surface and the normal to the slope in [°]

6.2.5.3 End-effect loads

If the elongation of the support surface in the level line is limited (e.g. at the end structures or in the gaps in between the structures), additional *end-effect loads* occur (Figure 6.7). The snowpack can flow sideways around a limited surface. Thereby, the back-pressure action extends sideways. This is particularly the case when there is severe snow glide. At the end of snow supporting structures, slope-parallel snow pressure forces may appear which are about two to five times greater than for endlessly long support surfaces (Figure 6.8a). The snow pressure increases continuously in the direction of the edge and reaches a maximum there. For the practical calculation, assume a constant end-effect load S'_R with a length of applied load Δl, which is equivalent to the edge forces. The end-effect loads S'_R are applied as additional slope-

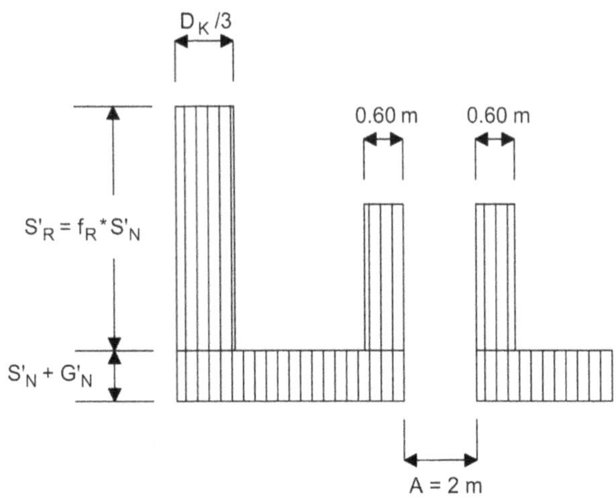

Fig. 6.7 End-effect load at the free end of the structure (A > 2 m) and with gaps in the structures of 2 m (according to [163])

(a)

(b)

Fig. 6.8 (a): End structures destroyed by end-effect loads (© Forest Office of canton Bern); (b): fortified end structures with double girder (© Margreth)

parallel load per metre with the length of applied load Δl (no additional slope-normal end-effect loads are assumed):

$$S'_R = f_R \cdot S'_N \quad [\text{kN/m}] \tag{6.8}$$

with end-effect factor f_R:

$$f_R = (0.92 + 0.65 \cdot \text{N}) \cdot \frac{\text{A}}{2} \leq (1.00 + 1.25 \cdot \text{N}) \tag{6.9}$$

where:

A Distance between structures in [m].
N Glide factor according to Table 6.3

The upper limit value for the right term in the Equation 6.9 is applicable for the separated structure (A > 2 m) and may not be surpassed (Figure 6.7). The length of applied load Δl of end-effect load S'_R is calculated dependent on the lateral distance between structures A or the effective supporting surface or net height D_K in accordance with Equation 6.10.

$$\Delta l = 0.3 \cdot A \leq \frac{D_K}{3} \tag{6.10}$$

6.2.5.4 Resulting snow pressure and load arrangement

The size of the resulting snow pressure R' is calculated from the vector addition of the slope-parallel and slope-vertical components according to the Equation 6.2 (as well as Equations 6.5–6.8. For an endlessly long support surface, the following is applicable (Figure 6.9a).

$$R' = \sqrt{R'^2_N + R'^2_Q} = \sqrt{(S'_N + G'_N)^2 + (S'_Q + G'_Q)^2} \quad [\text{kN/m}] \tag{6.11}$$

Within the length of applied load of the end-effect load, the end-effect load S'_R must be added to the slope-parallel components. The slope-parallel snow pressure component S'_N or S'_R contribute the most to the resulting snow pressure R'. The direction of the resultant

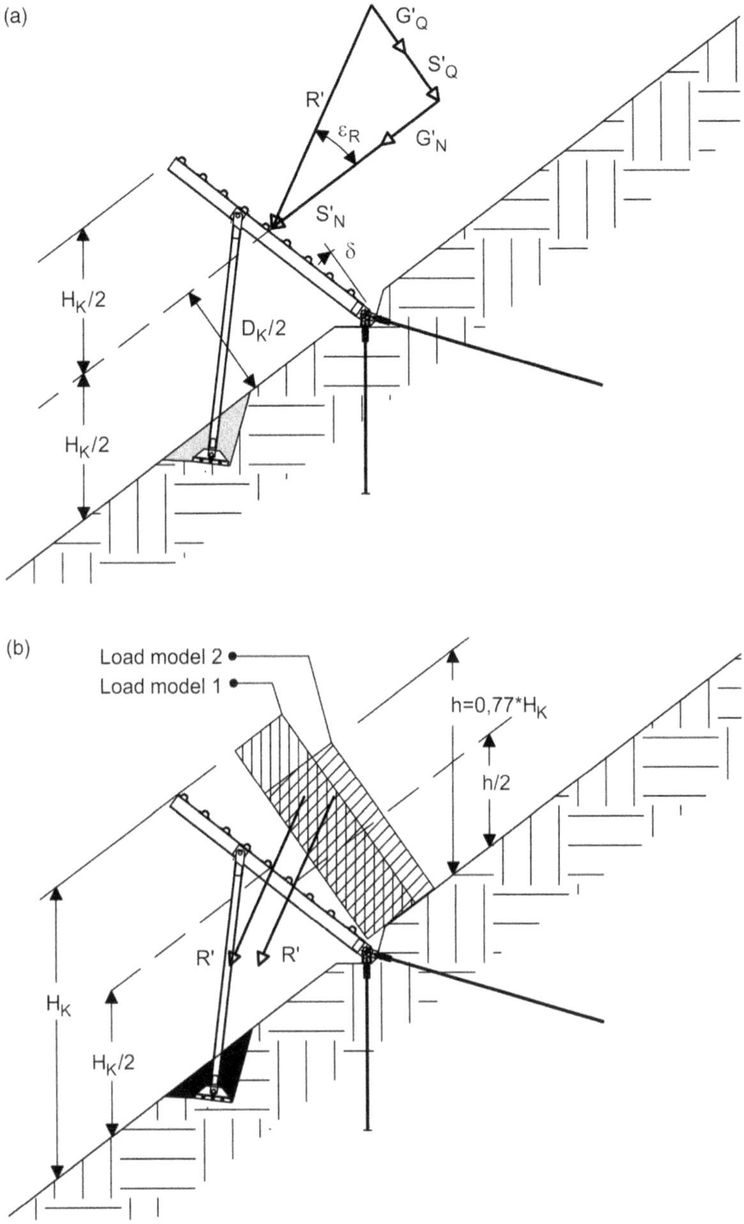

Fig. 6.9 (a): Resultant of the snow pressure; (b): attack height of the resulting and specific snow pressure distribution for load cases 1 and 2 (according to [159] (Herbert)

in the plane normal to the contour line is calculated using:

$$\tan \varepsilon_R = \frac{R'_Q}{R'_N} \qquad (6.12)$$

where:

ε_R Angle between the resultant and the line of slope (Figure 6.9b)

Note that in the area of the end-effect, load ε_R is smaller. When dimensioning the snow supporting structure, the direction of the resultant must be determined proportionally from the supporting surfaces in question. When dimensioning a snow supporting structure, differentiate between the following two load cases in the Swiss Guideline [163] (Figure 6.9b):

For *load case 1*, it is assumed that the snow supporting structure is fully snowed in. To simplify matters, it is assumed that the snow pressure is distributed evenly. The point of action of the resulting R' consequently lies at half height of structure in the load case.

Load case 2 takes into account the partial filling of the snow supporting structure by a settled snowpack with a snow height of 77% of the structure height H_k. In load case 2, the resulting R' acts at a height of 0.385 H_k and the specific snow pressure (Section 6.2.5.5) is increased by a factor of 1.3. The size and direction of the resultant R' does not change relative to load model 1. End-effect factors f_R and the length of applied load Δl (Equations 6.9 and 6.10) are in both load cases assumed to be the same size.

6.2.5.5 Snow pressure on grate

The specific snow pressure p_h normal to the grate is calculated using the resulting R'. Since the pressure distribution over the grate often is naturally irregular, the specific snow pressure from load case 2 is used for both load cases. The specific snow pressure p_h is calculated as follows (Figure 6.10):

$$p_h = \frac{R' \cdot \cos(\delta - \varepsilon_R)}{0.77 \cdot B_K} \quad [\text{kN/m}^2] \qquad (6.13)$$

where:

R' Resulting snow pressure [kN/m]
δ Grate inclination [°]
B_K Grate or net height: $B_K = D_K / \cos \delta$
ε_R Angle between R′ and the line of slope, to be calculated with a = 0.35 acc. to Equation 6.12

Since the specific snow pressure p_h can be greater close to the soil because of snow gliding, the specific snow pressure p_h is increased by 25% from the soil surface to 1/4 of the grate height and across the entire length of the structure (Figure 6.11). For grates where end-effect forces can occur, the following two load models must be taken into account:

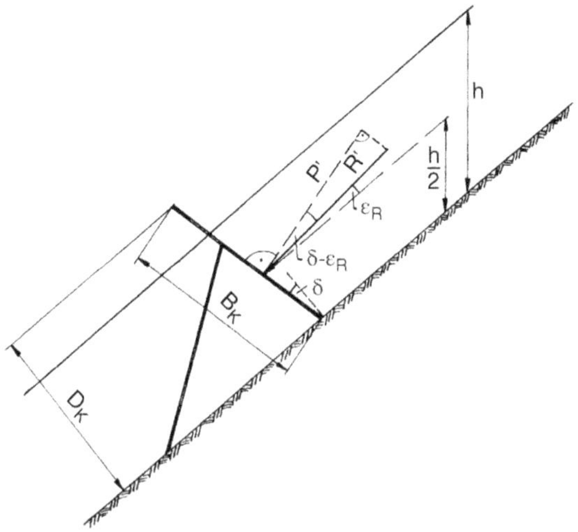

Fig. 6.10 Load P' normal to the grate (according to [163])

– Load model with end-effect forces according to Equation 6.8
– Load model without end-effect forces.

Because of the deformation of the snowpack, transverse forces also affect the grate. For snow bridges with horizontal crossbeams, the specific transverse load q_h is calculated as follows:

$$q_h = \frac{R' \cdot \sin(\varepsilon_R - \delta)}{0.77 \cdot B_K} \geq 0.2 \cdot p_h \quad \left[\mathrm{kN/m^2}\right] \tag{6.14}$$

Fig. 6.11 Distribution of the specific snow pressure p_h on the grate for load model 1 (according to [163])

The transverse load q_h is a minimum of 20% of p_h. This minimum value is almost always applicable for larger glide factors and slope inclinations. The transverse load acts at the outermost upslope edge of the crossbeam.

6.2.5.6 Snow load on slim components (structures)

Greatly increased end-effect forces act on slim, free-standing parts of objects such as supports. The width of action on slim objects can be much larger than their actual width, because the snowpack can flow around the component from both sides. Haefeli [89] introduced the influence factor η (Equation 6.15), which compares the relation between the effective snow pressure, including end-effect forces, with the snow pressure, which acts on an endlessly long supporting structure.

The influence factor η can be calculated as follows for free-standing objects like, for example, chairlift pylon that is subjected to the complete snow pressure [163]:

$$\eta = 1 + c \cdot \frac{H}{W} \tag{6.15}$$

where:

c Varies between 0.6 for sites with little snow gliding and 6 for sites with extensive snow gliding and great snow heights.
H Snow height in [m]
W Width of the component in [m]

If the snowpack hangs underneath a snow supporting structure, the support can be subjected to a transverse load by snow pressure forces directed downhill. This has already in some cases led to a sideways buckling of the supports (Section 7.2.3.4).

In order to better quantify these actions, the WLS Institute for Snow and Avalanche Research SLF (Switzerland) instrumented two supports in an area controlled with snow supporting structures where such support damage had occurred. During the winter 1998/99, a maximum strain with a bending moment of 10 kNm or 20 kNm was measured at a normal force of 34 kN or 36 kN. The corresponding transverse load on the supports was 7 kN/m or 13.5 kN/m [246]. Measurements on snow net supports in Iceland showed a max. bending moment of 19 kNm [126]. The transverse load q'_s can be calculated as an evenly distributed line load q'_s, which acts on the support normally:

$$q'_s = \eta_R \cdot S'_N \cdot \frac{\textit{Support diameter}}{\textit{Support length}} \cdot \sin\alpha \quad [\mathrm{kN/m}] \tag{6.16}$$

where:

η_R Reduced influence factor of the supports
which normally can be assumed to be 1.0, since the snow pressure is reduced by the support grate. At extreme creeping snow sites, however, factors of influence of up to 5 must be anticipated.
– Support diameter and length in [m]
α Angle between the support axis and the ground surface in [°].

Greatly increased factors of action act on thin steel cables or rods, which are subjected to the complete snow pressure (e.g. lateral guywires of snow supporting structures or retaining cable of rock fall protection barriers). There are only rough estimates for the determination of the influence factor since measurement data has not yet been collected. For a snow thickness of 200 cm and a cable diameter of 1 cm, one must anticipate an influence factor action η of approx. 50.

6.2.5.7 Lateral loads

Because of the irregularities in the terrain and in the snow height distribution, the resultant of all forces acting on the support surface in the ground plan is no longer perpendicular to the support surface. For this reason, a lateral force S_s parallel to the level line should be assumed. Note that you should expect an increased lateral force in the area of influence of the end-effect forces. In order to obtain sufficient lateral stability in the snow supporting structure, one must take a lateral force S_s acting on both ends which runs parallel to the level line into account. If the snow pressure result does not act normally on the support surface in the unprotected end of structure in the ground plan (e.g. in hollows), the lateral force S_s must be multiplied by the end-effect factor f_R. Per single structure for length l this is:

$$S_S = 0.1 \cdot S'_N \cdot l \quad [\text{kN}] \tag{6.17}$$

The attack point of lateral force S_s may be assumed at half structure height (equally distributed load across the height). The transfer of lateral force S_s from the grate to the support structure and the foundation must be ensured. This can in particular be critical for free-standing single structures or short rows of structures < 10 m). For longer rows of structures, lateral forces can be ignored.

6.2.5.8 Dead weight

The dead weight from snow supporting structures must only be taken into account if it accounts for a significant load (e.g. snow supporting structures made of concrete) in relation to the actions from snow pressure.

6.2.5.9 Wind load

Snow supporting structures may be stressed by upslope wind forces during snow-free periods. Sufficient transfer of these lifting powers from the bearing construction to the foundation and the substratum must be ensured. The wind forces can be determined according to EN 1991-1-4 (corresponding NAD) or the standard SIA 261 Chapter 6 'Wind'.

6.2.5.10 Other actions

The following different actions (loads) may have an effect on snow supporting structures and must be taken into account in individual cases:

- *Earthquake forces*: Usually earthquake forces are not taken into account when assessing snow supporting structures in the Alps.
- *Avalanche impact pressure*: Snow supporting structures must be arranged in such a manner that they cannot be actioned by avalanches that occur further upslope

(Section 6.2.6). If necessary, control measures such as deflecting dams must be implemented. Since the avalanche force from a large avalanche is many times larger than the snow pressure forces, assessing the avalanche force does not make sense and is not financially viable.

- *Rock fall*: When planning, it must be determined whether there is a starting zone for falling rocks or boulders in or above snow supporting structures. Using rock fall simulations, it is possible to estimate action energies. Rigid snow supporting structures are destroyed if the action energies are larger than approx. 50 kJ, and flexible snow nets are destroyed at action energies of approx. 200 kJ. If necessary, the snow supporting structures must be protected using rock fall barriers.
- *Landslides and solifluction*: In such situations, it is recommended to carry out a detailed preliminary study. There are two possibilities to counter such actions: On the one hand, one can try to stop the movement using anchored snow supporting structures; on the other hand snow supporting structures can be shaped in such a manner that they become somewhat resistant to soil displacement and erosion (Section 6.2.10).

6.2.6 Layout and configuration of snow supporting structures in the starting area

6.2.6.1 General rules for layout

The correct layout of the snow supporting structures in the terrain is essential for ensuring the protective effect of the technical construction and the safety of the snow supporting structure. The content of this section generally follows the Swiss Guideline [163] in shortened form and expanded for some aspects.

6.2.6.2 Slope Inclination suitable for snow supporting structures

Slopes fit for snow supporting structures have a slope between 30° and 50°. On slope inclinations below 30°, the formation of avalanches is very rare. For slope inclinations above 50°, one must expect a very frequent natural release so that snow packs generally cannot develop with a thickness from which destructive avalanches can form. In practice, it is very rare that avalanches form below 30°, especially when the snow is very dry or when there is a higher level of water content.

6.2.6.3 Vertical extension of defense area

Usually, the highest situated rows of snow supporting structures must be placed at a maximum distance of 2 to 3 times the structure height measured from the upper fracture line. Thereby sufficient support is ensured. In case of a formation of cornices, the distance must be kept even smaller, and one must take care that the structures are not put under excess pressure by the cornices.

If there is a steep bedrock area adjacent to the highest situated row of structures, it must be checked if there is a possible danger of rock fall, and if necessary, the second row of structures must be assessed and constructed accordingly. At such sites, one must ensure that a sufficient structure height is selected, since continuous snow deposits can be expected from the rock faces. Separate rock fall defense systems above the technical constructions may be required to protect the structures. Depending on the circumstances, rock face stabilizing systems may be required.

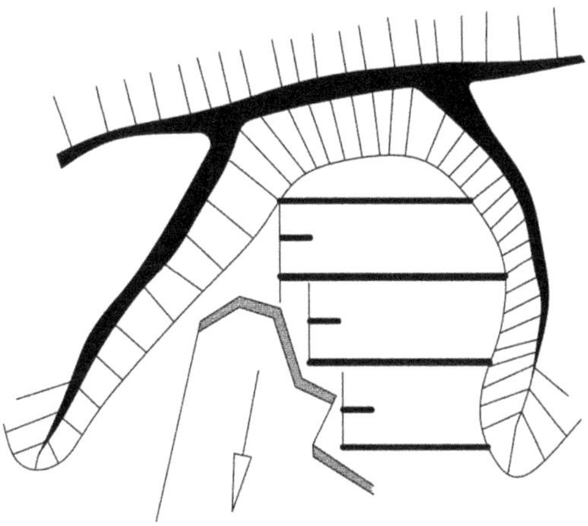

Fig. 6.12 Partially secured terrain unit (© Schilcher)

The lower delineation in snow supporting structures is preferably to be where the slope gradient is less than 30°. However, generally, for financial reasons, it is not possible to construct on the entire starting zone. In such cases, the project leaders must assess how the necessary protection can be attained with which defense area (Section 6.2.3). One way of proceeding is to construct on a large part of the starting zone, and then observe if further additions need to be made. If the terrain is suitable, it is possible to erect a financially viable avalanche catching dam downslope the residual starting zone. Avalanche simulation models (Section 3.2) can be used to evaluate the effect of avalanche formations from the sub-area in the starting zone on which no construction has been made yet.

6.2.6.4 Horizontal extension of defense area

Snow supporting structures are meant to cover an entire terrain unit with lateral limitation through ridges or terrain borders. This means the use of complicated measures to protect the end of structures can be avoided. If a defense structure does not cover the entire terrain unit, the row of structures must be arranged at the open side, tapered back in the downward direction (Figure 6.12). Thereby this avoids extreme loads that can cause damage because of avalanche formations and creeping snow at the edge structures. In case of large snow height and high glide factors, separating walls in the fall line or additional structures in the edge area can prevent damage at the open construction edge (Figure 6.12).

6.2.6.5 Concepts for arrangement of snow supporting structures

The preferred arrangement method in permanent avalanche defense constructions is a continuous structure. It ensures that rows of structures only are interrupted when so caused by terrain conditions. Figure 6.13 shows *continuous*, *separated-interupted*, *separated-staggered* and *combined arrangement methods*. The advantages and disadvantages of the different arrangement methods are listed in Table 6.4.

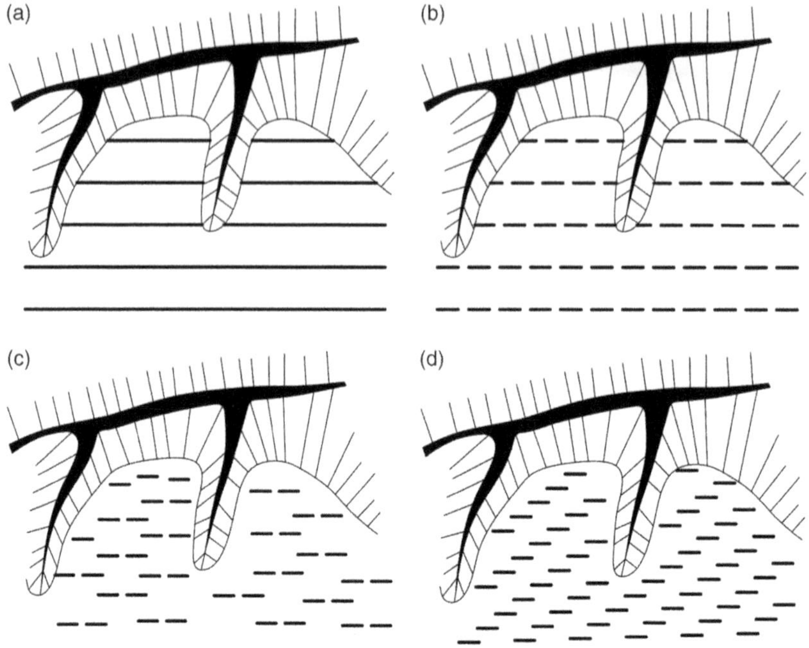

Fig. 6.13 Construction methods for the defense systems: (a) continuous, (b) separated-interrupted (c) separated-staggered and (d) separated-combined method (according to [163])

Table 6.4 Overview of the usage as well as advantages and disadvantages of the construction methods for avalanche defense systems according to [163]

Placement	Advantages	Disadvantages	Usage
Continuous	Propagation of shear fractures in the snowpack upslope and downslope across the rows of structures is increasingly complicated. Continuous obstacle for snowslides. Extensive avoidance of tensile stress in the snowpack. Strain on the structures caused by end-effect	Large and connected width development of existing shear and traction zones in the snowpack. Potentially lateral propagation of destructive effects on structures. Limited adaptation capability in very hilly terrain and frequently changing local snow conditions (depending	Normal Case

(*continued*)

Table 6.4 *(Continued)*

Placement	Advantages	Disadvantages	Usage
	forces, only at the ends of row of structures (minimal total snow pressure stress).	on the used construction type there is a big variation in weight).	
Separated, interrupted	Good adaptation capacity on horizontal terrain configuration and locally changing snow conditions. Localization of damage effects on single sections. From case to case, cost savings (compared to continuous structures).	Partial drift through of snow through structures intermediate spaces possible. Stress on the structures from end-effect forces, which correspond to the lateral gaps between structures. Easier propagation of shear fractures in the snowpack upslope and downslope beyond the structures series (compared to continuous construction).	On occasions in zones with for example bedrock ridges or local terrain steps.
Separated, staggered	Good adjustment capacity to the terrain configuration in all directions. Distribution of remaining traction and shear stress zones. Average decreased snow gliding compared to continuous and dispersed non-continuous arrangement method.	Stress on the structures from end-effect forces, which correspond to the independent structure. Increase of costs per metre (compared to continuous and dispersed non-continuous arrangement method). Propagation of shear fractures possible in all directions.	On exceptional occasions in very steep and very structured terrain, also in case of a compaction of older constructions that do not conform to the standards.

6.2.6.6 Height of snow supporting structures

Structure height H_K is calculated on the basis of the design snow height according to the method defined in Section 6.2.2.2. In this case, note that the snow supporting structures attain the optimal effect when they break through the entire snowpack. For this reason, the structure height (H_K) must be defined using an additional surcharge to H_{ext} (in Austria: 20 cm).

In many cases, the snow height varies within the defense area because of the topography, so that different structures heights may be required for a defense area.

6.2.6.7 Distance between (rows of) structures in the line of slope

The calculation of the distance between (rows of) structures in the line of slope is most often carried out according to the Swiss Guidelines [163]. In Austria, the slope-parallel distance is partially - deviating from this standard - calculated according to fixed height differences (according to ONR 24806 [203]).

The slope-parallel distance between structures L according to the Swiss Guidelines [163] is calculated as follows:

$$L = f_L \cdot \mathrm{H_K} \quad [\mathrm{m}] \tag{6.18}$$

where:

f_L Distance factor
H_K Structure height in [m]

The distance factor f_L depends on the slope gradient, the structures height, the creep factor, and the angle of friction tan ϕ between the soil and the snow. It can be calculated using Figure 6.14 with the help of the slope inclination.

The leeway in the distance measurement must be used to adjust the supporting structures to the requirements for the properties to be protected. Normally it is recommended that f_L lies in the area between tan $\phi = 0.55$ and tan $\phi = 0.5$. In order to attain sufficient safety against avalanche release, the climate must be taken into account when choosing the structures distances. Particularly on northern-facing slopes, or in precipitation-prone pre-alpine areas, one must assume even smaller values than for tan $\phi = 0.5$. If the slope inclination changes within a structure's distance, one must, when calculating L, take the inclination of the connection perpendiculars between the relevant structures foundations points into account.

When calculating the slope-parallel distances between rows of structures according to fixed height differences (method according to ONR 24806 [203]), the calculation of the slope-parallel distance for different structure heights must have a similar level of safety for the release of surface-layer avalanches. The conditions should be met to the extent possible by calculating the distances between rows of structures according to fixed height differences, independently of the structure height and, to the extent possible, independently from the slope inclination.

In Figure 6.15, the diagram with the curves for calculating the structures distances according to fixed height differences is shown. The measured slope inclination can be used to read off the slope-parallel distances between rows of structures on the ordinates based on the height lines (four curves for height differences from 11 m to 16 m).

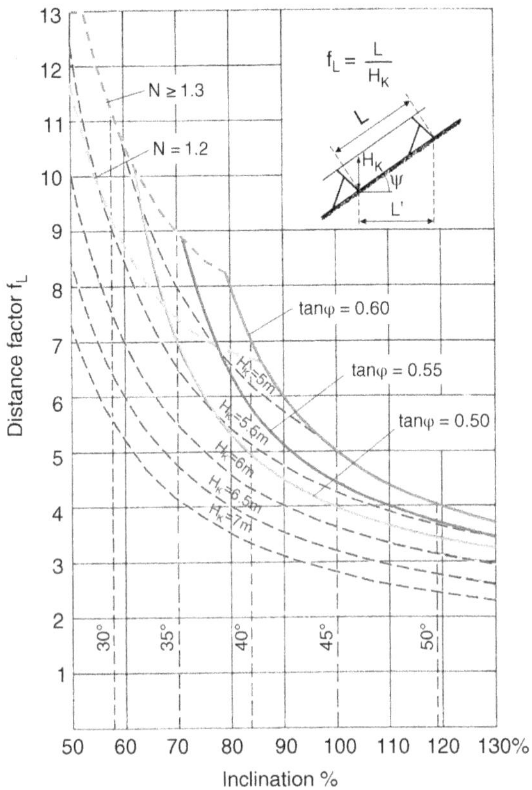

Fig. 6.14 Slope-parallel distances between rows of structures according to the Swiss Guidelines [163].

For smooth terrain (N > 2) or in case of high safety requirements, the values between tan $\phi = 0.55$ and tan $\phi = 0.50$ are applicable. For rough terrain (N < 2), if there are no high safety requirements, the values applicable are tan $\phi = 0.6$. The curve tan $\phi = 0.60$, N > 1.3 and $f_L = 13$ constitute the highest acceptable values for f_L. If the structures are designed at glide factor $N = 1.2$, the chosen distance factor may not lie above the corresponding indicated curve. If the vertically measured structure heights H_K are above 4.5 m, the max. allowable values for f_L are on the corresponding indicated curves

6.2.6.8 Lateral distance between structures

For non-continuous arrangement methods, lateral intermediate distances (Figure 6.16) between neighbouring structures that are not dependent on avalanche-safe terrain sections must be limited to max. 2 m. They must be fully covered from above by structures at distance *L*. If neighbouring structures are displaced into the fall line, they must be positioned as shown in Figure 6.16. For the different arrangement methods, the length of the connecting supporting grates must not be below 16 to 22 m, if the terrain conditions so allow. For practical reasons, the maximum length is set at 50 m.

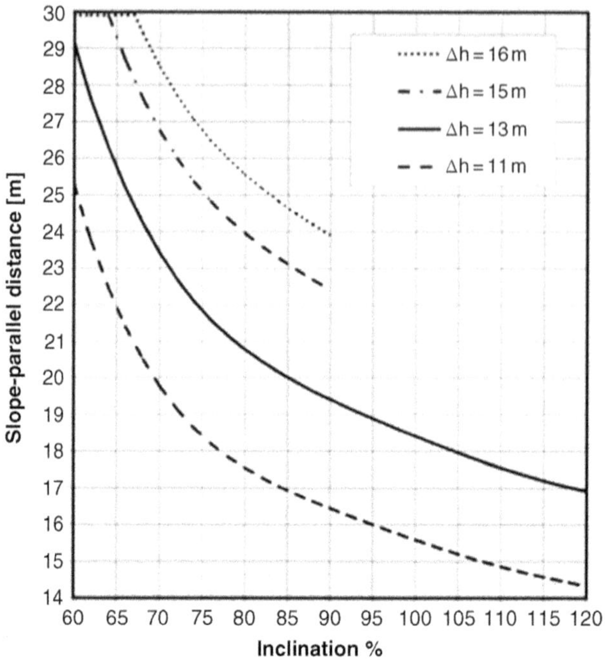

Fig. 6.15 Slope-parallel distance between rows of structures according to fixed height differences pursuant to ONR 24 806 [203]: The slope-parallel distance between rows of structures is read off with the measured slope inclination on the x-axis in Figure 6.15 from the ordinates in the slope-parallel distance between structures. Note: four lines for height differences from 11 m to 16 m are shown. For slope gradients below 75%, the curves were adjusted empirically to the practical experiences, and therefore deviate from the height differences by up to 2 m.

In order to use the diagram, the following conditions must be met:

- The max. distance between rows of structures may not exceed 30 m.
- The lines Δh 15 m and Δh 16 m may only be used at a slope inclination of max. 90% for snow thicknesses $D \geq 3$ m in order to prevent the structures from being overloaded. Because of the low assessment snow pressure, smaller structures can only support a smaller snow layer.
- The lines Δh 15 m and Δh 16 m are central alpine areas, in which the new-snow increments are lower, and there are therefore better conditions for snow supporting structures than in alpine areas that lie on the rim of the Alps.
- The line Δh 13 m is indicative for positions on the rim of the Alps and may be used for all structures $\geq$ 3 m snow thickness. For slope inclinations $\leq$ 90% the line Δh 13 m can also be used for structures $\geq D = 2.7$ m.
- The line Δh 11 m is meant for high safety requirements (e.g. protected objects in the avalanche path) and for smaller structures. Since small structures, because of the low assessment snow pressure only can support a small snow layer an unlimited utilization of the line Δh 11 is only possible for structures $\geq D = 2.5$ m. For slope inclinations $\leq$ 80% Δh 11 m can also be used for structures $\geq D = 2.0$ m. For smaller structures, either glide snow protection (e.g. glide snow tripods, berms) to reduce the layer pressure is required, or the structures must be assessed for a higher snow pressure

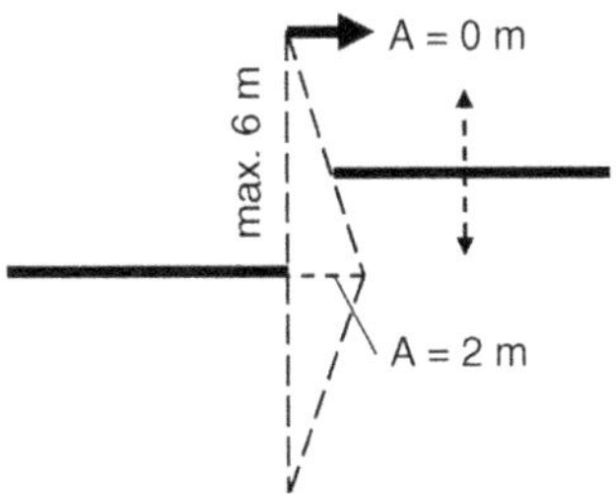

Fig. 6.16 Lateral structures intermediate distances (according to [163])

6.2.6.9 Combination of snow supporting structures with snow glide defense structures

Since defense structures often are constructed in the timberline region of a forest, or in the area of the potential forest limit, afforestation is often envisaged as a sustainable control measure (Figure 6.17) in such cases. In order to improve the conditions for future young stands, it is recommended to equip the areas between the defense systems with glide snow protection measures (Section 5.2.5).

In practice, it has made sense to set up small glide snow tripods in groups, for example, in order to create a foundation for group-based afforestation. When glide snow protection measures are used consistently, they can reduce the glide factor, and thereby create the conditions for weaker snow supporting structure construction types.

Fig. 6.17 Combination of snow supporting structures in steel with glide snow tripods (© Rudolf-Miklau)

6.2.7 Building materials for avalanche defense structures

6.2.7.1 General fundamentals of building materials

The following section gives a short overview of the building materials used in snow supporting structures. Compartmentalized snow supporting structures are for the most part manufactured from construction steel or wood or a combination of the two materials. In the foundation, pre-stressing steel, steel cable, concrete, and anchor grout are also used.

6.2.7.2 Construction steel

Steel is used for compartmentalized snow supporting structures. For supports, girders, and locks, hot-rolled standard profiles are used in most cases (Table 6.5).

The input values needed to carry out an assessment (e.g. dimensions, section moduli) of standard profiles are taken from tables. Such tables can be found in all standard component directories (e.g. [141, 248]) or in the manufacturer's product descriptions.

Cold-pressed omega profiles or hat-profiles (Figure 6.18) are used as crossbeams. These are not standard profiles. They have a total width of 210 to 242 mm, a width of the upper flange of 58 to 97 mm, and a height of 80 to 100 mm. The inclination of the web is between 10° and 25°. The metal sheets are between 5 and 10 mm thick.

Usually the construction steel used is types S235JR to S355JR. In Switzerland, pursuant to SIA 263:2003, table 18, one should usually choose quality group JR or higher.

6.2.7.3 Construction wood

Wood is predominantly used for glide snow protection or avalanche protection underneath the timber line. Predominantly, *round wood* and *half-round wood* with diameters up to 25 cm are used. For defense systems, *squared timber* (posts, boards) are also used. Timber for supporting structures with rectangular transverse cuts must conform to EN 14 081-1, and round wood must conform to EN 14 544.

Depending on the required durability, untreated fir, larch, pine, oak, sweet chestnut, and groundnut wood or pressure-impregnated spruce, larch and pine wood must be used. Untreated spruce is now only used rarely in practice (immediate measures following catastrophes), as it does not have a long lifespan.

Table 6.5 Frequently used steel profiles on separated snow supporting structures

Component	Profile	Dimensions	Steel Elements
Girders	IPE	140 to 270	S355
Supports/Locks	ROR	133/4 to 219.1/6.3	S235 and S355
	HEA	120 to 180	S235 and S275
	TPS	100/100/4 to 120/120/6	S355
	VHP	120/120/6 to 150/150/8	S275

Fig. 6.18 Crossbeams made from cold-pressed omega profiles (© Rudolf-Miklau)

The material characteristics relevant for the assessment are listed in EN 338. For spruce, fir, pine, and larch, the strength classes for coniferous wood are used. The characteristic values for oak, groundnut, and sweet chestnut are taken from the strength classes for deciduous wood.

6.2.7.4 Fasteners and connecting means

Depending on the components that need to be connected, and the required rigidity of the connection, wire cable clamps, screws, shackles, and bolts are used as connection materials. Wire cable clamps must conform to EN 13 411-5. Shackles are used in the sizes 5/8″, 3/4″ and so on according to the US Federal Specification RR-C-271: Shackles, completely rigid, in a curved shape, sixfold safety, with eyebolts, with payload and nominal size, conforming to EN 10 204-3.1. The design is rigid, welded steel; bending stresses are not accepted.

Joint connections are executed as bolt connections. Bolts used have a diameter of 40 to 60 mm, predominantly with steel strength class S355 (Figure 6.19a).

Screws used usually have a strength class of 8.8 ($f_{yb} = 640\,N/mm^2$, $f_{ub} = 800\,N/mm^2$) (Figure 6.19b).

6.2.7.5 Ropes and reinforcing steel

In tension piles, strand or spiral wire (injected with anchor grout) are utilized. Strand wire must conform to EN 12 385-4. Predominantly, wires with the dimensions 6 × 7, 6 × 19 and 6 × 36 with wire strength class 1770 N/mm^2 are used. These strand bundles have diameters of 6 mm to 28 mm. Spiral wire must conform to EN 12 385-10. Wires with spiral strand construction 1 × 19 and 1 × 37 and wire strength class 1770 are used. These cables have a diameter of 12 mm to 24 mm. The rigid steel wires have a characteristic stretch limit of $f_y = 1770\,N/mm^2$.

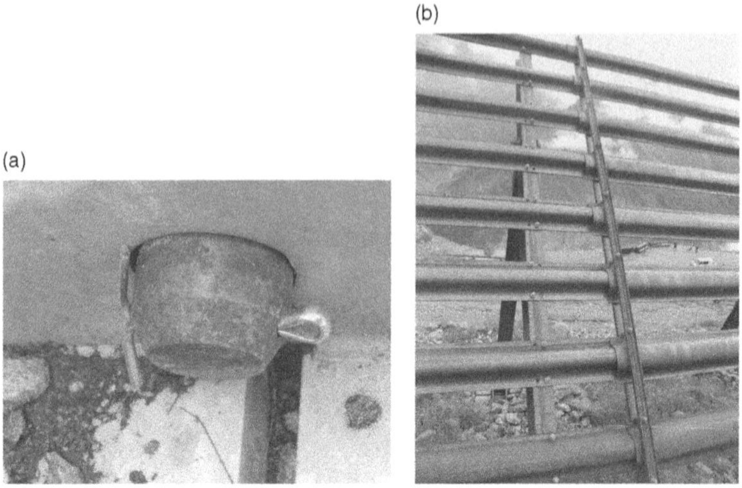

Fig. 6.19 (a): Joint bolt connections on the girder with the foundation of a snow supporting structure (© Rudolf-Miklau); (b): Screw connection of the girder with crossbeams (omega profile) and the clamping rail (connecting crossbeams) (© Rudolf-Miklau)

In tensile or pressure piles, profiled rods or threaded rods made from concrete steel of strength class BSt 500 B ($f_{yk} = 500\,N/mm^2$) or BSt 550 B ($f_{yk} = 550\,N/mm^2$) are also used. The concrete steel in ductility class B also has an increased ductility.

6.2.7.6 Anchor grout

The steel rod is embedded in anchor grout on tensile and pressure poles. Via the connection between steel rod and grout, as well as grout and the surrounding soil, the pile pressure is transferred into the soil. Furthermore, the grout protects the steel rod from corrosion (Section 6.2.9.7).

The largest loads acting on grout (compression stress up to 40 N/mm^2) occur in the case of wire rope anchors because of the deviation force at the drilling hole edge. The largest shear forces along the steel rod occur when there are relatively short rock anchors. The shear forces cause tensile and pressure forces in the grout. Small shifts between the grout body and the steel rod cause fissures. If the grout adheres well, there will be more fissures, however they are distributed evenly and are smaller. The adhesive stress increases linearly with the compressive strength. In order to attain a compressive strength of 5 N/mm^2, the compressive strength of the grout must be approx. 35 N/mm^2. Furthermore, the grout must, even after a short time (approx. 3 days), have a certain strength before one can carry out further construction work without long waiting times.

Anchor grout must have a good deformability, and not exhibit characteristics of brittleness. In order to keep the fissures due to volume decrease as small as possible, the anchor grout may only have a small degree of shrinkage. The resistance to frost is particularly important for avalanche defense systems, in order to minimize the damage risk to the grout in the head area of the anchor caused by the frequent interchange of

Table 6.6 Requirements for bolt grout in avalanche defense systems - tests and limiting values, according to [163]

Characteristic to be tested	Suitability test	Conformity test
Compressive strength 7 d	$Fc \geq 22\ N/mm^2$	$Fc \geq 22\ N/mm^2$
Compressive strength 28 d	$Fc \geq 35\ N/mm^2$	$Fc \geq 35\ N/mm^2$
Frost resistance (according to Swiss standard SIA162/1, laboratory test method Nr. 7) SIA 162/1, test no. 7) SIA 162/1, test no. 7)	High $fs \geq 1.5$	High $fs \geq 1.5$
Elasticity measurement after 28 d	$E \leq 25\,000\ N/mm^2$ (tensile step 5.0 . . . 0.5 N/ mm^2)	—
Shrinking after 28 d	$Ecs \leq 0.2\%$	—
Test frequency and test intensity	Every 3 yr	Dependent on the processed grout amount, the experience from the project, and the utilized grout/mixer combination
Manufacturing test baskets	Laboratory	Construction site

frost-dew. The resistance to frost can be determined according to different methods. In the Swiss Guideline [163], it is recommended that the resistance to frost be checked using the method of porosity. In that way, the relation between the measured air pore content and the free pore space is determined at critical water saturation limits. In permafrost soils, that is soils or bedrock with a temperature below 0 °C (Section 6.2.10), special anchor grout must be used.

In the Swiss Guideline [163], the anchor grout requirements are regulated through a two-step process. The checks and limit values are listed in Table 6.6.

- In order for a new product to be approved, a suitability test must be carried out successfully in the laboratory. The test has a validity of 3 years.
- During the mortaring work at the construction site, the conformity of the product applied must be checked continuously. Since it takes 1 to 2 months to receive the test results, it is recommended to carry out pre-tests before commencing construction, and to carry out mortaring of fresh grout characteristics with air pore measurements during the mortaring work. Using the air pore content, it is possible to determine the frost resistance level directly. In Switzerland, a frost resistance index of $FS \geq 1.3$ is tolerated for construction site tests.

6.2.8 Structure assessment and design

6.2.8.1 General fundamentals of structure assessment and design

The fundamentals for dimensioning steel constructions are listed in Section 6.2.5. This section deals with the particularities when assessing separated snow supporting structures in the STR ultimate limit state. In general, it is sufficient to prove existence of the ultimate limit state of payload for snow supporting structures.

It is not necessary to submit proof for the ultimate limit state for the serviceability (GZG) for the support structure. Such an ultimate limit state is applicable when the snow supporting structure can no longer fulfil its planned use, that is the prevention of avalanche occurrences and the deceleration and catching of smaller avalanches due to service load. This may happen because of extensive deformations, terrain displacement, or filling of the supporting grate (Section 7.2.3.4). The serviceability must be ensured by monitoring and maintaining the construction created.

6.2.8.2 Action combinations

In order to assess avalanche defense systems, it is recommended to use the action combinations listed in Table 6.7.

Table 6.7 Recommended action combinations (AC) for snow supporting structures

Design Situation	AC	Actions combined with another		
		Permanent action	Changeable action	Exceptional action
DS1	AC1	Dead weight	—	—
	AC2a	Dead weight[c]	Snow pressure[a]	—
	AC2b	Dead weight[c]	Snow pressure[b]	—
DS2	AC3	Dead weight	Wind	—
DS3	AC4	Dead weight[c]	—	Earthquake
	AC5	Dead weight[c]	Snow pressure	Avalanche action
	AC6	Dead weight[c]	—	Rock fall
	AC7	Dead weight[c]	—	Snow pressure[d]
	AC8	Dead weight[c]	—	Foundation soil deformation

a) Snow pressure (as a result of 100–150 yearly snow height) as a result of load model 1, pursuant to Table 6.8.

b) Snow pressure (as a result of 100–150 yearly snow height) as a result of load model 2, pursuant to Table 6.8.

c) May be ignored for steel and wooden supporting structures.

d) As a result of > 300 yearly snow height.

As rule combinations for snow supporting structures, AC2a, AC2b and AC3 are indicative. AC2a and AC2b only differ in the load model, which is used as a basis for the upslope snow pressure distribution (Section 6.2.5.4). For snow pressure, all snow pressure components pursuant to Section 6.2.5 are to be used with their maximum value and simultaneously, as they are directly coupled to the snow height and the gliding conditions. The lifting forces caused by wind (AC3) must be positioned without snow forces. The dead weight must be applied as a beneficial continuous action in this combination.

If indicated for a specific site, AC5 and AC6 are also examined as exceptional situations; however, a test of AC 4 (earthquake) is not carried out in practice. A snow pressure as an extraordinary action is generally not proven, as, if one takes the standard load case (snow pressure with 100 or 150 years snow depth as a changeable action) into account, the extraordinary action (snow pressure with > 300 years snow height) is covered.

The characteristic values of the individual actions are indicated in Section 6.2.5. Snow pressures is – corresponding to the guidelines valid at the time – calculated in Austria for a snow height with a return period of 150 years, and in Switzerland for a return period of 100 years (Section 6.2.2). All snow pressure components must be classified as variable actions that are dependent on each other. The dead loads from steel and wooden support structures – except for AC 3 (wind) – can be ignored.

6.2.8.3 Support reactions and internal forces

For the constructive design of snow supporting structures, it is important to reach an optimal balance between stress tolerance, total weight, size of the support reactions, and mounting ease. In general, snow supporting structures have a statically specific supporting. The support reactions and internal forces are to be calculated from the equilibrium conditions. Table 6.8 lists the action combinations, structure configuration and geometries to be used when calculating the indicative internal forces and support reactions for the supporting structure. The different snow pressure components (S'_N, S'_Q, G', S'_R and q'_S) have a simultaneous and maximum effect. As load distribution, load cases 1 and 2 with coefficients a of 0.35 and 0.5 are to be taken into account. According to the Swiss Guideline [163], in order to calculate the slope-normal snow pressure components, a is a coefficient between 0.35 and 0.5, and the least wanted case should always be taken into account. The size of the snow pressure remains constant in the middle area along the level line, and increases in the end areas.

In the arrangement of structures, there is a middle area, a single structure with lateral distances of 2 m, and an end structure with a lateral distance of more than 2 m, which must be examined (Figure 6.20a). Frequently, the support reactions and the internal forces of the single structure (with A = 2 m) are relevant for the assessment. The support reactions and internal sizes are further assessed for variable supports and bar lengths (index length: +/− 0.5 m), since length adjustments are required dependent on local topography (Figure 6.20b).

The maximum pressure force in the support is in load model 1 ($a = 0.35$ or $a = 0.5$ and $s = +0.5$ m). For snow supporting structures with bars the pressure force in the bar, load

Table 6.8 Action combinations and load models dependent on the arrangement of structures and geometries for the calculation of the relevant internal forces and support reactions on avalanche snow supporting structures

		AC2a Load case 1		AC2b Load case 2	
		a = 0.35	a = 0.5	a = 0.35	a = 0.5
Actions from snow pressure	S'_N	S'_N		S'_N	
	S'_Q	$S'_{Q,a=0.35}$	$S'_{Q,a=0.5}$	$S'_{Q,a=0.35}$	$S'_{Q,a=0.5}$
	G'	G'		G'	
	S_R	$S'_{R,a=0.35}$	$S'_{R,a=0.5}$	$S'_{R,a=0.35}$	$S'_{R,a=0.5}$
	q'_S	q'_S		q'_S	
Arrangement of structures	Middle field for continuous row of structures with A = 0 m	(X)	(X)	(X)	(X)
	Single structures with A = 2 m	X	X	X	X
	End of structure with A > 2 m	X	X	X	X
Structure Geometry: Support Length *s*	Max.: s + 0.5 m	X	X	(X)	(X)
	Normal: s +/− 0.0 m	(X)	(X)	(X)	(X)
	Min.:s − 0. 5 m	(X)	(X)	(X)	(X)
Structures Geometry: pressure bar Lengths *r*	Max.: r + 0.5 m	(X)	(X)	X	X
	Normal: r +/− 0.0 m	(X)	(X)	(X)	(X)
	Min.: r − 0.5 m	(X)	(X)	(X)	(X)

X = Generally applicable for the assessment, (X) = i. Gen. not relevant for the assessment.

model 2 ($a = 0.35$ or $a = 0.5$ and $r = +0.5$ m) is decisive. For the girder, load case 1 is for the projection moment and load case 2 for the span moment.

In Figure 6.21, the support reactions are given for a middle field of different types of snow supporting structures. For snow supporting structures with pressure beams, you have the largest pressure forces, for flexible snow nets, the largest traction forces.

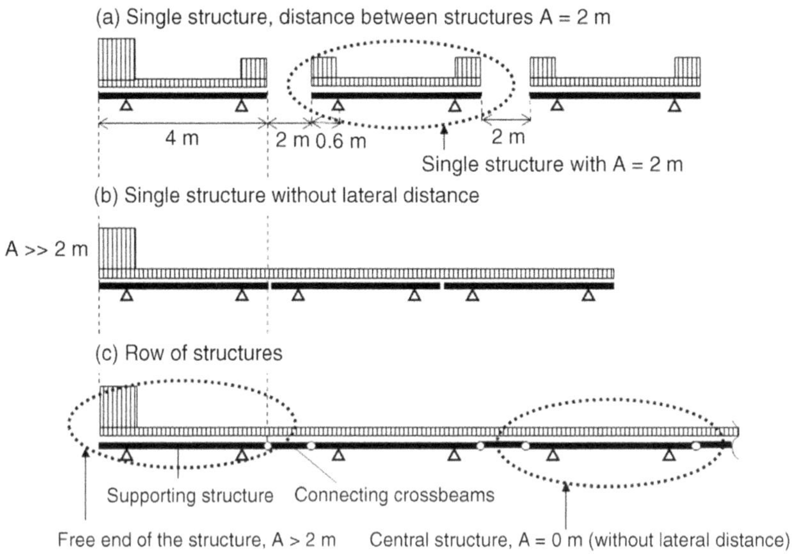

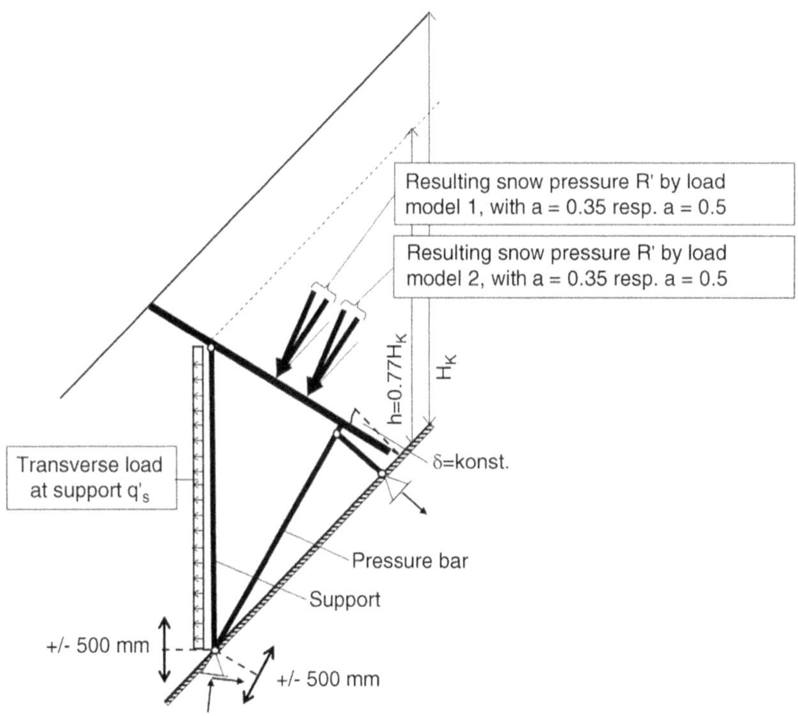

Fig. 6.20 (a): Possible arrangement of structures with snow pressure action, which are increased in the end areas (© Margreth); (b): relevant snow pressure action and snow supporting structure geometry (© Margreth)

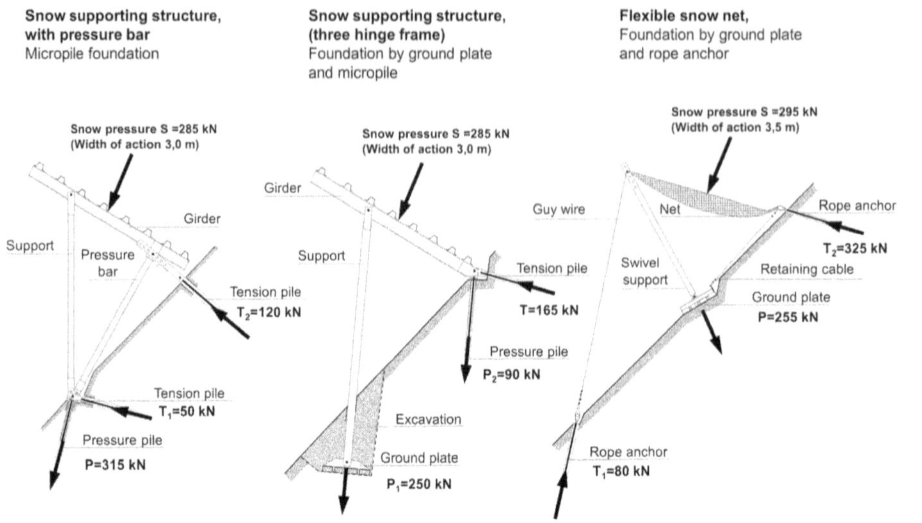

Fig. 6.21 Overview of the snow supporting structure types used today: The listed support reactions were calculated for a middle field and a structure height H_K of 4.0 m and a glide factor N of 2.5 (© Margreth)

6.2.8.4 Dimensioning of supporting constructions of snow supporting structures in steel

The supporting construction for the snow bridges consists of the crossbeams, the girders and the downslope supports (Figure 6.4).

Girders from I and C profiles must be dimensioned on lateral torsional buckling. Since the connected crossbeams (crossbeams) have a stabilizing effect on the drilling of the girder, in practice there is only action due to bending stress. In this connection, it is recommended to have a bearing reserve of approx. 20% (usage degree $\mu \leq 0.8$). Depending on the static system (Figure 6.4), the girder is modelled as a single-field beam with a cantilever on both sides or an articulated beam.

When dimensioning the supports on rigid and flexible snow supporting structures, in addition to the centric pressure force with action line in the support axis, one must take the transverse load from the snow pressure into account (Equation 6.16). Both actions are simultaneous and in full. A support must be dimensioned for flexural buckling with a pure pressure load (Section 6.2.5.3), or lateral torsional buckling with a stress equivalent to a pressure normal force and a bend moment (linear load from the attached snow).

Single-part supports (see static system Figure 6.4c) can be modelled as swivel supports (joint connection on both sides) with an Euler coefficient of $\beta = 1.0$. For the other static systems in Figure 6.4, the procedure is the same. However, on all joint-connected rods with an Euler coefficient to calculate the buckling length, $\beta = 1.0$ may be assumed.

The girder and the support, or the support and the foundation are connected jointly with bolts and connection straps. On the pressure-strained connection straps, a flexural buckling analysis must be carried out besides the tension analysis.

6.2.8.5 Dimensioning of grates of snow supporting structures in steel

The crossbeam profiles transfer the snow pressure forces to the supporting structure. They are screwed together with the girders and thereby ensure that the supporting structure is stiffened.

The crossbeams must be dimensioned according to the effective load widths incumbent on them, with the exception of the uppermost beam, which because of the possibility of an overload if the snow supporting structure is over-snowed, may not be dimensioned to be weaker than the neighbouring beams. The load width of the lowermost beam must always reach down to the soil (Figure 6.22b). The line load p'_B or q'_B, which acts on a beam, is calculated as follows:

$$p'_B = p'_h \cdot b \text{ or } q'_B = q_h \cdot b \geq 0.2p'_B \quad [\text{kN/m}] \tag{6.19}$$

where:

p_h Specific snow pressure perpendicular to the grate (Equation 6.13) in [kN/m^2]
b Effective load width crossbeams in [m]
q_h Specific transverse load (Equation 6.14) in [kN/m^2]

Since the transverse load q'_B acts on the outermost upslope edge of the beam, and not on the gravity point of the profile, there are torsion moments that bend the profile. The cross beams are modelled as single-file girders, the crossbeams at the end structures as single-span beam with a cantilever. Since a linear load acts on these beams, they must be dimensioned for torsional flexural buckling. Since the torsional flexural buckling test for these open crossbeam profiles can be quite complex with 2-axial torsion, the flexural torsion can be taken into account to an extent, by doubling the transverse load q'_B and assuming the action point in the gravitational point of the profile (Figure 6.22a). The dimensioning can hence be carried out for double-axis bending without torsion effect.

6.2.8.6 Dimensioning of snow supporting structures in wood

Wooden components must be dimensioned in accordance with EN 1995-1, taking the relevant NADs into account. The crossbeams, which predominantly are stressed by the buckling, and the upslope girder must be dimensioned in relation to buckling and transverse force. There may be no torsional flexural buckling on round cross-sections (ideal torsion cross-section).

The downslope support must be dimensioned in terms of torsional buckling (at centric normal force), and in terms of a combined pressure torsional stress with additional acting linear load (attached snow). All connections are assumed to be jointed. One-part supports (static system: see Figure 6.4c) can be modelled as swivel supports (joint connection on both sides) with an Euler coefficient of $\beta = 1.0$.

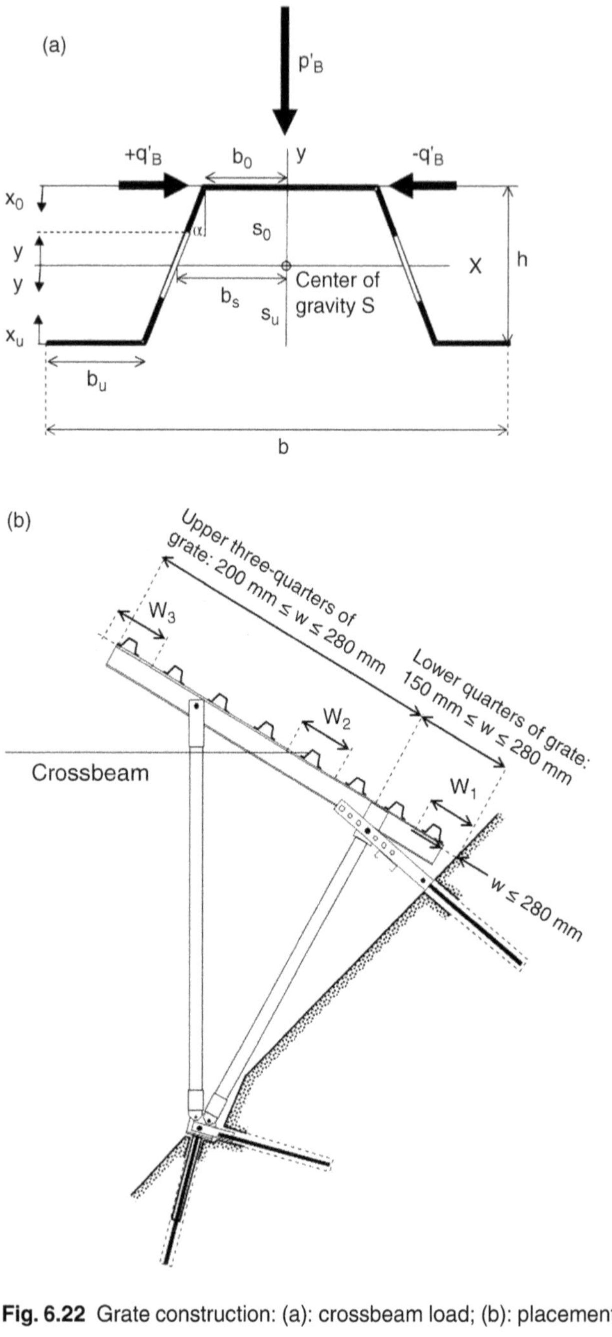

Fig. 6.22 Grate construction: (a): crossbeam load; (b): placement of the crossbeams (© Margreth)

A pressure analysis must be carried out transversely to the fibre on the supports of the crossbeams and on the support-girder connection.

The modification factor k_{mod} must be calculated for all proofs on the basis of utilization class 3. Actions from snow pressure on snow supporting structures are assigned to the class of the load action duration 'medium'.

6.2.8.7 Dimensioning of snow nets

The calculation of the support reactions and internal forces of snow nets is more difficult compared to rigid steel snow bridges. Because of the flexible support surface, spatial forces build up in the snow net, associated with large geometry changes. A snow net cannot stop the creep and gliding movement of the snowpack completely.

The theoretical basis for a simplified two-dimensional static of snow nets was made by Haefeli [90]. To check the calculated values, snow nets were instrumented in Switzerland [246], Iceland [126] and Austria [214] in order to measure the different forces. In Iceland, in the upslope anchor, maximum traction of 150 to 350 kN was measured. In Switzerland and Austria, the maximum upslope traction was 110 or 170 kN. The interpretation of the measurements showed that the snow pressure assumptions for the Swiss Guideline [163] and Haefeli's theory [90] are rather conservative.

The snow pressure forces are calculated as for rigid snow supporting structures. The slope-parallel snow pressure components S'_N (Equation 6.2) are, however, according to the Swiss Guideline [163] reduced by a factor of 0.8, in order to take the reduced snow pressure resulting from the flexible support surface into account. Furthermore, the slope-perpendicular snow pressure components S'_Q and the lateral force S_S are not taken into account during the dimensioning. Hence the weight G' of the snow prism, which is formed because of the slope-perpendicular and the net surface, is considerably greater compared to steel snow bridges. Load case 2, where the resulting snow pressure acts at a height of 0.385 H_k is applicable for the dimensioning of the support structure. A net sag has proven to be effective, which corresponds to 15% of the net cord. Note that for smaller sagging, the force increases greatly, and that the nets can touch the supports in the case of excessive sagging. In order to design the nets and their components for the force transfer to the support and foundation, the specific snow pressure p_H in accordance with load case 2 across the entire net height must be taken into account. The direction of p_H is parallel to the direction of the resulting snow pressure.

Because of their easy usage, snow nets even today are calculated using the *Haefeli* theory (Figure 6.23a). In accordance therewith, a section across the snow net is examined and the net is replaced by a parabola. Furthermore, it is assumed that the sag of the net is carried out in such a way that at the support head the tangents to the net parabola is perpendicular on the ground. Thereby, using the resulting snow pressure, it is possible to calculate the sizes of the wire end forces and the direction of the wire end force on the upslope anchor. It is then possible to use the wire end force on the support head to calculate the support force and the traction in the downslope guywire. The tractions S in the net can be calculated as follows using the specific snow pressure

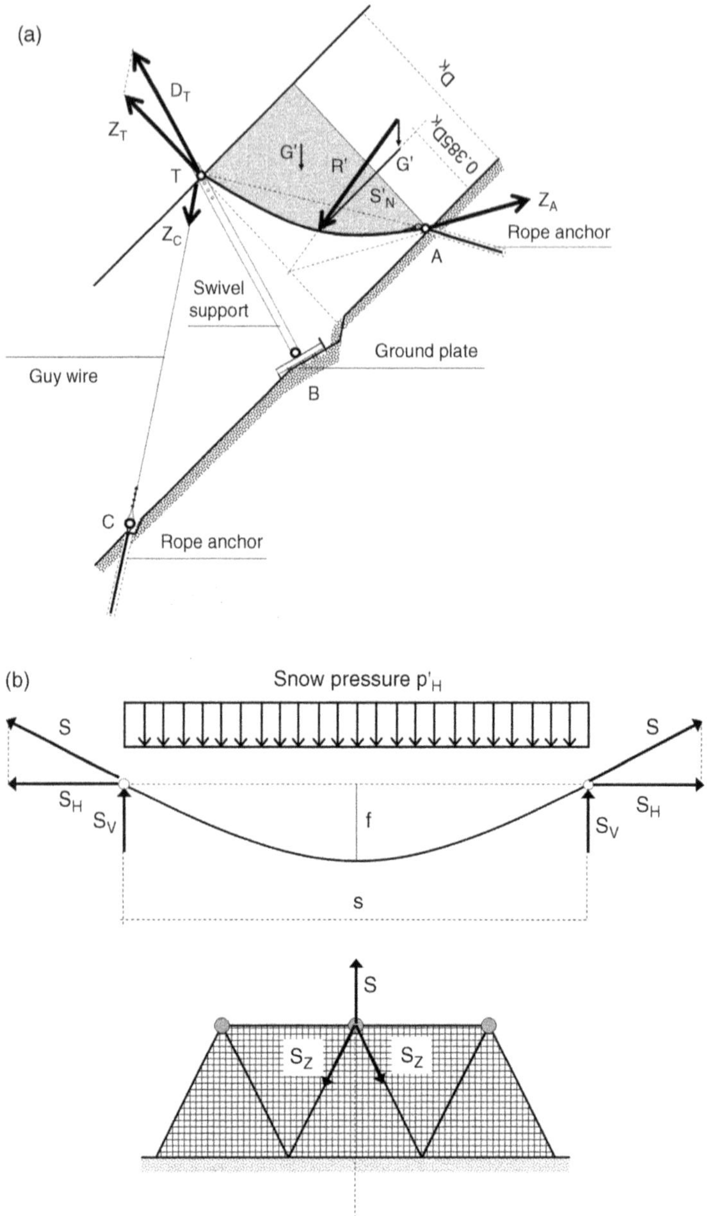

Fig. 6.23 Snow net: (a): Calculation of the support reactions according to *Haefeli*; (b): Calculation of the wire tension and internal forces (© Margreth)

p_H and the assumption that the net forms a parabola-shaped cable line:

$$S = \sqrt{S_H^2 + S_V^2} = \sqrt{\left(\frac{p'_H \cdot s^2}{8 \cdot f}\right)^2 + \left(\frac{p'_H \cdot s}{2}\right)^2} \quad [\text{kN}] \tag{6.20}$$

where:

p'_H Linear normal load of the snow pressure, which is calculated using the specific snow pressure p_h (Equation 6.13) and the corresponding net width, in [kN/m′]
s Length of the wire cord, in [m]
f Wire sag, in [m]

For triangular-shaped nets, the traction S must be distributed onto the four edge wires on the support head. For rectangular nets, S is the maximum traction in the mesh (Figure 6.23b). The forces in the mesh wires can also be calculated using S. In the edge areas of a net, the increased end-effect forces must be taken into account. To this end, additional reinforcement wires can be mounted. For triangularly shaped nets, the outermost upslope end anchor is strained less than the outermost anchor because of the smaller action surface. In practice, however, it is common to make the two outermost anchors equally strong. Guywires that are not covered by the net surface are subjected to the entire snow pressure.

In order to have sufficient braking and retaining capacity, the net surface must fulfil the following requirements:

- If no wire mesh is placed on the net surface, the gap between the wires should not be more than 100 mm.
- When placing a wire mesh with a mesh size of 50 mm, it is sufficient if the net wire has a mesh size of 200 to 250 mm.
- If covering a large area of the net, there may be a problem with early snowing-in of the net. However, in practice this problem has been solved by using a crosswise covering of the nets with a narrow-meshed wire mesh with strip widths of 250 to 500 mm.

6.2.8.8 Dimensioning of snow rakes

Snow rakes are in principle dimensioned in the same way as snow bridges (Section 6.2.8.4). The rake bars typically lie on two purlins. In order to dimension the bottom purlin, the load model 2 with the 25% increase in the specific snow pressure p_H is applicable (Figure 6.24). The rake bars must be dimensioned according to their effective load widths. The outermost crossbeam is measured for a load width that is equivalent to the axial distance to the neighbouring bar. In the same way as for the snow bridges, for the rake bars a transverse load must also be taken into account, the size of which is 10% of the maximum normal load p_B with

$$p'_B = p_h \cdot b \quad [\text{kN/m}] \tag{6.21}$$

where:

p_h Specific snow pressure perpendicular to the grate (Equation 6.13), in [kN/m^2]
b Effective load width, in [m]

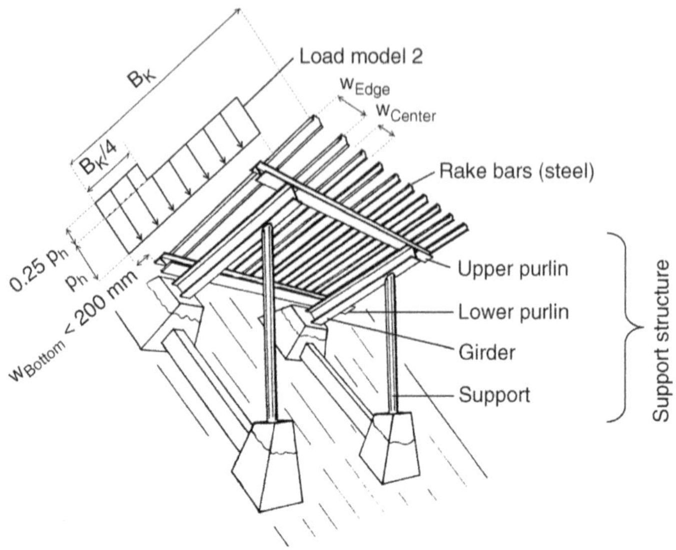

Fig. 6.24 Principle sketch snow rakes with load case 2 (© Margreth)

This acts at the outermost, upslope edge of the rake bar. For rake bars, the field moment for load case 2 can be indicative. For the diameter *w* between the rake bars, the ideal value is 300 mm.

6.2.8.9 Corrosion protection for steel structures above ground

According to the Swiss Guidelines [163], corrosion protection is not required for a steel structure over the ground. There must be a corrosion protection-compatible, constructive design. For components close to the soil (i.e. up to 40 cm above the soil surface), the remarks in Section 6.2.9.7 are correspondingly valid.

The lifespan of wood structure over the ground can be increased by choosing wood species that are not susceptible to fungus (ÖNORM EN 350-2) (e.g. sweet chestnut, groundnut, oak), or pressure-impregnated wood (ÖNORM EN 3802, DIN 68800). If a lifespan of less than 20 years is expected, larch (without sapwood) from slow-growing sites can also be used. In order to obtain a viable structure over the ground, moisture should not reach the wood on a continuous basis.

The steel wires from exchangeable steel cables and nets must have a zinc coating, class B, pursuant to EN 10 264, or be galvanized according to DIN 2078, or have corrosion protection of an equal value. Normally, avalanche nets are thickly galvanized by the manufacturers (zinc coating 250 g/m^2).

6.2.9 Geotechnical design of the foundations of snow supporting structures

6.2.9.1 Principles of geotechnical design

The first principle of design states that excessive deformation and failure in the ground below the construction must be excluded. Proof of safety against structural failure of any part of the construction (STR-proof) and against loss of stability in the ground (GEO-proof) must be given, as well as the ultimate limit state of loss of equilibrium (EQU). Eurocode 7 states that the system requires sufficient ductile behaviour, that is if the limit equilibrium is exceeded stresses are redistributed between the ground and construction. The following explanation are mainly based on [105, 107–110, 125].

ÖNORM B 1997-2 [196] (Table 6.9) describes requirements for the design, execution and evaluation of geotechnical investigations. Minimal requirements for planning the investigation are defined in Chapter 4 and depend on the assigned geotechnical category, which is based on the subsoil at the construction site and the size and complexity of the construction. The classification has to be verified during the geotechnical investigation and adapted as necessary by a geotechnical expert.

Foundations of snow supporting structures are defined as GC1 in the ONR 24 806 [203] (Table 6.9).

6.2.9.2 Design of foundations of snow supporting structures

Since the introduction of Eurocode 7 [195] the design of snow supporting structures in Austria follows verification procedures 2 and 2* of the ÖNORM EN 1997-1-1. The national appendix to the Eurocode 7, ÖNORM B 1997-1-1 (published in March 2010), integrates national requirements for calculations given in the Eurocode ÖNORM EN 1997-1 (Table 6.10).

When the concept of partial factors of safety was introduced for the design of mitigation measures against avalanches, the aim was to maintain the safety level given by the former global safety concept (this was also done in other geotechnical fields), in part to ensure stable construction costs. Therefore, partial factors of safety were defined accordingly and thus differ from those of other pile foundations included in the Eurocode 7, Part 1.

Table 6.9 Geotechnical category (GC) according to ÖNORM B 1997-2:2012 Appendix A. The classification is based on the complexity of the construction and the characteristics of the construction site

Geotechnical category 1:
Small and relatively simple constructions
Geotechnical category 2:
Well established, standard foundations and constructions
Geotechnical category 3:
Constructions and construction components not belonging to GC1 or GC2

Table 6.10 Standards for geotechnical proof of safety for snow supporting structures in Austria

Partial factor of safety according to ÖNORM EN 1997-1 with ÖNORM B 1997-1-1		
		Design approach
Analysis of base and slope failure	B 1997-1-5	3
Pile foundations	B 1997-1-3	2
Spread foundations: analysis of bearing capacity and settlement	B 1997-1-2	2

Concretely, this means that characteristic loads acting on snow supporting structures are increased by the respective partial factor of safety to receive design loads and the resistance of the foundation is diminished by the respective partial factor of characteristic safety to receive design resistance (compare Section 6.1.2.1). To prove that the ultimate limit state (ULS) is met it shall be verified that the following equation is fulfilled:

$$E_{\mathrm{d}} \leq \mathrm{R}_{\mathrm{d}} \tag{6.22}$$

where

E_d Design value of the effect of actions on the foundation
R_d Design value of the resistance of the foundation to an action

The design value of the effect of actions (E_d) is calculated as follows:

$$E_d = \gamma_F \cdot F_k \tag{6.23}$$

where

γ_F Partial factor of safety for an action
F_k Characteristic value for an action (e.g. snow pressure)

The design value of the resistance (R_{d}) is calculated as follows:

$$R_{\mathrm{d}} = R_k/\gamma_{\mathrm{M}} \tag{6.24}$$

where

R_d Design value of the resistance of the foundation to an action
R_k Characteristic value of the resistance
γ_M Partial factor of safety for the resistance of the foundation

6.2.9.3 Design situations

Proof of the foundations of snow supporting structures generally falls under design situation 1 (DS 1) in ONR 24 806 [203] (Section 6.3.1.1); DS 2 or DS 3 are considered for special cases only.

Table 6.11 Partial factors of safety for the resistance of spread foundations for different design situations (DS) according to ÖNORM B 1997-1-1 [193]

Resistance	Symbol	Value		
		DS 1	DS 2	DS 3
Ground failure	$\gamma_{R;v}$	1.4	1.3	1.2
Sliding	$\gamma_{R;h}$	1.1	1.1	1.1

6.2.9.4 Partial factors of safety for pile foundations

According to ONR 24 806 [203] and the Swiss Guidelines [163], load assumptions are considered variable loads, Q_k. For proof of structural safety, the factor of safety, γ_Q, for an unfavourable variable load is 1.5 (Table 6.11); the partial factor of safety (GEO) for the resistance of the pile foundation, γ_R, is 1.35 according to the Swiss Guidelines [163]. Notably, applying both of these factors, results in the old global factor of safety for the proof of safety for outer resistance of foundations (2.0). In Austria, the value of γ_R is 1.4, which corresponds to the values recommended in the ÖNORM B 1997-1-1 (Table 6.11).

Proof of the Serviceability Limit State (SLS) is not required for foundations of snow supporting structures, or can be guaranteed by regular inspection and maintenance.

6.2.9.5 Design of foundations for supports

The most common type of supports for snow supporting structures in Austria are the *three hinge frame* (Figure 6.4c), typically with a ground plate (*pressure plate*). Ground plates vary in size (400 × 400, 600 × 600 or 800 × 800 mm) depending on the quality of the subsoil and should have a zinc coating for protection against corrosion (Figure 6.25).

Experience with these structures has shown that ULS cannot be verified with standard geotechnical design methods, but that foundations are generally capable of bearing the acting loads (Figure 6.26). Proofs for foundations are calculated with an observation method according to EN 1997-1:2004, chapter 2.1, which is acknowledged as one of four effective methods to examine ultimate limit states. Spread foundations with ground plates are sufficiently stable if they fulfil the following criteria: a ground plate subject to pressure loading must be within the surface zone, defined in the Swiss Guidelines [163] as a minimum distance of 50 cm perpendicular to the slope and in the Austrian Standard ONR 24 806 [203] as 30 cm horizontal distance to the slope surface.

The surface zone applies only if the angle α between the force acting along the post perpendicular to the ground plate, $F_{\mathrm{St,k}}$, and the slope-parallel line is smaller than 75° [163]. For supports, it is assumed that the resistance of the subsoil depends on the resulting direction of force (Table 6.12). According to [163], an average resistance of the soil in a perpendicular direction to the slope, $q_{k,90°}$, ranges between 500 and 1000 kN/m^2. The ground plate has to be placed perpendicular to the axis of the post to

(a)

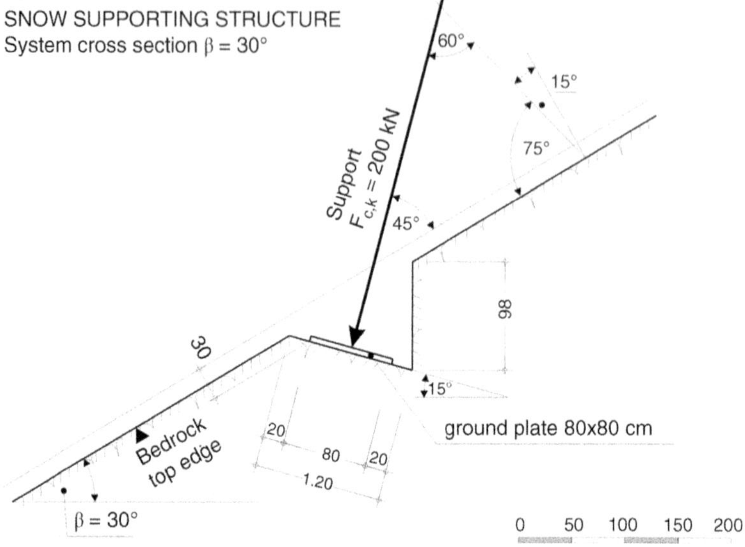

(b)

Fig. 6.25 Support foundation (© Hofmann); spread foundation with a 600 × 600 mm ground plate (© Rudolf-Miklau)

create uniform pressure on the ground. It is recommended to refill the excavated material on the ground plate and compact well.

$$q_{k\alpha} = f \cdot q_{k,90^\circ} \left[\mathrm{kN/m^2}\right] \tag{6.25}$$

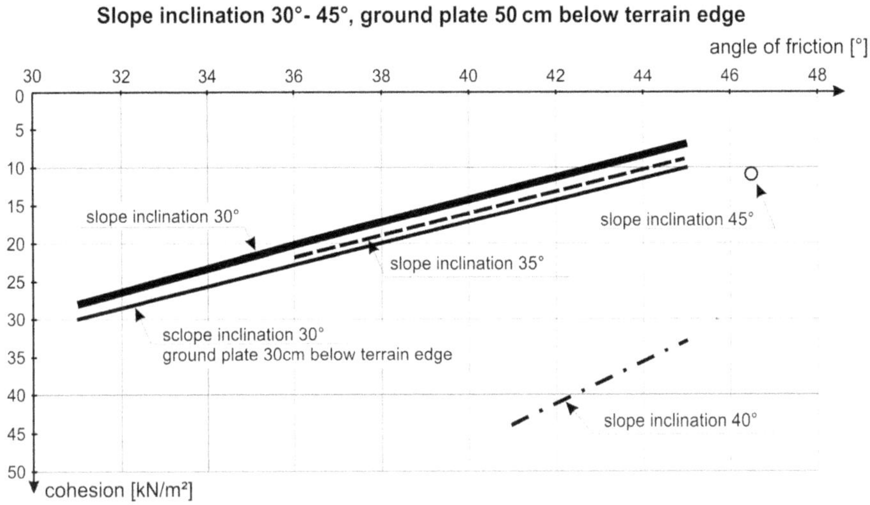

Fig. 6.26 Characteristic soil parameters required for standard geotechnical design methods (© Hofmann)

where

f Reduction factor according to Table 6.12
$q_{k,90°}$ Characteristic resistance of the ground perpendicular to the surface according to [163]

The size of the ground plate is calculated as follows:

$$F_c \geq \frac{F_{St,k} \cdot \gamma_Q}{q_k / \gamma_R} \quad [m^2] \tag{6.26}$$

where

$F_{St,k}$ Characteristic value of the force parallel to the support, perpendicular to the ground plate
γ_Q Partial Factor of safety (1.5)
$q_{k\alpha}$ Characteristic value of the ground resistance depending on angle α [kN/m^2]
γ_R Partial factor of safety for structural resistance according to ÖNORM B 1997-1-1 (1.40)

The tangential force, $T_{St,k}$, has to fulfil the following condition:

$$T_{St,k} \cdot \gamma_Q \leq \frac{F_{St,k} . \tan \varphi_k}{\gamma_R} \tag{6.27}$$

where

$T_{St,k}$ Characteristic value of the transverse force at the foot of the support, parallel to F_c [kN]

Table 6.12 Determination of the ground resistance depending on the angle α according to the Swiss Guidelines [163]

α°	0	15	30	45	60	75	90
f	0.40	0.53	0.66	0.80	0.90	0.97	1.00

$\tan \varphi_{Ek}$ Assumption for the characteristic angle of friction for transfer of compressive forces $\varphi_{Ek} = 35°$ ($\tan \varphi_{Ek} = 0.7$, according to [203])

If the foundation is placed in solid rock, rock anchors or micropiles are used. The design is based on expected loads. It is important that the axis of the micropile corresponds to the axis of the support to avoid an eccentric force.

Ground plates are also used for swivel supports on snow supporting nets. The angle α between the direction of the compressive force and the slope is significantly more favourable than for other structures (Figure 6.27). Depending on the recommendations of the producer, ground plates should be enforced by an additional micropile or steel wire rope between the support and the upper foundation of the net. Loose and humus-rich soils have to be removed before placing the ground plates.

6.2.9.6 Design of girder foundations

Foundations of the girders of rigid structures are usually made with *micropiles*; use of explosives or concrete for foundations is no longer common.

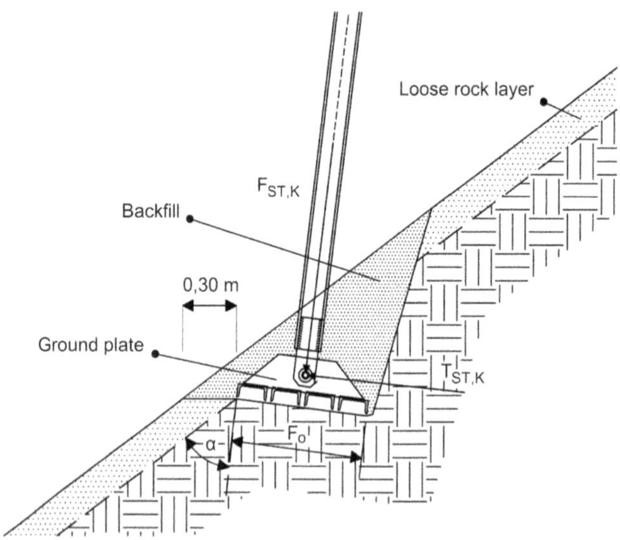

Fig. 6.27 Design of support foundations according to [203] (modified from [163])

Table 6.13 External specific bearing capacity of micropile under tensile loads in compact to slightly fractured rock depending on the diameter of the drill hole, according to [162]

Drill hole diameter	Characteristic value of shaft resistance $q_{s;k}$	Pull-out failure resistance per metre anchor length $R_{a,k'}$
mm	kN/m^2	kN/m′
30	100	95
45	200	280

According to ÖNORM EN 14 199 micropiles are foundation elements that transfer forces (pressure and tensile) into the ground. Micropiles are suitable in solid rock, scree and loose material. Loads have to act along the axis of the micropile, and transverse forces should be accounted for by an additional micropile or by the surrounding ground. The Swiss Guidelines [163] describe the use of rock and ground anchors in this case.

According to [163], micropiles in rock require a minimum drilling diameter of 1.5 times the diameter of the micropile and a minimum cover of 10 mm of grout. The length of the micropile is determined according to the rock quality, with a minimum length of 1.5 m. An estimation of the necessary length of the anchor (micropile) in solid rock is given in Table 6.13.

Micropiles in soil (loose material) are also called *ground anchors* if subject to tension forces in the Swiss Guidelines [163]. Contrary to foundations in solid rock, these are not effective under transverse forces since the surrounding soil cannot support the construction. A micropile that is subject to tensile loads is typically supported by a vertical micropile to withstand transverse forces. An array of two micropiles in the form of an A-rack requires that the resulting force of the snow cover acts perpendicular to the support plane of the construction, which is not necessarily the case with short snow supporting structures or irregular snow distribution. Sometimes a textile stocking (*net anchor* [163]) is placed around the micropile to prevent loss of grouting material into the surrounding soil or open joints in the rock (Sections 7.1.4.2, 7.1.4.3).

For micropiles in loose material the Swiss Guidelines [163] and ONR 24 806 [203] define the following mandatory requirements:

- Minimum drill diameter in loose material is 90 mm, in solid rock 50 mm;
- Textile stockings have to have a maximum mesh of 10 mm, if possible the use of stockings should be avoided;
- The steel rod should be centred in the drill hole;
- Minimum length of the micropile is 3 m;
- Minimal inclination of the micropile to a horizontal reference plane is 15°;
- Minimal cover of grout over the steel rod is 20 mm.

For the preliminary design of foundations, the pull-out failure resistance of micropiles, $R_{t,k}$, is estimated as 40 kN/m in loose material and 100 kN/m in solid rock.

However, the first metre in loose material and first half metre in solid rock should not be considered.

The load-bearing resistance of a micropile under compressive forces is approximately 50% higher than under tensile forces (estimated based on in-field testing and calculations). Only the diameter of the steel rod is needed to calculate the ULS (STR). To guarantee safety against buckling, micropiles subject to pressure loading should be reinforced with an extra tube around the steel rod in the uppermost 1.5 m of the pile or a concrete foundation at the head [163].

The final design of the micropiles requires results from pull-out tests (Section 6.2.9.8). Proof of load-bearing capacity is done in Austria according to ONR 24 806 [203] using the following equation:

$$R_{t;k} = min[(R_{t;m})_{\text{mean}}/\xi_1, (R_{t;m})_{min}/\xi_2] \tag{6.28}$$

where

$R_{t;k}$	Characteristic value of tensile resistance [kN/m]
$(R_{t;m})_{mean}$	Mean tensile resistance, measured by pile load tests at the respective site
$(R_{t;m})_{min}$	Lowest tensile resistance, measured by pile load tests at the respective site

$$R_{\text{t};d} = R_{t;k}/\gamma_{s:t} \tag{6.29}$$

where

$R_{t;d}$	Design tensile resistance of a pile [kN/m]
$\gamma_{s:t}$	Partial factor of safety for tensile resistance of a pile according to Table 6.14 for pile-skin friction with tensile load

In general, testing of specific properties should be done for every construction site, for example capability of the ground to absorb forces via skin friction of micropiles (Figure 6.28). During such tests, forces are applied incrementally until failure takes place, which determines typical pile-skin frictions for the tested pile type. Performing a series of tests will account for any variation in the ground within a homogenous area. With an increase in the number of tests, n, standard values of so-called scattering factors

Table 6.14 Partial factors of safety (γ_R) for the resistance of rammed or drilled piles according to ÖNORM B 1997-1-1 [195]

Resistance	Symbol	Value
Partial factor for peak pressure	γ_b	1.10
Partial factor for skin friction (pressure)	γ_s	1.10
Partial factor for total resistance (pressure)	γ_t	1.10
Partial factor for skin friction for tensile loads	$\gamma_{s;t}$	1.15

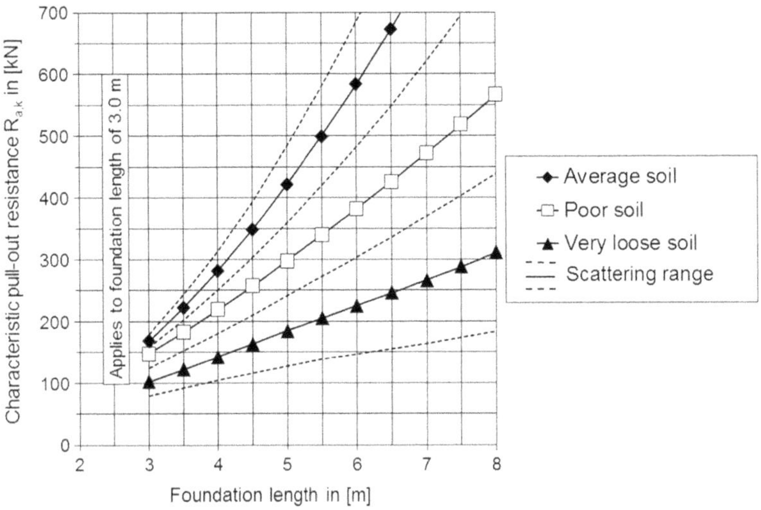

Fig. 6.28 Characteristic pull-out resistance, $R_{a,k}$, as a function of the foundation length and soil category for anchors (excluding blasted anchors and anchors with textile stockings) according to the Swiss Guidelines [163]. Average soil refers to a consolidated, close-grained material with cohesive, fine-grained content (e.g. coarse scree deposits with a content of cohesive fine material or dry, gravelly sand). Poor soil refers to loosely compacted material with non-cohesive fine content and a lower angle of friction (e.g. moist, fine-grained, weathered material that results in poor interlocking between the grout and wall of the bore hole) as well as scree material. Very loose soil refers to loose or humus-rich material with high porosity

can be reduced since these factors are in fact used to account for inhomogeneity of the site (Table 6.15).

If pile tests are not performed and the design of the foundation is based solely on values derived from tables, a fixed value of 2.5 is used to compensate for variations at the site.

Steel wire rope anchors are used primarily for foundations of snow nets. These anchors are not effective against transverse forces that can result due to leverage on the upper foundation along the construction; therefore, steel wire rope anchors require either an

Table 6.15 Scattering factors for pile foundations according to ÖNORM B 1997-1-1, chapter 4.5, Table 7 depending on the number of tests

ξ for $n = \ldots$	1	2	3	4	≥5
ξ_1	1.40	1.30	1.20	1.10	1.00
ξ_2	1.40	1.20	1.05	1.00	1.00

n number of test piles.

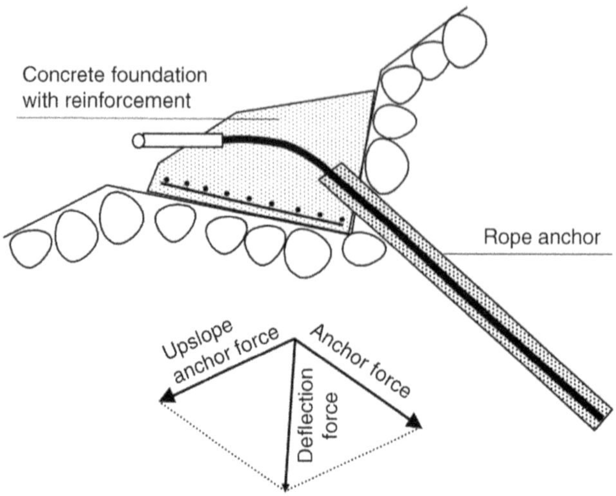

Fig. 6.29 Concrete foundation with reinforcement to account for transverse forces and reduce soil pressure on the steel wire rope anchor [163]

additional tube, if placed in loose material or a concrete foundation, if placed in loosely compacted material (Figure 6.29).

6.2.9.7 Corrosion protection for foundations

Corrosion protection is required for any part of the steel construction close to the ground (distance < 40 cm) and all steel parts of foundations. Either stainless steel should be used or a specified buffer diameter of steel according to Table 6.16. Zinc coating of the steel rods of micropiles is not recommended. If aggressive soils exist at the construction site, corrosion protection has to be increased accordingly.

Any part connected to the micropile has to adopt the requirements for the service life/lifespan of the micropile.

Wire rope anchors require class-A zinc coating according to EN 10 244-2 or a thick zinc coating according to DIN 2078. The anchor head of the wire rope anchor also requires corrosion protection and/or a sealed steel tube in the grout [162]. In aggressive soils, extra corrosion protection of these anchors should be considered.

6.2.9.8 Testing of micropiles

According to ONR 24 806 [203] *micropile tests* (suitability tests) are required for each homogenous part of the construction site (i.e. areas with similar properties). Depending on the number of tests performed, the characteristic pull-out failure resistance of the ground shall be reduced by the scattering factor ξ (Table 6.15).

Table 6.16 Reduction in the diameter of steel rods by corrosion in soils with/without groundwater according to ÖNORM EN 14 199, appendix D

Intended service life in years (a)	5	25	50	75	100
Undisturbed natural soils (sand, silt, clay, etc.)	0.00	0.30	0.60	0.90	1.20
Contaminated natural soils and industrial sites	0.15	0.75	1.50	2.25	3.00
Aggressive natural soils (swamp, peat, gypsum, etc.)	0.20	1.00	1.75	2.50	3.25
Unconsolidated and non-aggressive landfills (sand, silt, clay, etc.)	0.18	0.70	1.20	1.70	2.20
Unconsolidated aggressive landfills (ash, slag, waste products, etc.)	0.50	2.00	3.25	4.50	5.75

REMARK 1: Values are indicative; local conditions must be considered and appropriate values should be available in national appendices.
REMARK 2: The rate of corrosion in compacted landfills is lower than in non-compacted landfills and values can be divided by 2.
REMARK 3: Values for 5 and 25 years are based on measurements; the other values are extrapolated.

According to the Swiss Guidelines [163], a minimum of three test piles should be installed per homogenous area. If the number is reduced to 1 or 2, the characteristic value of the bearing capacity, $R_{a,k}$, will be determined by the lowest value measured in the test, R_a, reduced by 40% (one test pile) or 20% (two test piles). According to ONR 24 806 [203] acceptance tests have to be done for a minimum of one of every 25 installed micropiles and at least three piles for each homogenous area.

The force applied for the acceptance test is calculated as follows:

$$P_f = 1,25 \cdot P_{t,k} \tag{6.30}$$

where

P_f Proof load [kN]
$P_{t,k}$ Characteristic load value [kN]

These tests are not designed to evaluate forces due to friction between the steel rod and the grout, but to evaluate tensile resistance along the contact plane between the grout and soil (pile-skin friction). Therefore, the reacting forces are not applied directly beside the steel rod; forces are applied at a minimum distance of 5 times the diameter of the drill hole. One method to achieve this uses a beam structure (Figure 6.30).

Tests should involve at least two load cycles with a minimum duration of 5 minutes. The direction of the force is parallel to the axis of the micropile.

(a)

(b)

Fig. 6.30 Pull-out devices for testing micropiles (© Mölk; Tartarotti)

6.2.10 Design of snow supporting structures on permafrost sites

Frozen soil is designated *permafrost*, which is subjected to negative temperatures throughout the year, and is therefore a purely thermic phenomenon: only the temperature, and not the ice content, is decisive [21]. The soil layer between the soil surface and the so-called permafrost table, which makes up the upper limit of the permafrost body, is designated the active layer. The active layer, which can be anything from a few centimeters to several meters thick, thaws during the summer and freezes in winter. For this reason, permafrost soil has a tendency to creep downslope, and to settle when heated. Furthermore, there are often cases of rock fall. Permafrost may occur in the Alps above approx. 2200 meters above sea level.

Fig. 6.31 Aerial photograph of sliding steel snow bridges in ice-oversaturated block debris in the defense area Wisse Schijen above Randa (Wallis/Switzerland). Measurements have shown that the foundations move up to 0.1 m per year (© Margreth)

Permafrost subsoil may have different ice content. If snow supporting structures are erected on subsoil containing large amounts of ice, the stability and serviceability can be compromised by creep movements and settlements (Figure 6.31). The evaluation of possible creep movements on a slope makes up the final basis for the planning of defense structures. In the Swiss Guideline [163], possible defense structures were defined dependent on the creep rates (Table 6.17). Construction on permafrost slopes with creep rates of less than 5 cm must be carried out using flexible systems, for example snow nets with swivel supports, based on ground plates (Figure 6.32). Snow nets are less sensitive to creeping movements than rigid structures, and can relatively simply be set up again following any geometrical changes. On stable, non-creeping permafrost subsoil (predominantly bedrock), it is also possible to use rigid systems.

A possible permafrost occurrence is clarified in steps. The first phase is a general evaluation using aerial photograph analyses or models to calculate the spread of the

Table 6.17 Creeping rates and possible defense structures in permafrost

Soil conditions	Estimated creep rate	Measures
Stable (predominantly bedrock)	0 cm/year	All defense systems suitable for the site can be used.
Slightly creeping (ice-rich loose stone, predominantly stable talus)	0.5–2 cm/year	Flexible systems such as snow nets must be installed: Increased maintenance efforts are probable.
Moderately creeping (ice-rich loose stones, unstable talus)	2–5 cm/year	Flexible systems such as snow nets must be installed: construction with snow supporting structures should only be carried out if it is extremely cost-effective. There will be increased maintenance efforts, since damage will be expected in the medium-term. Reduced service life of the snow supporting structures is to be expected. Alternatives to constructing snow supporting structures must be tested in the starting zone.
Greatly creeping (ice-rich loose stone, block glaciers, very unstable talus)	>5 cm/year	Supporting structures only makes sense if there are no other alternatives: medium- and long-term life expectancy of the snow supporting structures is not ensured. Alternatives to constructing snow supporting structures in the starting zone should be used.

permafrost. Then, geological terrain evaluations are carried out, as are measurements in the terrain (e.g. soil surface temperature or geo-physical measurements). Creep movements can be determined using terrestrial surveying. In order to obtain concrete proof of the presence of permafrost, the soil temperature and ice content must be determined during a second phase. To this end, 8 to 10 m deep drillings have to be carried out. As a result of the drillings it is possible to evaluate the drill possibilities of the foundation soil. The load capacity of anchors in permafrost loose material can be determined using pull-out tests. Placement of the micropile and their injection should be carried out as soon as possible after the drilling, at least on the same day, in order to prevent the drilling holes freezing because of in-flowing water. Since micropiles usually must be installed in permafrost bodies, a special anchor grout, which also is set at negative temperatures, must be used.

Fig. 6.32 Snow nets with swivel supports, which are mounted on ground plates on a permafrost slope. The rough blocked debris layer, which is in a state of unstable equilibrium, is typical (construction Wisse Schijen at 3100 m above sea level, Randa, Wallis/Switzerland) (© SLF)

6.3 Design of snowdrift protection structures

6.3.1 Design of snowdrift fences and wind baffles

6.3.1.1 Principles of design

Snowdrift fences act as semi-permeable current obstacles in the wind field, and lead to snow deposits because of turbulence. For this reason, the permeability and filling degree of the fence, based on the slope inclination [179], is of decisive importance for the design of this construction type. The optimal fill ratio (relationship between effective table area and total area, including intermediate spaces, in [%]) is 50% on even terrain. As the slope inclination increases, higher fill ratios are recommended. The height is on average 3–6 m, a ground gap of 50–70 cm [254] or 10% of the fence height [266] is recommended.

Snow is deposited both in front of and behind the fence. The deposit area behind the fence must be large enough to prevent an additional influx of snow in the slope's lee area. Different measurement formulas for the distance L between the wind fence and the terrain edge are used in literature.

$$L = \frac{5 \cdot H}{f} [\text{m}] \tag{6.31}$$

according to [7]

$$L = 20 \cdot H \text{ to } 30 \cdot H [\text{m}] \tag{6.32}$$

according to [254]

where:

H Height in [m]
f Fill ratio [–]

6.3.1.2 Structural systems of snowdrift fences and wind baffles

Since snowdrift fences and wind baffles are always orientated as normally as possible towards the main wind direction, the highest stress normally occurs with regard to the fence level (table level). The structural systems normally on the fence or table level on snowdrift fences and wind baffles are very similar.

There are four structural systems shown in Figure 6.33. For system (a-a), there is a post mounted in the substratum, for system (a-b), the post is supported by a rigid tension/pressure strut. For system (a-c), the support is carried out by two rigid tension/pressure struts, whereby the support forces are distributed onto more support units. In system (a-d), the post is tightened to both sides with tension rods. In systems (a-b) to (a-d), all nodes are modelled like joints.

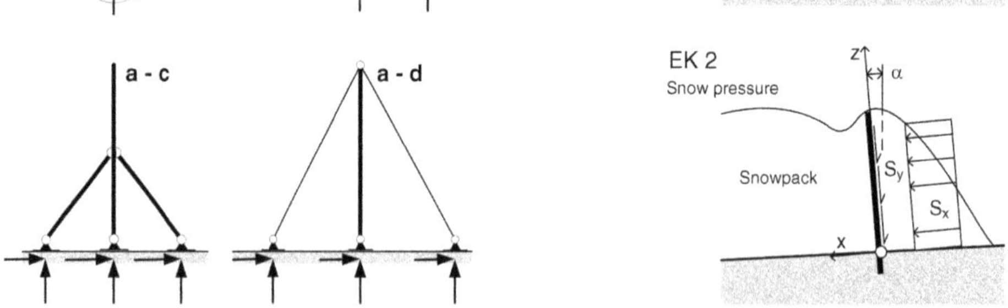

Fig. 6.33 (a): Structural systems (a-a to a-d) on snowdrift fences normal to the fence level: the arrows indicate the possible support reactions (© Suda); (b): action combinations (AC) for snowdrift fences (a section through the supporting structure is shown, © Suda)

On wind fences, the rigidity must not only be ensured transversely to the fence level, but also in the fence level. The rigidity in the fence level is usually ensured by positioning wind braces at the ends of the fence, and for longer fences also in between. Wind braces are usually manufactured using steel strip crosses.

6.3.1.3 Action and action combinations

Snowdrift fences and wind baffles are stressed in a free-standing condition by the wind, and in a snowed-in condition by snow pressure. For this reason, the action combinations EK1 and EK (Figure 6.33) must be taken into account when dimensioning the structure. The action combinations are assigned to DS 1. The wind and snow effects are to be categorized as changeable effects.

In EK1, the characteristic wind effects w_k, dependent on the fill ratio in accordance with EN 1991-1-4 [191], must be set for free-standing fences, or for free trusses and grid constructions. According to [7] snowdrift fences, and wind baffles must be dimensioned with a characteristic wind pressure of at least 3.0 kN/m^2. In [229], 3.3 kN/m^2 is recommended. One-third to one-half of the wind pressure that normally acts on the table level must additionally be assumed to be acting on the fence level, in order to measure the rigidity elements in the fence level.

The EK considers the situation where one has a snowed-in snowdrift fence or wind baffle. In order to avoid damage due to the settling snow pressure, snow pressure must at least be applied to the bearing elements (posts, struts, guy wires, foundations) on the snowdrift fences and wind baffles without ground fissure.

6.3.1.4 Construction principles

Snowdrift fences and wind baffles are to be constructed as free-standing walls or fences.

The dimensioning of the construction is, depending on the material, carried out according to EC3 or EC5. The posts are designed for torsion as wood or steel beams. Pressure struts are to be designed for flexural buckling as centrically stressed supports, traction strips in relation to the centric standard force.

6.3.2 Design of wind roofs (jet roof)

6.3.2.1 Principles of design

Wind roofs (jet roofs) are to be designed as free-standing pent roofs. The support structure is dimensioned according to EC3 or EC5, depending on material. The cladding elements and the girders are measured as wood or steel beams, the supports and strips for flexural buckling.

6.3.2.2 Structural systems

A wind roof (jet roof) is a spatial structural system. This spatial system can be divided into a transverse and lengthwise structural system (Figure 6.34). In terms of transverse, there are three common systems. For system (a-a), the supports are fastened to the subsoil, for system (a-b), the supports are supported by a pressure struts, in system (a-c) by a cross element with traction rods. In system (a-b) and (a-c), all nodes are modelled like joints. In the lengthwise direction, there are two possible systems. In system (a-d),

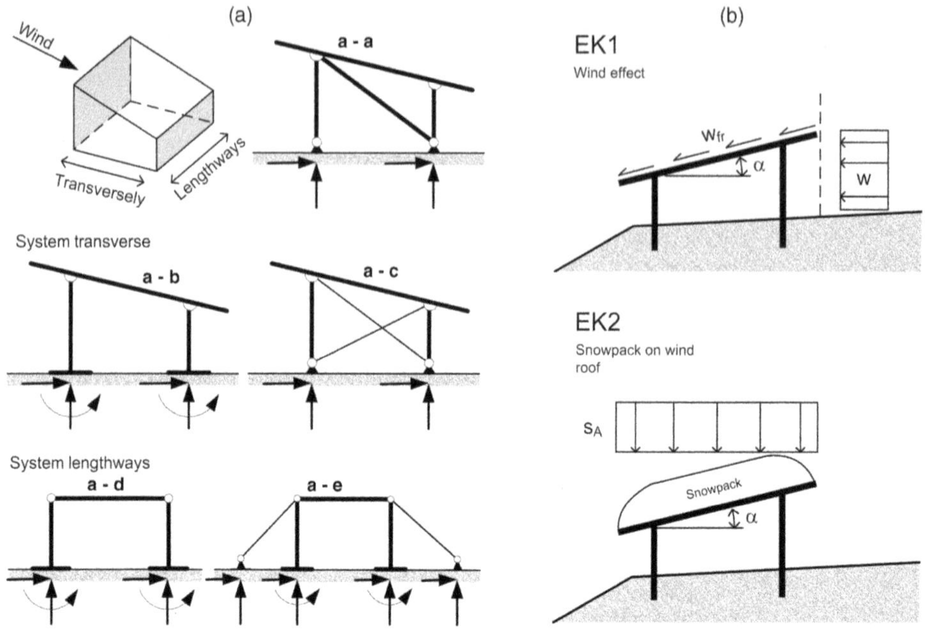

Fig. 6.34 (a): Static wind roof systems (a-a to a-e): the arrows indicate the possible support reactions (© Suda); (b); action combinations for wind roofs (a section through the support structure is shown) (© Suda)

the supports are fastened, in system (a-e), swivel supports are fastened laterally with tension rods.

6.3.2.3 Action and action combinations

The decisive action combinations are shown in Figure 6.34b. In normal conditions, the wind roof is strained by the wind (EK1). The wind generates positive and negative pressure, and friction forces on the roof surface and supports. The characteristic wind actions w_k and w_{fr} (friction) must be applied according to the regulations for free-standing podium roofs pursuant to EN 1991-1-4 and ÖNORMB 1991-1-4 [190]. In order to design the support structure (supports, girders, foundations, and fastenings), the total wind power F_W is used. In order to measure the cladding elements (roof), the total pressure coefficients (c_p) are used. The action on the supports is generated from the projecting surface of the supports, normally to the main wind direction.

If there is snowfall without wind, the snow can accumulate on the roof. This situation is illustrated in EC2. The characteristic action from the deposited snow $s_{A.k}$ is taken pursuant to EN 1991-1-3 and ÖNORM B 1991-1-3. In this connection, the regulations for podium roofs must be applied in a meaningful manner.

The actions from wind and snow are to be classified as variable actions. The action combinations are to be assigned to DS 1.

6.4 Design of avalanche catching, deflection and retarding structures

6.4.1 Determining of the required height of catching and deflection dams (classical approach)

The height is the most critical factor determining the effectiveness of catching and deflection dams, only second to the actual design of the dam and construction of the upstream side. Several approaches exist to estimate the required height, H_{DE}. Whereas newer methods are more physically based [15, 127], the classical method that is still commonly used is shown in Equations 6.33 to 6.35.

$$H_{DE} = h_u + h_f + h_s \tag{6.33}$$

where

h_u energy height [m]
h_f flow height [m]
h_s snow height on the ground [m]

For catching dams, the energy height of an avalanche is given by:

$$h_u = \frac{v^2}{2g\lambda} \tag{6.34}$$

where

v avalanche velocity [m/s]
g gravitational acceleration, constant [9.81 m/s^2]
λ empirical parameter [–]

For deflection dams, the energy height of an avalanche is given by:

$$h_u = \frac{(v \sin \varphi)^2}{2g\lambda} \tag{6.35}$$

where

φ deflection angle [°]

One challenge with this method is that the empirical factor, λ, is determined solely by expert judgement. Values range from 1 to 3, where 1 is assumed to hold for large, dry flowing avalanches and 3 for small avalanches. A value of 1 is generally used for deflection dams, implying that the energy loss is negligible for dams with a small deflection angle (Section 5.3.2.3).

6.4.2 Determining the required height of catching and deflection dams by a more physically based approach

6.4.2.1 General principles of design

The state of flow of an avalanche is the main factor that determines the interaction between an avalanche and a dam at action. The avalanche flow is generally assumed to be supercritical in the path, in which case the Froude number, Fr, of a dry-snow avalanche in the path is:

$$Fr \equiv \frac{v}{\sqrt{gh\cos\phi}} > 1 \tag{6.36}$$

where

v avalanche velocity [m/s]
h flow height [m]
g gravitational acceleration, constant [9.81 m/s^2]
ϕ local slope angle of the path [°]

This implies that an avalanche makes a transition from supercritical to subcritical (Fr <1) in the runout zone before deposition. The function of breaking mounds and catching dams is to artificially initiate or enhance this transition and thus to shorten the runout distance of the avalanche. Breaking mounds, catching and deflection dams have little or no effect on pure powder snow avalanches.

The Froude number upstream of the structure, Fr_1, determines the placement and design of breaking mounds, catching and deflection dams. It is usually assumed that the flow of dry-snow avalanches in the upstream area is still supercritical, that is Fr_1 >1, and typical values range between 2 and 6. The dimensions of catching and deflection dams are calculated based on two criteria: (i) supercritical overflow over the dam must be prevented and (ii) the flow height downstream of where the hydraulic jump is initiated should be lower than the freeboard (Figure 6.35).

As the Froude number is the determining factor for the design of catching and deflection dams, it is decisive to have reliable estimates of the avalanche flow velocity and flow height upstream of the structure.

Design criteria for catching and deflection dams and their physical principles can be found in [127] and a slightly modified procedure for caching dams in [14]. Both approaches are based on shock theory. The following overview mainly follows the approach developed during the SATSIE projects [127].

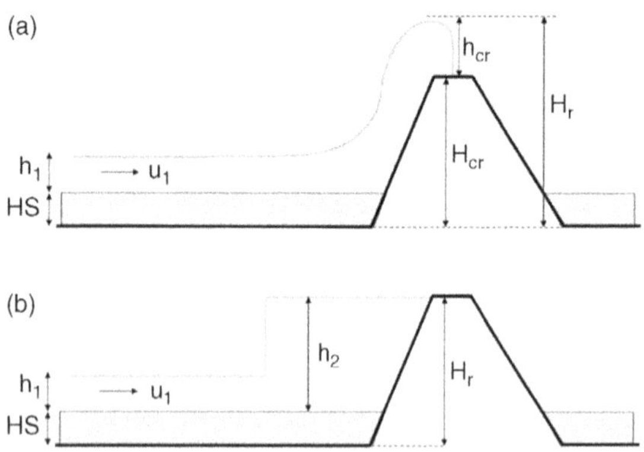

Fig. 6.35 Parameters required fulfilling the design criteria: (a) supercritical overflow and (b) hydraulic jump (© Gauer)

6.4.2.2 Catching and deflection dams

Although the purpose of a deflection dam differs from that of a catching dam, the bases of the design criteria are similar. The Froude number is defined relative to the dam axis:

$$Fr_{\perp} \equiv \frac{v_1 \sin \varphi}{\sqrt{g h_1 \cos \phi_b}} > 1 \tag{6.37}$$

where

v_1	avalanche velocity upstream of the structure [m/s]
h_1	flow height upstream of the structure [m]
g	gravitational acceleration [9.81 m/s²]
ϕ_b	local slope angle upstream of the structure [°]
φ	angle between avalanche flow direction and dam axis [°]

To fulfil the first design criterion the dam height required to prevent supercritical overflow is given by:

$$\frac{H_r - HS}{h_1} \geq \frac{(H_{cr} - HS + h_{cr})}{h_1} = \frac{1}{k} + \frac{1}{2}(kFr_{\perp})^2 - \frac{1}{2}(Fr_{\perp})^{2/3} \tag{6.38}$$

where

H_r	required dam height [m]
H_{cr}	critical dam height [m]
h_{cr}	critical flow height $\left(h_{cr}/h_1 = (Fr_{\perp})^{2/3}\right)$ [m]
HS	snow height [m]
k	momentum loss factor $k = 0.75 + 0.1 \cdot (60° - \alpha)/30°$ with angle of upstream side of the dam $30° < \alpha < 60°$; no momentum loss k = 1.0

To fulfil the second design criterion, the height of the hydraulic jump is given by:

$$\frac{H_r - HS}{h_1} \geq \frac{h_2}{h_1} = \frac{2\sqrt{(6Fr_{\perp}^2 + 4)}\cos\delta + 1}{3} \tag{6.39}$$

where

h_2 flow height downstream of the hydraulic jump [m]

$$\delta = \frac{1}{3}\left(\frac{\pi}{2} - \tan^{-1}\left(\frac{9Fr_{\perp}^2 - 8}{Fr_{\perp}\sqrt{27\left(16 + 13Fr_{\perp}^2 + 8Fr_{\perp}^4\right)}}\right)\right) \tag{6.40}$$

Figure 6.36 shows both criteria as a function of the Froude number. Preventing supercritical overflow is the determining factor for Froude numbers larger than approximately 3; the height of the hydraulic jump for Froude numbers less than 3.

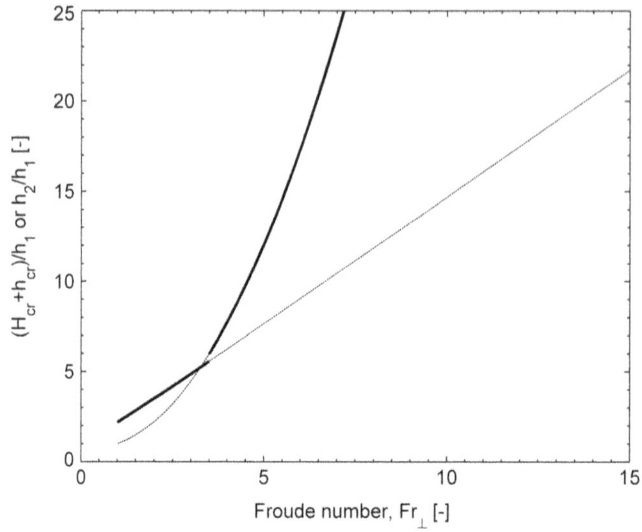

Fig. 6.36 Supercritical run-up, (Hcr + hcr)/h1, from Equation 6.38 and flow height downstream of the hydraulic jump, h2/h1, based on Equation 6.39 in relation to the Froude number. The supercritical run-up is calculated assuming no momentum loss (k = 1). Bold lines indicate the range over which each criterion is decisive for the dam height (adapted from [127])

If the terrain normal to the dam axis slopes towards a deflection dam and the criterion for the jump height is decisive, the required freeboard should be increased by:

$$\Delta H_{\phi\perp} = \frac{\sqrt{2}\tan\phi_\perp}{2\,Fr\cos\varphi}\xi \tag{6.41}$$

where

ξ distance along the dam [m] (measured from the upstream end)
$\phi_\perp$ local slope angle towards the dam perpendicular to its axis [°]
Fr $\equiv v_1/\sqrt{gh_1\cos\phi_b}$

Similarly, in the case of a curved deflection dam the freeboard should be increased by:

$$\Delta H_\kappa \approx \frac{\sqrt{2}(v_1\cos\varphi)^2}{2Fr\cos\varphi\, g\cos\phi_b\, R_\kappa}\xi \tag{6.42}$$

where

R_κ curvature of the dam axis [m]

It is not always possible or economical to prevent overflow over a catching dam, especially for the fluidized part of fast dry-snow avalanches (Section 3.1.2.4). Investigations on a catching dam at the avalanche test site Ryggfonn, Norway [72], have shown that the overrun length, l_{ovr}, measured from the crown of the dam can be estimated by:

$$\frac{l_{ovr}}{H_{De}} = b_1 \frac{v_1^2}{2gH_{De}} + b_0 \tag{6.43}$$

where

H_{De}	effective freeboard [m]
v_1	avalanche velocity upstream of the dam [m/s]
g	gravitational acceleration [9.81 m/s^2]
b_1	empirical parameter [≈ 2.56]
b_0	empirical parameter [≈ -1.41]

In the case of deflection dams, the deflection angle should not exceed a certain value to retain the effectiveness of the dam. If the deflection angle is too large, the hydraulic jump detaches and the jump height increases and overflow over the dam can result. In an extreme case, the dam acts as a catching dam. The maximum deflection angle is approximately (Figure 6.37):

$$\varphi_{\max} \approx \frac{\pi}{2} - \frac{2^{3/4}}{Fr^{1/2}} - \frac{2^{1/4}}{6Fr^{3/2}} \tag{6.44}$$

However, the recommended deflection angle is 10° smaller.

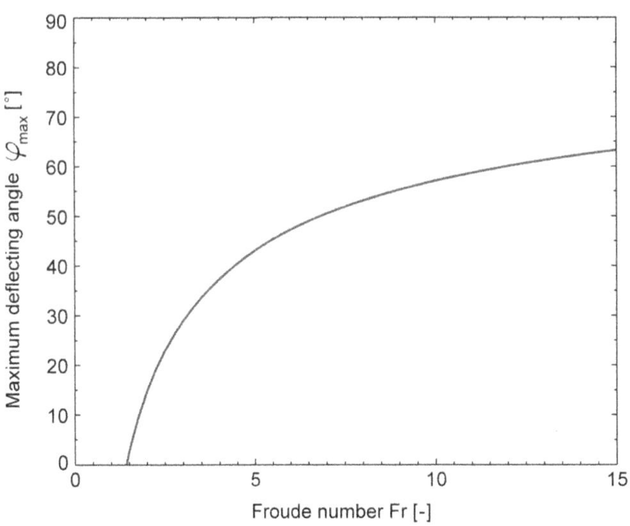

Fig. 6.37 Maximum deflection angle, φ_{max}, in relation to the Froude number, $Fr_{\perp}$, according to Equation 6.44 (adapted from [127])

6.4.2.3 Avalanche guiding dams

A guiding dam should prevent lateral spreading of the avalanche. The required height of a guiding wall, H_{LD}, depends on the avalanche velocity, flow height and height of the snowpack. Generally, two cases can be distinguished.

As long as the flow is undisturbed and no flow transition occurs, the flow height within the path can be estimated by:

$$H_{LD} = H_S + h_{1\uparrow} \tag{6.45}$$

where

H_{LD} height of the guiding dam normal to the ground [m]
H_S height of the snowpack and/or old depositions [m]
$h_{1\uparrow}$ flow height normal to the ground [m]

In the runout area or in areas in the avalanche path where deposition may occur the required height can be estimated by:

$$H_{LD} = H_S + h_{1\uparrow} \frac{\sqrt{1 + 8Fr_1^2} - 1}{2} \tag{6.46}$$

where

Fr_1 Froude number upstream of the dam (>1)

The second term on the right-hand side of Equation 6.46 accounts for the height of an expected hydraulic jump, which occurs during the transition from supercritical to subcritical flow. Constriction in the flow direction must be avoided.

6.4.2.4 Storage capacity

Storage capacity should be calculated carefully to obtain optimal effectiveness of deflection, guiding or catching dams. With guiding or deflection dams, deposition should not impair deflection or cause re-routing of the flow. With catching dams, storage capacity must be sufficient for repeated avalanche events per year. If necessary, intermittent removal of snow depositions in this area should be foreseen.

The maximum storage capacity, V_R, of a catching dam is given by the effective dam height and the surrounding topography. Traditionally, storage capacity upstream of a catching dam is calculated as the difference between the estimated surface of the avalanche deposit and the snow-covered terrain. It is assumed that the deposition height coincides with the dam crown and that the surface of the deposition has a slope, γ, of 5 to 10°. The deposition length is given by the distance from the dam to the intersection point between the snow-covered terrain and the surface of the avalanche deposit (Figure 6.38).

The storage capacity is given by the double integral of the estimated deposition height and length:

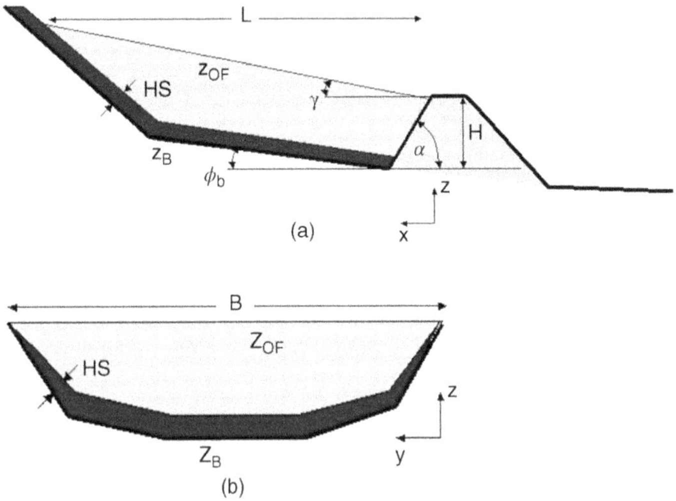

Fig. 6.38 Storage capacity upstream of a catching dam: (a) longitudinal section and (b) cross-section (© Gauer)

$$V_R = \int_{x_0}^{x_1} \int_{y_0}^{y_1} z_{OF}(x,y) - (z_B(x,y) + HS(x,y))dydx \qquad (6.47)$$

where

V_R	storage [m3]
z_B	height of the topography [m]
z_{OF}	height of the surface of the deposition [m]
HS	height of the snow height and/or old depositions [m]
$L \approx x_1 - x_0$	length of the deposition [m]
$B \approx y_1 - y_0$	width of the deposition [m]

Although this estimation neglects dynamic effects and probably overestimates the storage capacity for dry-snow avalanches, storage capacity is rarely the decisive factor for dam height.

6.4.2.5 Actions on avalanche deflection and retarding dams

In general, dynamic loads of avalanches are less important than geotechnical factors for dimensioning deflection, retarding and catching dams in the case of earth and stone constructions. If, however, the construction is designed as a wall (e.g. concrete wall), avalanche action pressures and resulting moments need to be considered. The force normal to a construction, F_{lx}, can be estimated by:

$$F'_{lx} \approx \rho h_1 v_1^2 \sin^2\varphi \left(1 + \frac{1}{Fr_\perp^2}\right) \quad [\mathrm{N/m}] \qquad (6.48)$$

where

F'_{Ix} normal force on to the wall per unit length [N/m]
h_I flow height of the avalanche normal to the ground [m]
v_I avalanche velocity [m/s]
ρ flow density [kg/m^3]
φ angle between the flow direction and wall axes [°]
$F_\perp$ Froude number normal to the wall (Equation 6.37)

For calculating the moment, a drift of the point of application needs to be considered. In the most extreme case, the point of application should be estimated as $H_r - h_I/2$. The static load of the deposition also needs to be considered. Forces parallel to the wall can be estimated by:

$$F'_{\parallel} \approx min(c_1 F'_{Ix}, \tau_y h) \quad [\mathrm{N/m}] \tag{6.49}$$

where

c_I friction coefficient for dry-snow avalanches [0.1–0.4]
τ_y yield strength of the avalanching snow [order of 10 kPa]
h height of the contact area [m]

6.4.3 Geotechnical design of avalanche deflection and retarding dams

6.4.3.1 Fundamentals of geotechnical dam design

In Eurocode 7 (EN 1997-1 (November 2004) + AC (February 2009) [193]) in the scope of application of EN 1997-1, 1.1.2, the design of supporting structures is explicitly mentioned. The topics handled are: the foundations of geotechnical design, geotechnical documents, construction monitoring, control measurements, and maintenance, filling, dewatering, soil improvement, soil reinforcement, surface foundations, pile foundations, anchor, supporting structures, hydraulically caused failure, total stability, and earth dams.

Figure 6.39 shows failure mechanisms for which proof of total stability or load capacity for surface foundations must be provided.

For proof of total stability (terrain fracture), proof process 3 from EN 1997-1 [193] was chosen. For this proof process, the partial safety coefficients are used for the subsoil characteristics. It must be proven that the fracture to the subsoil or too large deformations cannot occur. In that connection, the external actions on the subsoil (e.g. traffic loads and support structure loads) are coupled to partial safety coefficients.

When calculating terrain fractures, the proof is carried out via the comparison of resistance to actions. Appendix A of EN 1997-1 [193] includes recommendations for partial safety coefficients. This standard indicates, pursuant to EC 7, the partial safety coefficients dependent on the risk. In cases of exceptional risk or uncommon or extremely difficult subsoil conditions or loads, there should, according to EN 1997-1 (2.4.7.1 (4)) [193] be an indication of higher partial safety coefficients than in appendix A to this standard. However, a reduction of the partial safety coefficients must

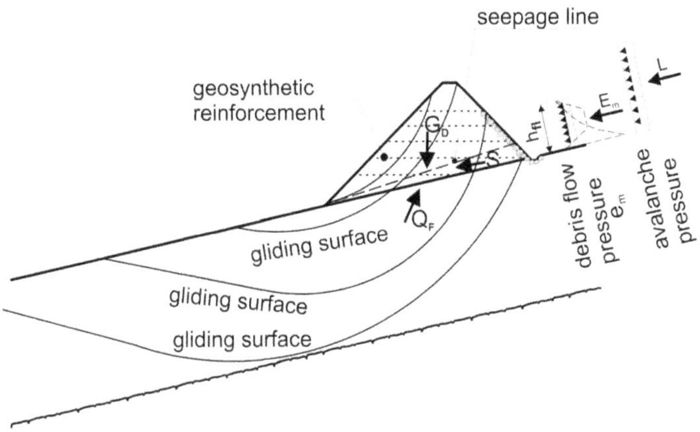

Fig. 6.39 Principal failure mechanisms in an avalanche catching dam (© Hofmann)

be carried out for support structures under time constraint, or in case of temporary design situations.

According to the new safety concept, the calculative proof is in principle carried out as according to the valid ÖNORM B 4433, slope fracture calculations [189] (pursuant to the global safety concept). The subsoil characteristics (tangent of the angle of friction and cohesion) are reduced by the same partial safety coefficient. However, the difference lies in the application of the changeable unfavourable outer effects according to a partial safety coefficient.

The settlements (progressive settlements) must be evaluated for size. The settlement differences must be indicated.

6.4.3.2 Rules of dam construction

To construct avalanche dams, generally coarse fill material (without organic particles) is used. The minimum required angle of friction for the material must be indicated so that the planned downstream slopes can be reached. This value must be confirmed through suitability tests for the fill material (Figure 6.40a and b). The proof must be provided via a framework shear test with a maximum particle size of 2.0 mm. A geotechnical construction site evaluation must be carried out.

Preparation of the dam bed is carried out in accordance with the following rules:

- Exchange of humus, rooted and swampy or soft soils, as well as very loose and moderately loose soils. If there are local landfills detected during the course of the excavation work, it must be checked if these are to be replaced at full depth.
- Generally, a compaction of the excavation bed must be carried out at a compaction degree of at least $D_{Pr} = 98\%$ of the simple proctor density and a load plate module of $E_{v1} \geq 15\,\text{MN/m}^2$. (Details can only be defined in connection with the construction).
- Depending on the soil conditions, higher E_{v1} values may be required. Corresponding information can only be defined in connection with the construction work.

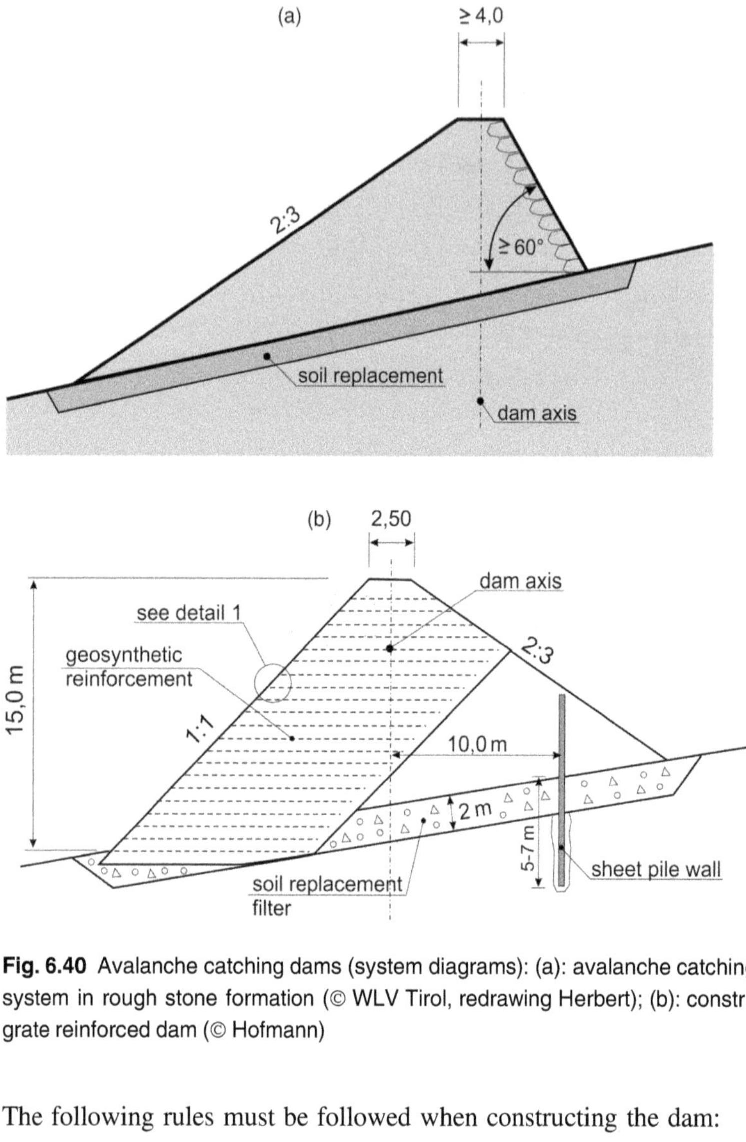

Fig. 6.40 Avalanche catching dams (system diagrams): (a): avalanche catching dams with safety system in rough stone formation (© WLV Tirol, redrawing Herbert); (b): construction model, geograte reinforced dam (© Hofmann)

The following rules must be followed when constructing the dam:

- The dam itself is to be constructed in layers with a compaction degree of $D_{Pr} \geq 98$–100% of the simple proctor density. Before the start of construction, soil mechanical suitability tests must be carried out with the debris.
- Fine debris must primarily be compacted statically. The debris height must be limited in accordance with the effectiveness of the compaction machine and grain distribution, plasticity, as well as inflow water level of the debris.
- Debris layers, which already have been tested in terms of soil mechanics, can however become soft from heavy rains, and must be reconstructed. For fine-grained materials, it may be necessary to pull off the uppermost zone and carry out compaction once again.

– In that connection, not only the permitted fill levels and permeability coefficients must be defined, but also the required minimum values for the deformation modules.

The corresponding earth construction measures must be defined by a geotechnical expert, and inspected continuously during the construction period. In this connection, EN 1997-1 [193] must be adhered to.

6.5 Design of avalanche breakers

6.5.1 General remarks

According to the classification of the static systems pursuant to ONR 24 800 [200], avalanche breakers are volume-active support structures. They consist of boards, discs and beams, and are exclusively constructed from steel concrete. The action of the breaker depends, on the one hand, on the turbulence of the snow powder-air mixture, and on the other hand on the back-pressure of the flowing component from the mixed-snow avalanche (Section 3.1.2.4).

6.5.2 Actions on avalanche breaker and structural systems

When measuring avalanche breakers, one generally must differentiate between those in a confined avalanche path (Figure 6.43) and those in an open avalanche path with an effect against unconfined avalanches (Figure 6.41). The latter are only stressed by dynamic avalanche pressures. Constructions in confined avalanche paths are often stressed by torrential processes (e.g. debris flow). In this case, pursuant to [20], the structural design also has to take into account the indicative torrent processes.

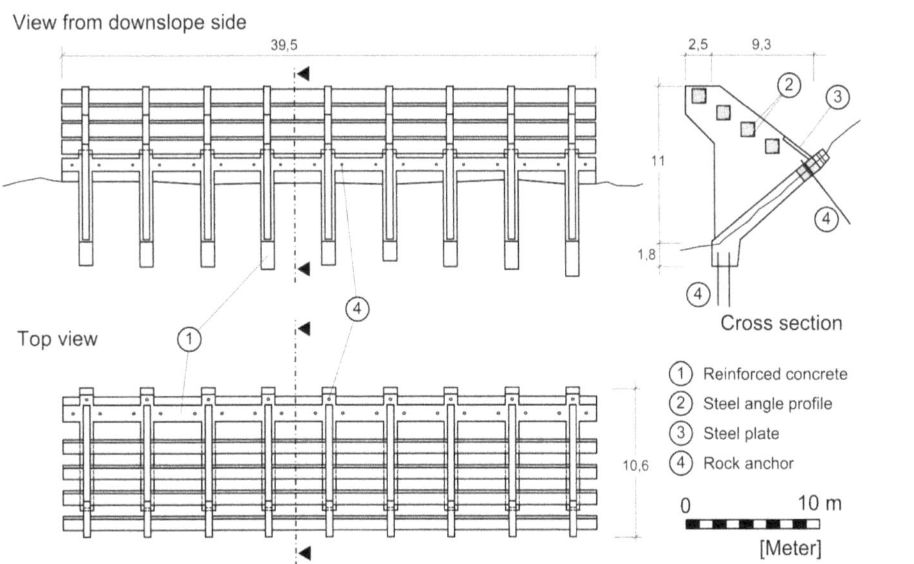

Fig. 6.41 Avalanche breaker in the unconfined avalanche Gamsleiten (Obertauern/Salzburg) (© Suda: Drawing made from detailed plans from WLV Salzburg)

The actions from earth, water, and avalanche pressure must be set on the reverse side of the projecting discharge section surface. If discs are simultaneously acting as lateral framing, earth pressure must be positioned on the disc surface (board bearing effect).

For the construction, proofs must be carried out for the GEO ultimate limit state (base failure, creeping, and terrain fracture), as well as the STR ultimate limit state. For constructions in torrent areas, the HYD and UPL ultimate limit state could also be indicative.

The design of solid constructions – and thereby also avalanche breakers – is carried out as for supporting structures pursuant to proof process 2 and 2* EN 1997-1 [193]. Details about the indicative actions and the design process in connection with the construction and functionality of constructions similar to torrent breakers can, for example, be found in [20].

Structures of this type are financially best dimensioned and optimized using FE models. In connection with manual pre-dimensioning or verifications of results from FE models, the individual parts of the constructions must be examined. A component's support force is positioned in each instance as an action onto the component below. Boards can be dimensioned using board tables or FE models, discs using suitable frame models or non-linear FE models.

6.5.3 Constructive design

The general construction of the steel concrete parts is according to EN 1992-1. The concrete covers follow the indicative exposition classes. Generally, the concrete covering will be between 25 and 40 mm. If the avalanche breaker is located in a torrent, there must be a nominal concrete cover of at least 60 mm, pursuant to ONR 24 802 [201].

Discs and beams made from steel concrete are executed in a steel-clad manner from the inflow side and along the flow sides at least to the height of the avalanche flowing component (Figure 6.42a, Figure 6.43). Heavy plates with a thickness of between 8 and 20 mm are used for steel cladding. They are connected by welded head bolts or reinforcement rods with the construction concrete. The sheets act as abrasive protection for the concrete from ice and stone blocks.
For constructions that also are stressed by torrent processes, the construction principles for torrent dams must also be taken into account. These can be found in [20] and in detail in ONR 24 802 [201], for example.

6.6 Design of avalanche galleries (tunnels)

In avalanche galleries, there is a differentiation between open and closed galleries. The avalanches flow over the gallery roof. Besides the actions from avalanches, there is also earth load, rock fall, and stresses from the road traffic (action) to be taken into account during the design. In special cases, the gallery roof must also be calculated for other uses (e.g. forest roads). In Switzerland, a guideline was published regarding the measurement of avalanche galleries [10]. The minimum action combinations to be examined are stated in Table 6.18.

Fig. 6.42 Examples of Avalanche Breakers: (a): upslope view of an avalanche breaker for confined avalanches; (b): avalanche breaker for unconfined avalanche; (c): downslope view of an avalanche breaker for confined avalanches (© WLV Salzburg)

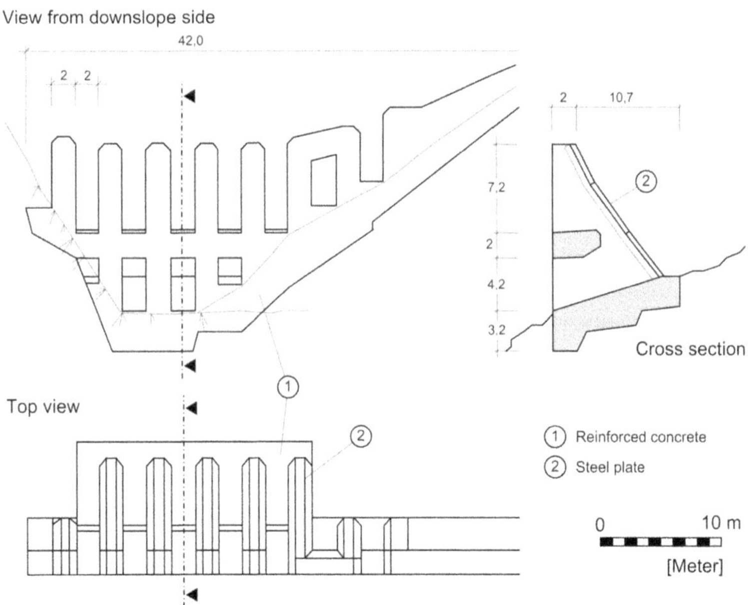

Fig. 6.43 Avalanche breaker in a confined avalanche (© Suda: Drawing made from detailed plans from WLV Salzburg)

Table 6.18 Recommended action combinations (AC) for avalanche protection galleries for the limiting state of the payload (ULS) and usability (SLS)

Design Situation	AC	Actions combined with another		
		Permanent action	Variable action	Extraordinary action
DS1	AC1	Deadweight Earth pressure		—
	AC2	Deadweight Earth pressure	Avalanche Pressure	—
DS2[b)]	AC3	Deadweight Earth pressure	Avalanche Pressure	—
DS3	AC4	Deadweight Earth pressure	—	Extraordinary traffic loads[a)]
	AC5	Deadweight Earth pressure	—	Extraordinary avalanche pressure
	AC6	Deadweight Earth pressure		Earthquake
	AC7	Deadweight Earth pressure		Rock fall, rock slide

a) Action types and models acc. to EN 1991-2 and ÖNORM B 1991-2 [192].
b) Construction conditions or repair work.

The load-bearing capacity (STR and GEO ultimate limit states) must be proofed for the indicated action combinations. Indicative design situations must also be taken into account (DS2). All actions from avalanche snow in movement (friction, normal loads, and deviation loads), together constitute either leading or accompanying actions. When setting up the action combinations, leading and accompanying actions must be defined, and varied if required. Eurocode-compatible combination coefficients can be found in the Swiss Guideline for avalanche galleries [10].

Actions from the *ordinary* avalanche result (design event) are to be classified as variable. The increase of the earth pressure must only be taken into account for deposited snow (natural snowpack or deposited avalanche snow). Horizontal pressure within the deposited snow need not be taken into account.

An *extraordinary* avalanche event is according to [10] an extremely rare (return period > design event) and statistically difficult to measure avalanche event with high intensity and great magnitude. For such an event to occur requires different, independent factors to occur simultaneously. This event must be treated as an extraordinary action.

For the action combination vehicle action, only the almost constant accompanying actions must be taken into account. Extraordinary traffic loads are the result of a vehicle

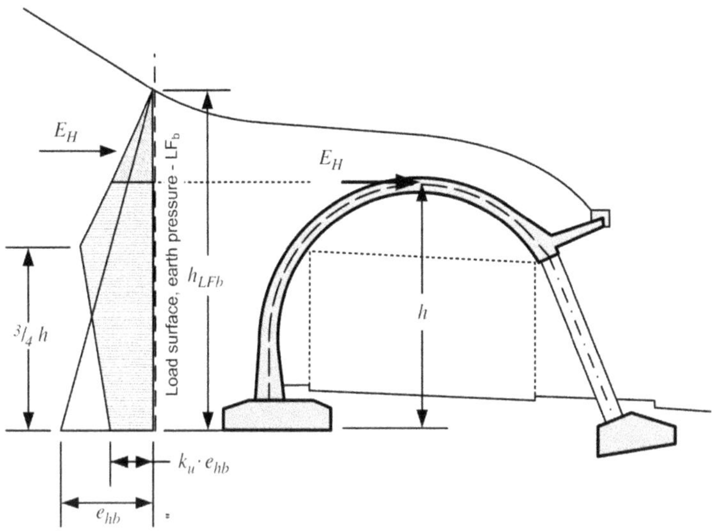

Fig. 6.44 Recommended earth pressure distribution according to *Schober*; pursuant to [56]

action. This must be set according to the rules in EN 1991-2 [194] (action loads on support constructions and top-builds).

Cross-sections of open galleries are stressed except for the superposition, predominantly because of the slope and backfill, while for two-sided closed galleries it is, due to the backfill, also possible to position a counter-pressure from the valley side. The actions on galleries are positioned on vertical load surfaces (replacement walls).

Vertical and horizontal loads occur on the support structure of the gallery due to earth pressure and avalanches. The vertical actions are a result of the cover and the avalanche pressures, the horizontal actions are a result of the earth pressures. When calculating the earth pressures from the backfill and cover, load redistributions must be considered. Schober (pursuant to [56]) recommends applying increased active earth pressure to calculate the horizontal earth pressure tension (Figure 6.44). For this, he brings in redistribution factors k_u dependent on the inclination angle of the slope of the excavation.

In order to calculate the basic earth pressure without redistribution, e_{ub}, a 50% partial activation of the zero pressure and active earth pressure is assumed on the mountain side. This gives the basic earth pressure (without redistribution) according to Equation 6.50, when no creep pressure needs to be applied.

$$e_{hb} = 0,5 \cdot e_0 + 0,5 \cdot e_{ah} \quad [\text{kN/m}^2] \tag{6.50}$$

where:

e_0 Earth zero pressure, in [kN/m^2]
e_{ah} Active earth pressure, in [kN/m^2]

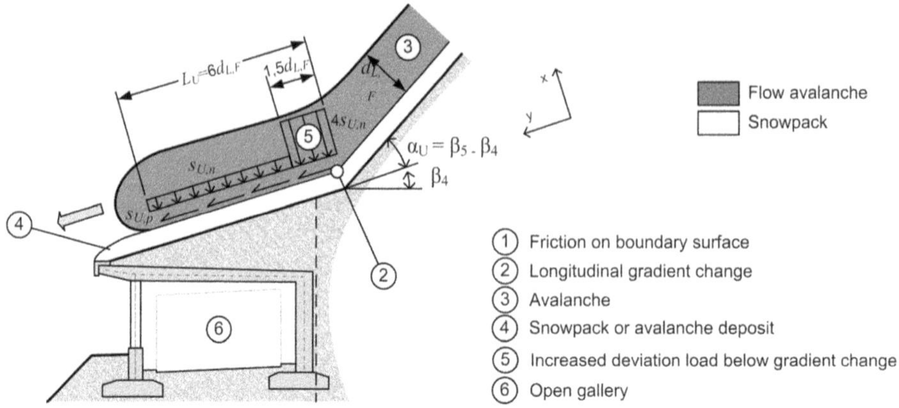

Fig. 6.45 Action of a flowing avalanche on a protective gallery, according to [10]

This earth pressure must be shifted squarely.

If the space for constructing the gallery is constructed from the bedrock, or if the construction walls are secured with a nail wall and the inclination angle in the excavation slope is larger than 70°, no gliding surfaces will occur in the backfill. In this case, the silo pressure according to ÖNORM B 1997-1-4 [198] is to be used for the earth pressure calculation. Details regarding the earth pressure approach on avalanche galleries can be found in [56] and [145], for example.

The indicative loads from avalanches occur because of the flowing component and act on the gallery roof, or the terrain surface on closed galleries. On the one hand, flowing avalanches generate a normal load on the terrain surface ($p_{n,k}$ in [kN/m^2]) and because of friction on the soil, a tangent (parallel) load on the terrain surface (p_t in [kN/m^2]) (Figure 6.45). The load that is normal on the terrain surface is dependent on the inclination of the terrain surface in the terrain area in question (ψ in [°]). For more about this model, see Section 3.3.1.

After a longitudinal gradient change, the deviation of the avalanche leads to a deviating force (Figure 6.45) in addition to the forces from the flow Table 6.19. This is separated, in the same way as the flow forces, into a characteristic load normally ($q_{un,k}$ in [kN/m^2]) and in parallel (q_{up} in [kN/m^2]) to the terrain surface. The calculation takes place using the flow velocity v_{Lf} in front of the longitudinal gradient change. The size of the vertical deviation force is dependent on the deviation angle (δ_u in [°]). It is the difference between the terrain incline before and after the longitudinal gradient change. The vertical deviation force is to be placed on the length of the deviation stretch (l_u in [m]). In [10], it is recommended to place 4 times the deviation force after the longitudinal gradient change (Figure 6.45). These load models are based on studies by Platzer & Margreth [165].

The geotechnical proofs (GEO limiting states) must take place as for stabilizing structures according to ÖNORM B 1997-1-1. For avalanche galleries, the limiting state for serviceability must be examined under all circumstances. In terms of usage

Table 6.19 Formulas for the calculation of the load ratios on the substratum caused by flowing avalanches (characteristic values); according to [10]

Action	Normal to the terrain surface	Parallel to the terrain surface[a)]	Unit
Flowing avalanche	$p_{n,k} = \rho_{L,F} \cdot g \cdot d_{L,F} \cdot \cos\beta$	$p_{t,k} = \mu \cdot p_{n,k}$	[kN/m²]
Deviation force[a)]	$q_{n,k} = \frac{\rho_{L,F} \cdot d_{L,F} \cdot v_L^2 \cdot \sin\delta_U}{l_U}$	$q_{p,k} = \mu \cdot q_{n,k}$	[kN/m²]

a) Acts on the gliding surface of the avalanche, as a result of friction.

suitability, according to [10], and if needed, the actual avalanche actions and winter-specific actions such as icicles, black ice, melt water, salty snowmelt, and spray fog, backflow of avalanches, and so on must be taken into account. The proof of usage suitability (fissure width limitation, limitation of the deformation, limitation of the tensile and compressive stress) must be carried out in accordance with EN 1992-1-1 (+NAD). When choosing the concrete covering and the concrete composition, exposition classes for chloride and frost-thaw-stress may be indicative. The standard values for permissible deformations can be found in the Swiss Guideline [10].

7 Construction work and maintenance of structural avalanche control

Florian Rudolf-Miklau, Wolfgang Schilcher, Stefan Margreth, Gebhard Walter and Jürgen Suda

7.1 Construction work (avalanche defense structures)

7.1.1 Fundamentals of construction work in Alpine environments

7.1.1.1 Conditions on avalanche control construction sites

Construction sites in the starting zone of avalanches are certainly in Alpine terrain. Exceptional, to a degree extreme, framework conditions therefore apply to the construction site equipment, the development, the worker protection, but especially to execution of the actual construction (Figure 7.1a). While avalanche braking and avalanche deflection construction measures are mainly performed in areas near valleys and are comparable to normal earth and dam constructions, the construction of the snow supporting structure is basically from a different construction sector and is to be seen as a special construction method. Further explanations in this section therefore essentially cover the construction methods for defense structures in the avalanche starting zones.

Avalanche construction sites in the starting zone are characterized by their exposed locations and steepness of the terrain, the specific framework conditions are shown in Table 7.1.

7.1.1.2 Demands for building methods and construction machines in avalanche control

Avalanche defense structures in the starting zone place considerable demands on the construction method used. Workflows can hardly be standardized and require individual adaptation to the framework conditions of the construction site.

Avalanche defense requires specific machines, equipment, and safety equipment. The development and supply of the construction site in the rugged mountain terrain, the establishment of the construction site as well as the accommodation and catering of the construction workers are particularly challenging. Due to the steep and narrow space constraints, the construction equipment and materials as well as operating resources can often only be used manually and this places high demands on logistics, weight-saving and intelligent methods. Characteristic for the elevation is a lower performance and more rapid wear on the equipment – compared to valley locations.

7.1.2 Construction site facilities and infrastructure

7.1.2.1 Construction site facilities: overview and requirements

The construction site equipment and facilities include all of the production, storage and transport equipment and workplaces, which are required for the construction, alteration or renovation of avalanche defense structures. The objective of the building site is to employ the resources with short hold-back times and a high utilization rates on the construction site. Contrary to the requirements, on avalanche construction sites the ideal

The Technical Avalanche Protection Handbook, First Edition. Edited by Florian Rudolf-Miklau, Siegfried Sauermoser and Arthur I. Mears.

Fig. 7.1 Work conditions on an Alpine elevated construction site for defense structures in avalanche starting zones: (a): Assembly work and material manipulation with helicopter (Wanne-Litze avalanche, Vorarlberg); (b): Construction site equipment (© WLV Vorarlberg)

Table 7.1 Framework conditions (environment) for the construction of the defense structures in avalanche starting zones

Topography (avalanche starting zone)	On open slopes between 30° and 50° steep; often rocky, strongly structured topography, but also homogeneous steep slopes; in certain cases also in the forest (forest avalanches)
Elevation	Between 1200 and 2700 m above sea level, in special cases in the permafrost region (over 2400 m), some even below the forest limit
Substratum	Often talus deposit or bedrock (with layers of debris), less often deep, fine-grained soil; occasionally steep mountain meadows or Alpine dwarf shrub heath
Accessibility	Long transport routes, long walkways, development through a difficult (steep) terrain
Weather conditions	Strong sunlight (UV exposure), extreme (short-term) temperature changes, exposure to rapid changes in weather, snowfall and freezing in the summer possible, strong wind (storm), thunderstorm
Risk factors	On steep slopes often dangers of fall and falling rocks; extreme and rapidly changing weather conditions: storm, lightning, heavy rain, frost, snow and ice; physical risks: physical overexertion, dizziness, dehydration, exposure to harmful sun rays
Construction season	Snow-free season (duration greatly dependent on altitude), in the Alpine areas between June and October

Table 7.2 Elements of equipment for an avalanche site (overview)

Social and office facilities	Rooms for breaks, changing, sanitation and meetings, crew quarters and canteens
Storage rooms and warehouse space	Storage infrastructure in steep terrain: Storage of drilling equipment, injection equipment, grout, supporting parts, anchoring parts, operating resources
Communication facilities	Radio, telephone, signal devices, satellite communication equipment
Elements for the protection of the environment	Fire, noise and water protection, waste disposal, water supply and sanitation services (generally no connection to public supply networks)
Accessibility	Construction access roads, ascent routes, handling areas, parking areas, conveyor and cable car installations, helicopter landing pads; platforms, scaffolding
Safety and protection devices	Devices for the prevention of accidents, protective gear, rescue equipment; protective equipment against rock fall, fall protection

situation is often not achievable even with optimal preparatory planning due to technical and local situations (Figure 7.1b).

Due to framework conditions at avalanche construction sites as shown in Section 7.1.1.1, special difficulties apply to virtually all areas of construction equipment and facilities. The establishment of the avalanche construction sites differs significantly from the typical construction site equipment for construction and civil engineering (Table 7.2).

7.1.2.2 Social and office rooms, housing for workers

Social and office rooms at the avalanche site were built exclusively as permanent buildings until the 1990s (wooden sheds, huts) (Figure 7.2). A permanent construction however is only lucrative for long-term projects and mainly on construction areas with rapidly changing weather conditions to protect the team, or as required for an emergency overnight stay.

Today, on the other hand, mainly transportable, multiple-use light-construction containers are used. Usually these are custom-made products, which are suitable for transport by helicopter and truck. The container should have sufficient thermal insulation, ventilation sliders for diagonal ventilation as well as a free-bearing welded steel frame that also acts as a load-bearing structure during transport and as lightning protection for the occupants. The weight of the container suitable for helicopters, including inventory (seats, heater, table, ladder, etc.), should not exceed 750 kg to allow for transport with any common type of helicopter. Special attention should be given to

(a) (b)

Fig. 7.2 Social and office premises: (a): Fixed team accommodation in the area of the construction site Kruckengraben avalanche (Styria) including the necessary supply infrastructure (© WLV Styria); (b) Office and social space of lightweight containers in the area of Kehm and Kenneckbach avalanches (Tyrol) (© WLV Tyrol)

the lightning protection because in the Alpine area, thunderstorms quickly form and the lightning risk on Alpine construction sites is very extreme due to the exposed location.

The question of accommodation in avalanche construction sites is closely linked to the accessibility and the mobility of the workers. Up to the 1980s, it was standard practice to build labour camps with a corresponding company kitchen on avalanche construction sites. The economic feasibility of these facilities was justified by the large size of the team and the high cost of mobility. Since the 1990s, the framework requirements of avalanche constructions have changed such that labour camps can only be built in the immediate area of construction on very remote construction sites. The reasons for this are the increased requirements for worker protection, the high expenses for supply and disposal (waste, water, energy) as well as the elaborate technical infrastructure. Another reason lies in the constantly declining workforce size (6–10 workers per site) as a result of increasing mechanization and rationalization of work processes. Furthermore, due to higher mobility and the increased requirements of social institutions, workers are less willing to sleep at the avalanche construction sites. Occasionally, transportation from the valley to the avalanche construction site is only possible by helicopter.

7.1.2.3 Storage and handling of construction material and equipment

Lightweight containers are mainly used for *storage* and *device warehousing* in the area of the avalanche construction sites (Figure 7.3a). Since these facilities in the vicinity of construction sites must be moved frequently, the transport weight of this container (max. 750 kg) is of significant importance. Storage and device warehousing should be locked from unauthorized access and must be anchored against extreme winds.

Storage and manipulation areas are required both in the valley and in the immediate area of the avalanche construction site. The storage areas in the valley are to be situated on a road that can be accessed by truck. Alternatively, these storage areas can also be at

(a)

(b)

Fig. 7.3 Storage rooms and warehouse space: (a): Storage rooms in the form of lightweight containers in the area of the avalanche construction site Hohe Munde (Telfs/Tyrol) (© Rudolf-Miklau); (b): Storage and manipulation platform on a suspended scaffold with stored fuel tanks, compressors and chemical toilet (Vordernberger Mauer, Styria) (© Rudolf-Miklau)

the end of an access road at the elevation. Storage areas are to be flat and should have a minimum size of 30 × 50 meters. Clear airspace (no steep terrain, trees or power lines) is essential for operating the helicopter. The classification and preparation of the construction materials and the construction of the avalanche structures for the transport flight (helicopters) takes place on these facilities. The intermediate storage of construction materials and resources is on the storage locations directly next to the construction area. In the absence of flat surfaces in steep terrain, these can also be erected as platforms on suspension scaffolds (Figure 7.3b). Helicopter landing pads can also be erected in the form of platforms.

7.1.2.4 Supply and disposal at construction zones

Supplies and the disposal of waste at avalanche construction sites is a special challenge due to the location, the loads to be transported, but also for reasons of environmental protection. The following are the supply and disposal requirements in the vicinity of avalanche construction sites: water (drinking water, service water), food, waste water and sewage, waste, energy (electricity, fuel storage).

To minimize the transport costs for the water supply to the construction site (1 m^3 of water ~1000 kg), water will be obtained on site at high altitude sites. To do this, either springs in the catchment area will be tapped or rain or melting water will be collected in a tarpaulin and stored in plastic tanks (Figure 7.4a). On diverse construction fields, with the corresponding weather conditions, the water needs for the entire summer season can be secured. To maintain crucial hygiene regulations, the water extraction on site is normally confined to industrial water. The drinking water supply is, on the other hand, to be brought in from the valley (bottled drinking water, drinking water container) by helicopter, cable car or truck.

Sanitary systems in the area of the avalanche construction sites will use mainly chemical toilets involving chemical substances to reduce putrefaction (odour) and for the disinfection of excretions. Faeces will be collected in a tank. Amongst other things, formaldehyde, glutaraldehyde or quaternary ammonium compounds, to degree also biodegradable products, are used as disinfectants. The faeces collected in chemical toilets must be transported regularly to the valley from the avalanche construction site. At avalanche construction sites with fixed team accommodation, small wastewater treatment systems are used in some cases. The systems usually consist of a sedimentation tank (for example three-chamber sedimentation tank), in which the solids and

(a) (b)

Fig. 7.4 Sanitation and disposal (industrial water): (a): simple collection of precipitation and melt water in the area of the avalanche construction site (© WLV Vorarlberg); (b): wastewater treatment system in the area of the avalanche construction site Vordernberger Mauern (© WLV Styria)

floating fats in the wastewater are separated. Subsequently, the wastewater is then cleaned in a biological step (Figure 7.4b).

Mobile diesel generators are generally used to supply the avalanche construction site with electrical energy (electricity). Most are devices with a frequency of 50 Hz, a rated speed of 1500 rpm, a nominal voltage of 400 V and a rated current of 18 to 45 A. Increasingly also photovoltaic (solar power) systems are used for the power supply to social and office rooms.

7.1.3 Transportation systems on avalanche defense construction sites

7.1.3.1 Principles of transportation

In the construction of avalanche defense structures, the accessibility of the construction site and the material transport are special cost factors. Not only the supply of the construction site but also the manipulation of the equipment and construction materials within the construction site are of importance. The site infrastructure (accessibility) also represents a central safety factor for rescue missions and a rapid retreat possibility in bad weather.

For the site infrastructure, there basically are three concepts to choose from in construction of avalanche defense structures that can be selected after evaluating the costs and safety factors:

- Goods road connection (Section 7.1.3.2)
- Material ropeway incl. systems for the cross transfer on the construction site (footbridges, platforms) (Section 7.1.3.3)
- Helicopter transport (Section 7.1.3.4).

7.1.3.2 Transportation road

If the topography permits, the avalanche construction sites will be connected with truck accessible routes (Figure 7.5a). The *goods road connection* for avalanche construction

(a) (b)

Fig. 7.5 Construction site accessibility and transportation: (a): construction of an access road for an avalanche construction site in Alpine terrain; (b): material ropeway used during transport (© WLV Tyrol)

sites is not only a transport road but is also a workplace and travel route for the workers as well as storage and manipulation area. When building access roads, all three functions have to be taken into consideration. Furthermore, the construction of access roads has important synergy effects for the forestry service in the catchment area and the protective forests [48] as well as for the future maintenance and care of the protective structures. Basically, development should be sought with a road navigable by truck, since both the construction site safety is increased and the cost for manipulation can be reduced significantly if the necessary material and equipment can be brought directly by truck to the immediate site.

The following principles – compiled in accordance with [48, 80, 81, 212, 251] and described in detail in Table 7.3 – are to be considered in particular in the construction of access roads in Alpine terrain:

Table 7.3 Construction methods for access roads in the Alpine area (in accordance with [48, 80, 81, 212, 251])

Planning and routing	Basic principles: variation of gradient, harmonious integration in the terrain. Maximum gradient: 12% (in exceptional cases: up to max. 22%), minimum lengthwise gradient: 3% (due to the risk of waterlogging the body of the track). Negative points are: unstable side slopes, spring horizons and rock-strewn slopes.
Landscape-adapted excavator construction	Construction equipment: *hydraulic excavator*; in steep terrain (more than 70% slope gradient) additional mass length compensation with trucks. Hydraulic excavators as universal construction equipment: forwarding of tree trunks in the track (savings of forwarding equipment), manufacturing, and compression of the raw formation including a valley-side catch ditch as an embankment base (depth: 1.0–1.5 m), production of the longitudinal drainage, production of cross drainage (pipe culverts, rinses), removal of rock (tearable rock, hydraulic breaker tools, excavator chisel for preparation of blasting) and production of artificial slope backups. Construction speed: 5–10 meters/hour General profile: Road width min. 3.5 m plus double-sided bank strip of 0.5 m each; longitudinal drainage: for example spiked ditch (0.5–0.75 m); mountain-side embankment: inclination 1:1 (for bedrock: 5:1); valley-side embankment (fill): inclination of 2:3.
Bedrock	Landscape-protective blasting, scraping or milling methods. *Blasting technique:* production of drill holes in head drills, blasting portions (smaller amounts of explosives). *Scraping technology:* hydraulic excavators with hydraulic chisel, cutting of the worn rock material on the spot.

(continued)

Table 7.3 (*Continued*)

Serpentines	For 2 axle trucks, a minimum radius of 8 meters is required, for trucks with trailers, a minimum radius of 10 m.
Channel crossing, road drainage system	*Channel crossing:* Fords or culverts. Hydraulic dimensions according to [48]. *Road drainage:* Create longitudinal drainage of the subgrade either by lateral incline (mountain side) of 2–3% or camber of the road; culverts (concrete pipes, corrugated pipes, plastic pipes) at regular intervals (Table 7.4).
Retaining structures, slope re-vegetation	*Securing of overly steep embankments* by means of retaining structures: Overview in Table 7.5. *Re-vegetation methods:* Hay flower seed (manual seeding), mechanical seeding or humus free ray seeding (on rock slopes), mechanical injection procedure; details on slope re-vegetation for example in [54, 181, 240].

- The planning and routing of the road should be adapted to the landscape.
- In the Alpine terrain, access roads are to be built exclusively in a landscape-friendly excavator construction.
- Serpentines are to be built in areas with relatively low inclination (maximum transverse gradient: 40%) and on stable ridges.
- On rocky substratum, landscape-protective blasting, scraping or milling methods are to be used for the construction of the raw track.
- Channel crossings are to be designed in a ford-like longitudinal profile and to be protected against erosion.
- The water drainage in the longitudinal direction is to be established in the form of an edge or a triangular or trapezoidal ditch. Ensure regular transverse drainage (Table 7.4).
- The rich topsoil is to be removed carefully when building the raw track, to be stored on the side and, after completion of the road, to be moved to the road embankment.
- Mountain and valley-side slopes in unstable or extremely steep slopes are to be secured as soon as possible by retaining structures (Table 7.5).
- Road embankments are also to be re-vegetated as quickly as possible to protect against erosion.
- Access roads are to be maintained regularly, but especially at the beginning of a construction season (grading, rolling, and clearing the water drainage).

Table 7.4 Recommended distances for underground lateral drainage on access routes (according to [80])

Street longitudinal tilt %	4	5	6–8	9–10	11	12–14
Spacing of the culverts [m]	120	110	100	90	80	70

Table 7.5 Overview of the supporting structures to secure unstable embankments (according to [54, 181, 283])

Construction method	Construction materials	Height up to . . .	Seeding and mulching possible	Areas of application
Wooden crib + Single walled + Double-walled	Wood	5 m	Yes	Mountain-side slope securing, stabilizing fill material, for deep loose material
Retaining structure from prefabricated concrete elements	Prefabricated concrete elements	7 (10) m	Yes	Most expensive stabilizing methods, anchoring for unstable bedrock
Back-anchored log crib wall	Wood, steel tracks, injection drilling anchor	Staggered over 5 m	Yes	Rock and loose soil; securing steep slopes; protection of old wooden cribs
Dry-stone pitching	Rough stones	12 m	Yes	Point-shaped or linear securing of the embankment; only with stable foundation
Reinforced earth	Geotextile steel wire mesh reinforcement belts Earth (filling)	10 m+	partially	Permanent stabilization of embankments and steep slopes as well as dams
Gabions (stone baskets)	Wire baskets Stones (filling)	10 m	partially	Linear slope retaining structure for soaked embankments
Retaining walls	Concrete Reinforced concrete	10 m+	no	Linear, vertical support of embankments at the foot of the slope as well as for securing piles; bridge abutments

Table 7.6 Ropeway systems in the avalanche defense in the starting zone (according to [48, 207])

Type	Properties	Range
Permanent cable car system (Material ropeways)	Permanently installed cable car system for transporting material, occasionally also for people.	up to 2000 m
Semi-mobile cable logging systems (long-distance, short-distance cable cranes)	Permanently mounted cable, stationary drive unit and carriage or wheel winch; mountain and valley-side anchoring of the bearing cable with dead man anchor in the loose material or rock anchors in the compact bedrock; supports from trunks or prefabricated truss supports (element supports)	Long-distance systems: 800–900 m; mid-distance systems: 400–800 m
Mobile cable logging system (tilting mast-cable cranes)	Drive unit, cable drum and tiltable or telescopically extendable main mast on a self-driven or drawn wheel or caterpillar track vehicle	

7.1.3.3 Material ropeway and cable cranes

For the direct connection of the construction site to a road, mobile *cable crane* are often used. Occasionally fixed *material ropeways* are also built (see Table 7.6 and Figure 7.5b).

The advantages of the cable crane and material ropeways are the permanent availability, weather independence and the independence from helicopter availability. In addition, transportation of heavier loads is possible. The cable crane is preferable due to the noise prevention for environmental protection. One disadvantage of cable cranes and material ropeways is the potential avalanche hazard of the system and the resulting damage, since this is usually limited to snow loads (snow slope-parallel pressure, ice attachment). Another disadvantage is the high physical strain of workers during the assembly (part weights up to 250 kg/piece.). Furthermore, the delivery by cable crane in strongly structured terrain is not the optimal solution (transverse transport) [48]. Another disadvantage is high personnel expenses for operation because the winch, as well as the valley station, typically needs to be staffed.

7.1.3.4 Heavy transport helicopters

Most heavy load transports in the area of the avalanche barriers are made today by *helicopter.*

The helicopter types (Table 7.7) are suitable in the high mountains for the transportation of loads between 900 and 1200 kg. The same applies to lifting and assembly work in avalanche control sites, where the helicopter is also used as a *construction crane* (Section 7.1.4.7). For all machines and constructions, which are required for the avalanche construction site, ensure that the total weight (load + slings) does not exceed 900 kg

Table 7.7 Comparison of helicopter types used in load flights for avalanche barrier construction in Austria and Switzerland (overview according to manufacturer's information – selection)

Helicopter type	BELL 205 A1 T53/17a	Eurocopter AS 350 B3 Equreuil	Eurocopter SA 315 B Lama	AS 332 Super Puma C1
Manufacturer	Bell/USA	Aerospatiale/ France	Aerospatiale/ France	Aerospatiale/ France
Engine	1 × Lycoming T53-L-13turbine	1 × Turbomeca Arriel 2B Turbine	1 × Turbomeca Artouste IIIB	2 × Makila 1A1
Performance	1305 kW	642 kW	649 kW	2800 kW
Max. net load	2000 kg	1200 kg	1000 kg	4500 kg
in 1000 m a. s.l. in 2000 m a. s.l. 3000 m a.s.l.	1650 kg 1500 kg 1350 kg	approx. 1150 kg approx. 950 kg approx. 750 kg	approx. 900 kg approx. 810 kg approx. 700 kg	Only for special transports (large loads)
Empty weight	2385 kg	1200 kg	1021 kg	4350 kg
Highest speed	210 km/h	287 km/h	210 km/h	300 km/h
Range	510 km with Reserve	680 km without Reserve	425 km	1060 km

(depending on altitude, Table 7.7). Thus, the use of cheaper helicopters is possible for most transport flights. The use of the helicopter for lifting and assembly work (load flights) is usually extremely noisy, the communication between the ground crew and the pilot is therefore by radio and the usual hand signals for landing [3] (Figure 7.6).

The advantages of using a helicopter for lifting and assembly work in avalanche defense structures in the starting zone are the lower costs for the construction site facilities and the possibility of rapid delivery to the construction site. It is also beneficial that the assembly work is relatively easy for the workers (manual manipulation and mounting of steel parts is minimized at the construction site) and the transport in autumn is still possible even after the first snowfall (longer construction season). A disadvantage is the dependence on the weather (wind and visibility), the high risk of working under suspended loads, the assembly under extreme noise and air turbulence and the increased risk of rock fall caused by helicopter down winds. Often it is best to use a combination of

Table 7.8 Characteristic performance metrics for drills in avalanche control (information sections Vorarlberg, torrent and avalanche control/Austria)

	Hydraulic Rotary engine	Pneumatic Rotary engine	Hydraulic top hammer	Pneumatic top hammer
Slewing torque	930 Nm	400 Nm	370 Nm	85 Nm
Power hydraulic aggregate compressor	17 KW	35 KW	26 KW	40 KW
Impact energy	—	—	85 J	approx. 45 J

(a) (b)

Fig. 7.6 Helicopter use in the avalanche defense construction works: (a): platform as a landing pad (passenger transport); (b): use for a load flight with mounting cable (© WLV Vorarlberg)

cable crane and helicopter, whereby light loads can be delivered with the cable crane and heavy loads or the installation of the snow supporting structures can be performed with the helicopter. The advantage of this combination is that, on the one hand, a cable crane is sufficient for light loads, and this is also available at any time for emergencies or equipment replacement and, on the other hand, the advantages of the helicopter can be exploited for transporting heavy loads such as steel snow bridges.

7.1.4 Special construction methods in avalanche defense in the starting zone

7.1.4.1 Principles of construction work in avalanche control

The methods for the construction of avalanche defense structures in the starting zone are characterized by the building operations in the remote, steep terrain.

The work for the construction of snow supporting structures can be divided into two sections. Section 1 includes the anchoring or foundation of the supporting structures, section 2 the assembly of the constructions. The most expensive part of the construction of snow bridges and snow nets is for the foundation. The assembly (establishment of supporting structures on the foundations) requires only a fraction of this expense. Depending on the foundation soil, the expenses for the foundation in comparison to the assembly is a ratio of 4 (8):1.

7.1.4.2 Construction of micropile foundations and drill technology

For *micropile foundations*, drill holes are drilled with a minimum diameter of 90 mm in accordance with [163]. The depth of the holes varies depending on the load between 3 and 8 m. Normally, the lightest possible hydraulically or pneumatically operated drilling equipment is used to make the drill holes. These must be manipulated by hand in the steep terrain, however, several-hundred-pound drilling units are also used, which are manipulated and secured with a hand winch (Figure 7.7a). In suitable site conditions, the holes can also be produced by a climbing excavator with drilling equipment.

To drill the holes and install micropiles, the following three methods can be used, depending on the subsurface conditions:

- *Method 1:*In the cohesive loose material and in the bedrock, the required holes (minimum diameter: 90 mm) are made with a drill carriage with rotary motor and deep hole hammer (Figure 7.7b). Then the anchor rods and spacers can be inserted into the hole. Pressure grouting is carried out via the injection channel using injection anchors or by an injection hose, ensuring the drilled hole is filled completely down to the deepest part of the drilling hole. If, while drilling, it is determined that there are cavities in the substratum, it is recommended that textile stockings be used to curb the loss of anchor grout. These however have the disadvantage of possible damage from sharp-edged material and large cavities that the desired effect is therefore not achieved. For a tear-resistant alternative, hoses made of fine geo grids can be used. Make sure that the permeability of the grout material is sufficient. In addition, this material does not have the elasticity of a stocking, consequently larger diameters must be selected for the hoses. In such cases, it is important to perform a relevant number of pullout tests (Section 6.2.9.8).
- *Method 2:*Drill equipment for colluvium to manufacture cased drill holes is used when the holes could collapse when drilling or when the anchor is inserted. With cased drillings, a casing tube is drilled in addition to the drill pipe. After drilling to the corresponding hole depth, the down-the-hole hammer is pulled with the drill pipe. Then the anchor rod (also with stocking or geotextile tube) can be inserted into the pipe without any obstacles. Once the anchor rod has been positioned, the casing tube is pulled. Then a pile cap is set up for the connection of the support structures and the borehole is grouted. Due to the slower drill progression and the increased expenditure of manipulation, the cased drilling is likely to require twice the time as for a normal hole.
- *Method 3:*Cohesionless, loose material can also be processed with drilling injection anchors. In this procedure, a top hammer drills directly on the drilling injection anchor with lost cutter (Figure 7.9). This is rinsed continuously during the drilling with cement slurry or grout. The cement or the grout settles around the anchor rod so that the pile is ready installed once the necessary depth has been reached. This method can be used only when no larger cavities are found, and this is also more expensive than method 1.

(a)

(b)

Fig. 7.7 Production of drill holes for micropiles: (a): drilling unit with a drilling rig (© Morath); (b): drill carriage with hydraulic motor and pneumatic down-the-hole hammer (© Schilcher)

In avalanche control, the drill holes necessary for the foundation (with micro piles) are manufactured almost entirely with portable light drill carriages. The basic philosophy of the portable drill unit is the dismantling of each device into portable individual parts.

This is based on a hydraulic drilling carriage (Figure 7.9a), which can be used for different drilling methods and drill hole diameters. Portable light drilling carriages are generally made of the following components: Mount body with hydraulic feed, saddle, rotary motor, down-the-hole hammer and fastener with support leg.

In geologically unfavourable conditions (*drill hole collapses*), a drilling technique is used with reinforced rotary motor and a casing of the bore hole. In fine-grained talus deposits, self-drilling anchors can also be used alternatively. In this case, a powerful top hammer is used (Figure 7.8). The drive can be either hydraulic or pneumatic.

In the past, mainly pneumatic drill equipment was used in avalanche defense construction works. Compressed air was used for the feed, the rotary engine, the hammer (stroke movement) and the flushing of the rock drill. Two compressors with a capacity of 5.5 m^3/min each were required.

In the course of the development of the foundation methods in snow supporting structures, drills with hydraulic drive prevailed (Figure 7.9a). With this technology, both the feed and the rotary engine for the drill are operated hydraulically. The support leg, which is required to adjust the proper height of the carriage, is also hydraulically operated. Generally, a hydraulic drilling system has a higher drilling capacity with the same energy consumption. Cased holes and drilling with self-drilling anchors are generally only useful with a hydraulic drive.

Since the summer of 2000, a specially adapted climbing excavator with drilling apparatus (Figure 7.9b) has been used in the Tyrol on certain avalanche construction sites to establish the foundation. A lightweight carriage with hydraulic drive has been mounted on the climbing excavator, which is powered via the oil pump on the climbing excavator and has its own controller. With the equipment, cased holes can also be drilled; an eccentric drill head is used for this. As a life-line for the excavator, two steel winches are assembled on the underside with 50 m steel rope each (breaking load: 24 t) and a tensile force of 7 t each.

7.1.4.3 Construction of wire rope anchors

The *wire rope anchor* was developed for the foundation of snow nets. The installation of the rope anchor is the same as for a micropile. After drilling the borehole, a galvanized spiral rope is fitted as a tension anchor instead of a rod.

The spiral rope is manufactured in the factory as a loop with spacers. The installation is carried out so that the two ends of the rope are in the deepest part of the bore hole and a loop to hold the load is on the soil surface. The use of a halved PE tube has proven successful for inserting the spiral cable link. This is inserted into the hole and permits an easier insertion of the rope anchor, to which the injection hose is attached. On a fissured substratum, the use of a stocking can reduce the anchor grout

(a)

(b)

Fig. 7.8 Production of drill holes for micropiles: (a): with drilling injection anchors using a hydraulic top hammer (© WLV Vorarlberg); (b): through drilling in colluvium (pipe drilling in soft ground) (© WLV Tyrol)

consumption when pressing. If the load is not axial, an additional reinforcement pipe is installed in the loose material near the top soil. On a poor substratum or when stronger bending stress is expected, the top soil is excavated and a concrete foundation is made for mounting the non-axial loads.

(a)

(b)

Fig. 7.9 Lightweight drill carriages: (a): with hydraulic drive; (b): climbing excavator with drilling apparatus (© WLV Tyrol)

7.1.4.4 Construction of ground plate foundation

Ground plate foundations are used today for transferring the pressure forces on the system supports into the ground and are used frequently for steel snow bridges and snow nets.

The angle of force transmission for snow bridge supports is significantly less favourable than for snow nets, for this reason, much larger excavations are required for the

Fig. 7.10 Manufacturing a ground plate foundation on a steep rock substratum (© Schilcher)

installation of ground plates for snow bridge foundations. On bearing substratum, the ground plates for snow nets can be installed without almost any excavation. The size of the ground plates is dependent on the load-bearing capacity of the soil and the load (Section 6.2.9.5).

The excavations for the ground plate foundations are made by hand in most cases. Much of the foundation excavations are made with shovel, pickaxe and pneumatic hammers (Figure 7.10). In rare cases, blasting operations are required. The material dug out is deposited to the side so it can be poured back without much effort after the establishment of the constructions. In the corresponding terrain and ground conditions, the excavations can be inexpensive with an excavator. After the excavations, the ground plate must be inserted immediately in the loose talus because the slopes are not stable and material constantly falls into the excavation area. On solid ground, it is possible to keep dug holes open for a longer period of time, so that after making a series of foundation holes, the supporting structures can be set in place with a helicopter, including the ground plates.

7.1.4.5 Construction of concrete foundation

Until the 1970s, most of the snow bridges built in Switzerland and Austria used concrete foundations. Concrete foundations have proven to be very durable and solid foundation methods, but also very expensive. For supports and girders in a snow bridge, 1.5 m^3 up to 2.5 m^3 of cast-in-situ concrete are required; this means a transport weight of 3–5 t. For larger sections using rope cranes, this type of foundation can be made at a reasonable cost. The cost of transporting the concrete increases sharply, with the increasing use of the helicopter. Furthermore the large amount of work for the production of the excavations, shuttering and reinforcement, the transverse transport on the construction site and the manufacture and placement of the concrete, the excess excavated material that often led to damage on the underlying plots, and the negative impact on the landscape are reasons why concrete foundations are rarely used anymore.

In Switzerland today, snow bridges have concrete foundations installed in combination with micropiles, allowing the amount of concrete to be significantly reduced. Small concrete foundations to absorb the pressure forces in the supports are also manufactured for snow nets as an alternative to ground plates. In coarse talus without fine material, where micropile foundations are very complicated and expensive, concrete foundations or combined concrete foundations can be an alternative.

7.1.4.6 Construction of micropile foundation in solid rock (rock anchor)

Micropiles according to ÖNORM EN 14199 [199] are also used in the rock for the foundation. These are also called *rock anchors* when used in bedrock. Rock anchors in accordance with [163] are to be fitted with a drill hole that has at least 1.5 times the diameter of the steel tensile rod. This kind of anchoring for snow bridge girders is used for outcropping bedrock or bedrock projected up to about 1 m. For the manufacturing of the foundation, either a part of the rock underground is to be excavated niche-like or the loose material is to be removed up to the outcropping rock. After that, the holes for the micropiles can be drilled. The use of machines for this foundation method is confined to a compressed air powered hand drill, a mining hammer and a compressor.

7.1.4.7 Mounting methods for snow supporting structures

The mounting of supporting structures includes assembling the snow bridges or snow nets at a manipulation site and transporting the structured or their individual parts from the site to the construction area. Today the most used method is mounting with a helicopter. If no suitable helicopter is available, the transport can also be performed with a rope crane. In favourably connected construction areas, supporting structures can also be installed with a truck crane or a telescopic crane.

Helicopter assembly has the advantage that most heavy steel construction parts for the supporting structure can be assembled at a suitable manipulation site with the help of a truck-mounted crane (Figure 7.11a). When the helicopter has sufficient lifting capacity, this flies the complete supporting structure into the construction area. The assembly team then attaches the supporting structure hanging from the helicopter (Figure 7.11b). The snow bridges are attached to the foundations with bolts. Because the individual steel snow bridges often differ by the length of their supports, a precise numbering of the structures and a clear assignment to the correct position are necessary to avoid unnecessary (expensive) flight times. For the helicopter transport of snow nets, supporting systems are offered by the manufacturers. The snow nets are fastened to the foundations with ropes and clamps. These workflows require great experience and the highest concentration of pilots, because the assembly teams of normally four to six men work under the helicopter and sometimes under the suspended load.

Cable crane assembly means the transport of components with the rope crane from the storage yard in the valley to the construction area. Transverse distribution of the components from the mountain station is possible with a narrow trolley on suspension scaffolds.

(a) (b)

Fig. 7.11 Assembly of snow supporting structures: (a): assembling a snow bridge with a truck-mounted crane; (b): installation of the snow bridge by helicopter (© WLV Vorarlberg)

7.1.5 Safety engineering in avalanche control

7.1.5.1 General principles of employee protection at construction zones in alpine environment

Avalanche construction sites are subject to the general provisions of occupational safety for the construction industry. These are in the interest of the personal safety of employees and third parties and are obligatory for both the employers and the workers. The principles of hazard prevention are that the establishment of the construction site, the construction processes, the equipment and materials are set up, designed and used to ensure the safety and health of workers.

There are no specific legal provisions for the protection of workers on construction sites for avalanche control in any country. The protection of workers is an essential part of the planning, because 60% of all fatal accidents at work can be traced to incorrect decisions made before starting work. This applies particularly to construction sites for the snow supporting structures [2].

The explanations in this section refer to the specificities of the protection of workers during construction of the avalanche defense structures in the starting zone, which must be performed regularly, in particularly exposed and steep terrain. These construction sites are typical summer construction sites and thus confined to the snow-free times. In the remaining time of the year, the risk of avalanches or falls due to icing are too high to carry out the construction work (Section 7.1.1.1).

7.1.5.2 Preventive employee protection (prior start of construction)

The organization and preparation of the avalanche construction sites according to the standards of worker protection places high demands on the logistics of the construction company. In addition to the special precautions in the scope of construction management (Section 7.1.2), specific regulatory permits for industrial safety in the avalanche control play a major role. These include approvals for the blocking of roads and railway lines (due to the risk of rock fall), approval of transport flights with helicopters (prohibition of overflight by public transport facilities with attached loads), blasting work and special weekly work time models.

Other important areas of preventive protection of workers include the provision of *personal protective equipment* (PPE), the planning of the safety devices (in particular, fall protection, lightning protection, rock fall protection), construction site coordination, the construction schedule, the safety and health protection plan, as well as contingency planning (alarm plan, special warning systems, for example, rock fall).

7.1.5.3 Requirements for employees on avalanche control construction zones

Working in extreme terrain, as regularly occurs in avalanche construction sites, may only be performed by persons with the necessary physical, mental and technical capacities. Construction workers in the avalanche area must be free from vertigo and sure-footed and have experience in Alpine terrain. Workers who tend to have dizzy spells, epilepsy, or anxiety, or have a hearing loss, strong visual impairments or physical disabilities are not suitable for employment on avalanche construction sites. Also, drug-addicted or alcoholic workers are not permitted. The construction workers employed in the area of the avalanche braking control must be familiar with the specific construction methods and safety requirements in high Alpine sites. They must have high physical and mental performance (capacities). The work presents a high risk of accidents; therefore, it is necessary to wear special personal protective equipment (PPE) (Section 7.1.5.4). The typical size of a working team on avalanche construction sites is between 6 and 10 workers.

The responsible site manager as well as the construction supervisor must have appropriate special training and adequate experience in the planning and implementation of construction operations in Alpine terrain and be familiar with the weather phenomena in the mountains, the requirements and techniques for the protection of workers on steep slopes and Alpine rescue processes. The technology of coarse and fine stakeout for snow supporting structures, drilling and foundation engineering (Section 7.1.4.2), the installation procedure for supporting structures (Section 7.1.4.7), the selection and the use of special construction equipment and the organization of construction management in Alpine terrain also require many years of experience.

7.1.5.4 Personal protective equipment (PPE)

The most important element of safety at work is the *personal protective equipment* (PPE) provided to the employee. The PPE required for avalanche construction sites varies from the standard equipment for building and civil engineering sites (helmet, gloves, safety shoes, protective clothing, dust mask, hearing protection at over 85 dB

(A), chainsaw protective trousers for chain saw work, safety glasses, warning clothing) in many areas and includes in particular pieces of equipment for Alpine use. These include an Alpine helmet (with chinstrap), weather- and cold proof clothing (rain protection), safety mountain shoes (with sole tread), fall protection for work with a safety harness (Section 7.1.5.5), sunglasses and sunscreen as well as UV-protective clothing (Table 7.9). Protective helmets must be worn at all times in the area of avalanche construction sites.

Table 7.9 Personal Protective Equipment (PPE) for use on avalanche construction sites (special features)

PPE avalanche control	Properties, standards
Alpine helmet	Protective helmets that also provide sufficient safety against rock fall: When working at high altitude, only helmets with a chinstrap according to EN 397, which opens at 25 daN to prevent strangulation, may be worn. In exceptional cases, helmets with a chinstrap sized to 50 daN are allowed to avoid a loss of the helmet in the event of a fall.
Weather- and cold proof clothing	Water and cold resistant work clothes: 100% waterproof, permeable to water vapour, according to EN 343, class II.
Safety mountain boots	Safety boots for avalanche construction sites must comply with EN 345 and be equipped with toe caps (steel, aluminium, or plastic), which must withstand a force test of at least 200 J. Additional properties: Sole profile (6.5 mm), cut protection liner (Kevlar).
Hearing protection	Hearing protection (ear muffs) is legally mandatory when working with a noise of more than 85 dB, in particular in the field of compressors, drills, pneumatic hammers, or when working with a chain saw.
Safety glasses	Protective goggles are to be worn for scraping or cutting work with the power saw and drilling work causing fine dust. These must comply with EN 166 and be certified by an independent test institute.
Sunglasses, sunscreen	In the Alpine area, special (optical) sunglasses with 100% UV protection and side protection are to be worn as well as sunscreen according to the personal skin type.
UV-protective clothing	Clothing as personal protective equipment: plenty of skin cover (such as sun protection shirt). Head cover and neck protection: A wide-brimmed hat with sufficient neck protection or special neck protection for helmets

Fig. 7.12 Work requiring a safety harness in the area of the avalanche construction site Hohe Munde (Telfs/Tyrol) (© WLV Tyrol)

In the area of avalanche construction sites, there is often a danger of falling. Protection against falling (the wearing of a safety harness) serves to protect people at work from a risk of falling when technical or structural protective measures (such as rails) are not possible or are not sufficient as well as when working in rock faces (Figure 7.12).

The personal protective equipment for work with a safety harness (PPE-sh) include the safety harnesses (harness), the safety rope, the anchorage point, a shock absorber and an arrester (tightness of long safety ropes) and, if necessary, fall arrest systems [4] (Table 7.10). The breaking strength of the PPE-sh is designed to provide multiple

Table 7.10 Personal Protective Equipment for work requiring a safety harness according to [4]

PPE work requiring a safety harness	Properties, standards
Fall arrest harness	Harnesses have designated fall arrest attachment points (A). Only the fall arrest attachment loops have the task of transferring the energy of the fall gently over the leg loops to the body.
Safety ropes	Safety ropes for high working heights must comply with EN 1891 (half-static ropes) and are used in connection with impact means, catch systems and rope shorteners. Attachment, extension or shortening of safety ropes using knots is not permitted in Austria.

(continued)

Table 7.10 *(Continued)*

PPE work requiring a safety harness	Properties, standards
Sling points	Sling points must withstand at least a load of 10 kN or, according to EN 795, twice the value of the force that arises when catching. The use of slings is allowed. Anchors are often used in the area of the avalanche site, in special cases, horizontal cable systems are used as sling points.
Fall arrest systems	Fall arrest systems must comply with EN 363. For work with safety harnesses on steep slopes, running safety devices including a flexible anchor line can be used in accordance with EN 353-2.

safety features, nevertheless these may only be used strictly according to the instructions. All components are tested in accordance with the European PPE Directive 89/686 EEC, marked and approved only for personal protection. Avoid, in particular, running the cables over sharp edges, using damaged or stressed PPE-sh or improper cleaning or storage. Before each use, visually check the PPE-sh, have the PPE-sh checked yearly by a person authorized by the manufacturer or a competent, trained person. Eliminate any damaged PPE-sh.

In addition to direct work with the safety rope, clean-up work is required on a case-by-case basis to protect avalanche construction sites from falling rock from overlying rock slopes. The use of a *two-rope technique* is legally mandatory for this, work while hanging freely in the rope also requires the use of a descent seat.

7.1.5.5 Fall protection equipment and scaffolding

In the area of the avalanche construction sites, there is a risk of falling in many areas, especially in steep rocky or grassy slopes. *Fall protection systems* are legally mandatory for fall heights of more than 2.0 m. As a means of fall protection, weirs (railings), suspended scaffolding and catch nets or abseil protection are generally available.

Hanging scaffolds have special importance in avalanche defense in the starting zone. In the past, the scaffolding on site was made completely from wood. The hanging scaffoldings used on avalanche building sites today consist of aluminium-steel brackets (Figure 7.13a). These have fall protections with removable railing supports (calculated for 600 kg/console). This allows quick assembly. In addition, the consoles are adjustable for various inclinations.

In Austria, the hanging scaffoldings for special applications of avalanche constructions have been further developed. For example, for the construction site at the Breitlehner avalanche (Telfs/Tyrol), an 'adjustable (telescopic) scaffold piece that can be disassembled into individual parts' was constructed for extremely steep and rocky starting

(a)

(b)

Fig. 7.13 Suspension scaffolds and fall protection in the avalanche barriers: (a): Hanging scaffolding with aluminium scaffolding console (© WLV Vorarlberg); (b): 'adjustable (telescopic) scaffold piece that can be disassembled into individual parts' on the Breitlehner avalanche (Telfs/Tyrol) (© WLV Tyrol)

zones (platform width: 1.25 m; field width of max. 2.0 m; calculated according to ÖNORM B 4007 for light work and MOT approved) (Figure 7.13b) or the frame piece *East Tyrol* (lightweight, very flexible scaffolding trestle made of wood, which has a low weight of approx. 13 kg, can be used for all inclinations, calculated according to ÖNORM B 4007 for light work).

7.1.5.6 Safety regulation for helicopter transportation

Special safety regulations apply for occupational safety during helicopter transport flights (lifting and mounting structures) (Figure 7.6a, b). The loads are generally to be limited, elevation dependent, to the maximum lifting capacity of the helicopter (Table 7.8) (for example, 900 kg). The loads are to be fixed securely, so that they cannot become dislodged nor fall out during the transport flight. Long loads are to be broken into multiple loads, whereby the loading tackle sling may not be at a larger angle than 60°. Crossing traffic areas (roads, cable cars) with attached loads usually requires obtaining official permits.

Cargo transport with a helicopter must follow a set schedule and is coordinated by a marshal who is in constant radio contact with the pilots and the crew at the loading and unloading point. Communication at the loading and unloading point between the pilot and the marshal uses predetermined hand signals [3] and radio communication.

7.2 Maintenance of avalanche defense structures

7.2.1 Principles of maintenance

Several decades of avalanche control in the Alps have brought about a substantial stock of protection structures, which ensures a high level of security against avalanche dangers. Maintenance measures for existing protection structures (systems) will therefore gain a great deal of importance in technical avalanche defense in the future.

The following section presents the maintenance of technical avalanche protection measures as part of a comprehensive lifecycle approach. The maintenance concept established in the construction industry is based on a *cradle-to-grave* principle, which considers the technical system (structure) comprehensively from the start of manufacture until the end of service life (removal, rot) and aims at optimal cost efficiency for the whole lifecycle (life-cycle costs). In addition to these systemic bases, this section also covers practical issues related to the analysis and assessment of damage and defects on avalanche protection structures as well as the technical and organizational standards and methods of inspection, service, repair and restoration of avalanche protection structures.

7.2.2 Maintenance management and condition assessment

7.2.2.1 Lifecycle of avalanche defense structures

After their completion, avalanche defense structures are exposed to rapid ageing and wear and tear due to the extreme environmental conditions (effects) and function, which results in a limited service life [221]. The defense structures can fulfil their protective functions over the entire scheduled lifecycle only under the condition that they are regularly monitored and maintained during the use phase. In comparison to other structures, avalanche defense structures – particularly in Alpine areas – are exposed to extreme environmental conditions (humidity, temperature fluctuations, Section 7.1.1.1) and exposed to dynamic impact from snow, avalanches and wind, which lead to great wear and tear (damage) and rapid physiological ageing. Extreme snow conditions, avalanche effects, rocks and movements in the substratum, which were not considered in the planning and design of the protective structures, can lead to serious damage to the equipment. An overview of the causes of damage and the classification of the typical damages to avalanche protection structures is in Section 7.2.3. Experience has shown that the snow supporting structures are particularly vulnerable for dynamic effects, while the risk of damage to avalanche braking and deflection structures is comparatively lower.

So far, there have been hardly any long-term studies on the effective service life of technical avalanche protection measures. Depending on the type of protection measures, in practice (on the basis of the observation of existing technical avalanche protections) a service life of 40 to 80 years is expected for snow supporting structures, the effective service life of avalanche braking and deflecting constructions may, in certain conditions, be much higher. The scheduled life depends considerably on the acting processes, the functioning of the system and the construction materials used, the maintenance strategy (frequency and time of maintenance) (Section 7.2.2.2) as well as the intensity and frequency of inspection and condition monitoring are also of great importance [1].

The service life of temporary, technical avalanche protections (Section 9.2.1) on the other hand depends first and foremost on the operation of the system, the maintenance expenditure as well as the technological lifecycle of the mechanical and electronic components. In these systems, therefore, the preservation status of the structural components can hardly be seen as a limiting factor. The same applies to avalanche monitoring systems (Section 9.3).

7.2.2.2 Functions and strategies of maintenance

According to the definition [163, 204, 249], maintenance is the combination of technical and administrative measures as well as measures of management (monitoring, service, repair and renovation) throughout the lifecycle of a protection structure (maintenance management). The objective is the maintenance of the functional state or return to this so that the structure can fulfil the required function. Therefore, on the one hand, maintenance management includes the monitoring of protection structures in terms of stability and serviceability, on the other hand, the maintenance measures (Section 7.2.4), which are required as a result of the preservation condition (Section 7.2.2.3) (Figure 7.14).

The regular maintenance of avalanche defense structures is therefore the basis for determining the serviceability. The maintenance of structures for permanent avalanche protection must, in accordance with ONR 24807 [204], follow a *maintenance strategy* laid down at the beginning, which defines the time and scope of the measures (Table 7.11). The maintenance strategy depends on the requirements for safety and protection, the number of protection structures, the category of the structures and the economic efficiency of the conservation measures and includes, in particular, the wear stock (according to DIN 31051:2003). This is defined for avalanche protection constructions according to [204] as a supply for the possible fulfilment of function under standard conditions, which are inherent in a protective structure based on manufacture, repair or improvement. The wear limit is reached when the wear

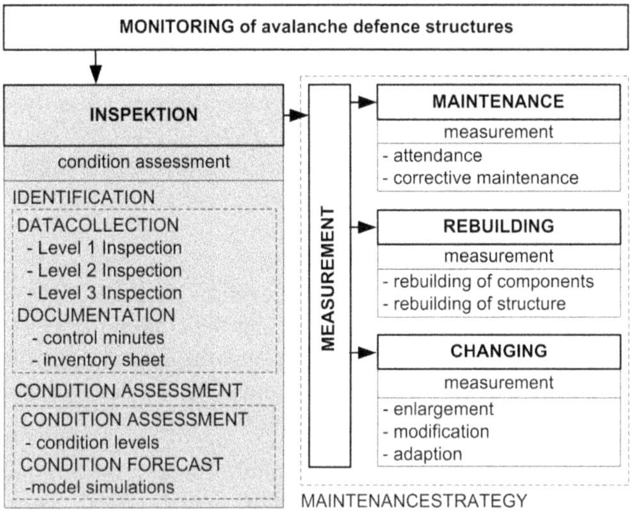

Fig. 7.14 Maintenance management in the avalanche defense structures: Principle functions of the maintenance management of the structures of permanent technical avalanche protection according to ONR 24807 [204]

Table 7.11 Possible maintenance strategies for avalanche protection structures according to [237] and ONR 24807 [204]

Maintenance strategy	Measures and effects
Predictive maintenance strategy	Smaller, predetermined measures are implemented on the basis of maintenance directives (operating guidelines). In certain circumstances, the *wear stock* (in accordance with DIN 31051) is not exploited.
Condition-based maintenance strategy	Preservation measures are determined based on the condition. The requirement is the continuous monitoring of the condition of the protective structures.
Medium/long-term corrective strategy	Preservation measures are only determined once protective structures reach a structurally and functionally critical condition. Wear stocks are fully exploited.
Short-term corrective strategy	The measures are only taken if there is an immediate loss of protection or a threatened failure of the structure (exceeding the limit of wear stock). This strategy can lead to an irreversible loss of protection function. A neglected maintenance usually leads to the same result.

supplies have been fully depleted. The repairs must be initiated at the latest at this point to exclude a breakdown or a failure of the system. From an economic perspective – at least in theory – the time period represents the *preservation cost optimum*, an earlier intervention can lead to relatively higher maintenance costs over the lifetime of the system (wear stock is not fully exploited), a later intervention may, however, lead to a temporary or complete loss of the protective effect and require a premature re-investment [237].

7.2.2.3 Inspection and condition monitoring

The *inspection* serves to ensure the reliability (stability, serviceability, durability) of the structure. This is divided into the condition documentation and the condition assessment. The inspection includes the operational instruments and the instruments for documentation, the condition assessment includes those of assessing the damage and defects on avalanche protection structures (see Section 7.2.3.3).

The inspection and condition assessment of protective structures is also determined by legal and technical standards (Austria: water act, ONR 24807 [204]; Swiss Guideline [163]). These standards are therefore of great importance for a universal, uniform quality of maintenance and inspection measures in avalanche protection since numerous public and private actors (state avalanche control services, communities, transport carriers, cable car operators, landowners, beneficiaries of protective measures) are involved. The preservation and periodic inspection of avalanche protection structures is to be regarded as a legally binding obligation of the owner (operator) or concession holder of the avalanche protection systems or the protected parties, in some cases, the property owner is liable for maintenance on the basis of the relevant legal standards.

The ONR 24807 [204] sets forth a structured and comprehensible procedure for carrying out the necessary inspections on avalanche protection structures, which includes recording the condition, documenting the structure condition and condition assessment. In the condition assessment, damage and defects to the avalanche protection structures are collected and the consequences for the stability and serviceability are assessed. In the spirit of efficiency, the concept provides for several levels of condition assessment, that is the first assessment, continuous monitoring, control and examination, which are based on different survey intensities (visual, engineering methods) and different qualifications of the recording staff (trained person, expert). The inspection of protective structures aims to use a comprehensive method to survey the preservation status of the facilities and then assign them to a condition level (Section 7.2.3.3). However, the severity of the impact of failure of the structure on the group measures (the associated measures) or the protected areas is decisive for determining the nature and urgency of preservation measures.

7.2.3 Damage and functional defects of avalanche defense structures

7.2.3.1 Overview and classification

Avalanche defense structures are subject to extreme effects, which are often difficult to quantify. Damage analysis can provide valuable information, to check the effects and hazard scenarios and, on the other hand, to improve structures. Since research has only provided few data about impact on avalanche protection structures (Section 9.3.6), damage analyses are often the only way to better understand the interaction between processes and structures.

In addition to the correct construction, the vulnerability of avalanche protection structures depends highly on the acting processes, the type and condition of the protective structures. The vulnerability of avalanche protection structures is generally low. For example, the total damage in Switzerland during the extreme avalanche winter of 1999 was less than 1% of the total investment in the structural avalanche protection [246]. Experience has shown that snow supporting structures are the protection structures most vulnerable to damage. Avalanche deflection, braking and catching structures as well as avalanche galleries, on the other hand, are robust structures and relatively insensitive to damage. Snowdrift structures are built mostly in flat terrain, greatly reducing the probability of snow pressure damage. On the other hand, damage may occur due to wind exposure.

The consequences of damage are very different for the various avalanche protection structures:

- The collapse of the ceiling of an avalanche gallery due to large avalanche impact can lead to a direct risk to persons (key structure).
- If several snow supporting structures are damaged – for example due to severe snow gliding – there are generally only spatially limited damages. Experience has shown that such damages do not lead to a total failure of an avalanche support structures, since they consist of many individual structures. Furthermore, the harmful event has no immediate effect on the protected area, because it is independent of the damaging avalanche process.

ONR 24807 [204] defines *damage* as changes in the structure, its foundation or the surroundings from internal or external causes that lead to a reduction in load-carrying capacity, serviceability or durability of the structure. According to the Swiss standard SIA 269, damage is an unpredicted weakening of the material substance of a structure occurring after approval. A *defect*, on the other hand, is the absence of a property that the structure should have during approval according to the contract or according to accepted engineering practice. Often concealed defects, such as an insufficiently injected anchor, lead to subsequent damage.

7.2.3.2 Causes for damages and functional deficits

The causes of damage to protective structures can be systematized according to [216] by differentiating between internal causes that directly affect the structure and external causes, such as environmental impacts (Table 7.12). A further subdivision is performed into typical, predictable, or atypical, often unforeseeable influences:

- A typical, internal cause for damages is, for example, ageing.
- Typical, external causes (normal actions), such as snow pressure or impact of small snow slides in supporting structures, are predictable and are by default taken into account for the calculation of protective structures and therefore usually cause no damage.

Table 7.12 Outline of causes of damage to snow supporting structures

Influence	Internal cause (Structure/material)	External cause (effects from the environment)
Typical (predictable)	Material ageing (Corrosion, embrittlement) load changes (material fatigue)	Snow pressure Impact of snow slides Erosion
Atypical (often not predictable)	Material defects, construction defects, construction faults, planning errors	Avalanche impact Cornice collapse Rock and block fall impact Landslide Strong erosion Dead wood (falling trees)

- Atypical, internal causes such as material errors, design errors or construction defects often lead to damage.
- Atypical, external causes (unusual effects) such as rock fall or avalanche impact on snow supporting structures are ignored consciously or unconsciously when designing a measure. Such influences are often unpredictable, but can cause great damage.

The probability of damage depends greatly on the respective location of a protection measure, in particular, how complex the hazard situation, the topography or the geotechnical situation presents itself. The vulnerability to damage increases with increasing age. To make matters worse, the design and execution of older measures often no longer correspond to the current state of technology.

7.2.3.3 Damage analysis and condition assessment

Recording of damage or loss of function in the scope of the inspection is the basis for a condition assessment. In the Swiss guideline [163] and the ONR 24807 [204] it is proposed to create the condition level of a protection structure as a function of the structural safety and serviceability, as well as a forecast for the occurrence of consequential damages (Tables 7.13 and 7.14). The classification of all protective measures in one condition level, and the ongoing review of changes allow an estimate of the future cost of renovation and maintenance measures. Generally, damage to support structures, dams or snowdrift structures can be recognized early through consistent visual observations. The load-carrying capacity of non-visible components such as anchors or the stability of earth dams, however, cannot be checked visually. This requires more in-depth condition assessments – such as with anchor pullout tests, geotechnical testing or engineering surveys.

The condition of a protection structure is essential for the assessment of its effectiveness in the context of risk assessments. A four-step procedure can be applied in accordance

Table 7.13 Mapping the condition levels for the structural safety and serviceability at the time of recording and the time horizon for the occurrence of consequential damages

Condition level		Structural stability[a)]	Suitable for use[b)]	Time horizon for occurrence of consequential damage	Examples of steel snow bridges
ONR 24807 [204]	Swiss Guideline [163]				
0	—	—	—	—	—
2	1	guaranteed	guaranteed	>5 yr	Bent crossbeams, erosion to foundation base, less than 10 cm up to 20 cm, collection of debris on the grate with a thickness of less than 50 cm, uniform corrosion (rust).
4	2	limited	limited	2 to 5 yr	Exposed anchors, between 20 and 40 cm (still intact), a pressed in micropile anchors, slightly bent supports.
5	3	very limited (risk of collapse)	very limited (no support function or support function highly limited)	1 yr	Buckled supports, severely deformed or broken girders, broken or pulled out anchors, bent micropiles.

a) at the time of recording (= security of reaching the limit condition of the load-carrying capacity, loss of the overall stability of the support structure)

b) at the time of recording (= serviceability of the protective structure)

Table 7.14 Condition levels of avalanche protection structures

Condition levels according to ONR 24807 [204]		Condition levels and assessment of repair urgency and need for action in accordance with *Swiss Guideline* [163]	
0	Structure is not necessary	—	—
2	Good state of conservation	1	Good, little urgency: watch.
4	Inadequate state of conservation	2	Defective, medium urgency, maintenance in 1–3 yrs.
5	Poor state of conservation	3	Bad, high urgency, immediate corrective maintenance.

with the Swiss Guide PROTECT [166] for assessing the effect of protection structures (Figure 7.15):

- In a first step (rough assessment), a preliminary assessment is made based on the existing knowledge of the process and the protection structure.

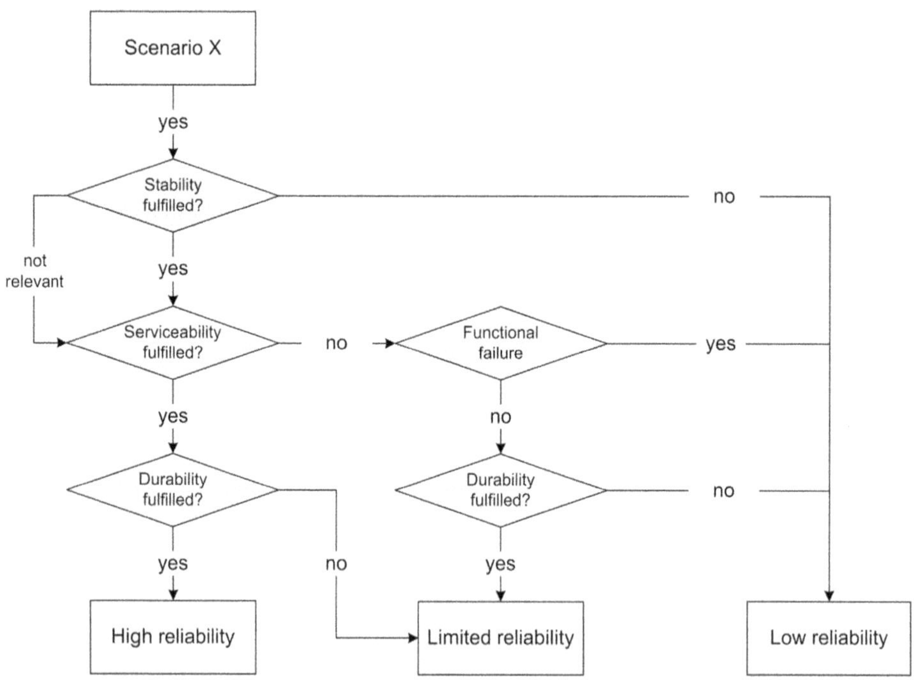

Fig. 7.15 Assessment of the reliability of a measure for a particular scenario X on the basis of the stability, serviceability and durability, according to PROTECT [212]

- In the second step (assessment of measures), the reliability of the structure is determined by assessing the qualitative and quantitative condition and behaviour for various hazard images, effects or scenarios, based on the stability, serviceability and durability.
- The reliability of the measures is used in the third step (impact assessment) to determine the relevant scenarios for the risk assessment. If the reliability of a structure (e.g. due to being in a poor state of repair) is restricted, only a partial effect can be assumed.
- The fourth step addresses the spatial planning implementation of the endangered area reduced by the protection structure.

7.2.3.4 Damages at snow supporting structures

The greatest damage to snow supporting structures occurs during dynamic avalanche effects, especially when avalanches enter the defense area from the top or the sides. Supporting structures can be destroyed completely. In the 1999 disaster winter, avalanches caused approximately 20% of the damage done to supporting structures [246] in Switzerland. Rock fall can be just as destructive as avalanche impact.

The most common forms of damage are deformation of the superstructure and the foundations due to great snow pressure as a result of severe snow gliding or when supporting structures are snowed over due to being built too short or from redistribution of snow by wind. The choice of site-adapted construction height is crucial for a sustainable use of snow supporting structures. On the superstructure, supports bent towards the valley are the most common damage. The hollow profiles used often are, compared to HEA profiles, significantly more susceptible to damage. Today, Swiss Guideline [163] requires that when designing the supports, a transverse load must also be taken into account.

A systematic survey of the damage in Tyrol by Klaus [139] showed that less than 1% of support structures were damaged. Approximately 62% of the damage affected the foundations and around 38% affected the superstructure. In the past, damage to the foundation was often attributable to an optimistically estimated load-carrying capacity of the anchors and micropiles. Only in recent years have anchor pullout tests (Section 6.2.9.8) been performed regularly for avalanche control. In Austria, a relatively high number of cases of damage have been recorded to the girder foundation with the parallel rod anchor (Section 5.2.3.3), this is no longer incorporated. In Switzerland, support foundations in the loose rock on only a micropile have not proven to be successful. A damage catalogue with the most common damage to permanent snow supporting structures and possible causes is listed in Table 7.15 (Figure 7.16: A, B, C), Table 7.16 (Figure 7.17: D, E, F), Table 7.17 (Figure 7.18: G, H, I, J), Table 7.18 (Figure 7.19: K, L, M, N, O) and Table 7.19 (Figure 7.20: P, Q).

7.2.3.5 Damages at avalanche dams

Avalanche dams are generally robust structures that are hardly vulnerable to damage. The most common cause for maintenance work is the deposition of rock fall and debris in the area before the dam. On the one hand, this reduces the effective height; on the other hand, the run-off can be hampered. Other criteria to consider when dams are

Table 7.15 Damages and causes on crossbeams for snow supporting structure (condition levels according to ONR 24807) [204] (Figure 7.16: A, B, C)

Index	Condition level	Damage and observations	Possible causes
A	2	Deformation of the crossbeams (dents), water does not drain, bending (beam centre, cantilever), rust formation	Upper crossbeam: excess snow Lower crossbeam: snow gliding, rockfall, surface corrosion
B	5	Missing beam or intermediate beams; defective fasteners on the crossbeam (broken bow, missing or loose screws)	Falling rock, storm, avalanche impact, overload from snow pressure
C	2	Filling the support grate with stones and earth material leads to a reduction of the effective height	Local erosion or landslide: Effective height is reduced (from approx. 0.5 m problematic)

checked are any deformations (settling) of the dam body as a result of terrain movements, stability problems or erosion as a result of heavy rainfall. The superstructure and the foundations of brake or deflection walls made of steel or concrete can be deformed and damaged as a result of excessive impact forces. A damage catalogue with the most common damage, and possible causes, to earth dams and deflecting walls is listed in the following Table 7.20 (Figure 7.21: R, S, T, U).

7.2.3.6 Damage at avalanche galleries and tunnels

Damage to the load-bearing construction of avalanche galleries is very rare. Damage may occur with old, poorly designed buildings or unforeseen effects such as vertical avalanche impact (snow masses fall over rock steps vertically on the roof of the gallery, Figure 7.22a), block impact or impact by vehicles. As a result of suction or impact forces, damage was determined on the valley side of the gallery front (Figure 7.22b). The most common damage to avalanche galleries is concrete spalling, corrosion of the reinforcement, water penetration or defective drainage systems. Causes are the high chloride concentrations as a result of the use of road salt, especially in the superficial concrete of the supports and the back wall, leaking dilatation joints or defective seals. The waterproofing systems and the concrete cover for the reinforcement of older galleries mostly no longer correspond to the current state of technology.

7.2.3.7 Damage to snowdrift structures

There are two prominent causes of damage in drift constructions:

- Dynamic pressure of the wind: The most common damage occurs when wind peaks (squalls) break boards or *vibrate* them out (Figure 7.23a). The boards are often

Fig. 7.16 Damage to the support grate: A, B, C. [A] Deformation of the crossbeams (© Margreth). [B] Missing crossbeams (© Margreth). [C] Filling the support grate with stones/earth (© Margreth)

Table 7.16 Damage and causes on load-bearing constructions for snow supporting structure (condition levels according to ONR 24807) [204] (Figure 7.17: D, E, F)

Index	Condition level	Damage and observations	Possible causes
D	5	Valley-side buckling or bending of the supports, damaged connection pieces (often at the structure edges)	Valley-side attachment of the snow cover creates transverse load, overload caused by oversnow (constructed heights too low), too much buckling length (extra-long support), increased border forces, undersizing
E	5	Deformation and lateral tilting of girders, bending/folding of the connector tabs, deformation of the crossbeams, breaking of the bolts (often at the structure edges)	Overloading due to over snowing (effective heights too low), poorly trained forces (construction defects) undersizing
F	5	Missing bolts, missing splints, missing screws, loose nuts	Construction defect, vibrated out as a result of strong wind

designed as a predetermined breaking point. Therefore, it is relatively rare that the load-bearing structure of a snowdrift fence will fail due to the high dynamic pressure of the wind and the structure collapses.

- Unexpected snow mechanic effects: Snowdrift structures are mostly blown free when in a correct, wind-exposed position. However, unfavourable winds or very large amounts of snow can cause these to lose their function and be completely snowed in. In flat terrain, lateral bracing supports or struts in a driving snow wall can be damaged by forces as a result of the settling of the snow cover. The result is a loosening of cables or bending of braces (Figure 7.23b). As secondary damage, supports can also be damaged. In sloped terrain, snowdrift structures can be impacted by snow pressure. Wind baffles are relatively sensitive to damage when they are completely snowed in.

7.2.4 Maintenance measures for avalanche defense structures

7.2.4.1 Methods of maintenance

The maintenance of avalanche protection structures includes service and repair. Measures that slow down the increase in the degree of age-related wear and tear include service (operational maintenance). This includes smaller measures that are required to protect the substance and function. This includes, for example, the removal of deposits on the protection structures. Repair (structural maintenance) includes

Fig. 7.17 Damage to the load-bearing construction: D, E, F. [D] Buckling of the supports in a snow bridge (© Margreth). [E] Deformation of the girder (© Margreth)

Fig. 7.17 (*Continued*) [F] Missing bolts (© Margreth)

Table 7.17 Damage and causes of snow nets (condition levels according to ONR 24807) [204] (Figure 7.18: G, H, I, J)

Index	Condition level	Damage and observations	Possible causes
G	5	Valley-side bending or break-off of the support, position of the support (turning)	Valley-side attachment of the snow cover produces transverse load, twisting of the support due to the uneven load
H	4	Geometry support foot and foundation, support is on the foundation or is dislocated, joint is blocked	Tilted foundation, lifting forces due to wind, soiling from debris or earth material
I	5	Status of the network boundary rope, the mesh ropes and net coverage or net attachment: wear, cable breakage, damaged galvanizing	Pressure overload due to snow pressure or falling rock, too much slack (caused by loosening of the guy ropes)
J	2	Sag (lack of tension) in the ropes, position and clamping action of the wire rope clamps, condition of the guy ropes: wear, cable breakage, damaged galvanizing	Overload caused by snow pressure or falling rocks, construction deficiency: for example clamps not tightened properly

Fig. 7.18 Damage to snow nets: G, H, I, J. [G] Buckling of the supports in a snow net (© Margreth). [H] Tilted foundation leads to inadmissible geometry on the support foot (© Margreth). [I] Nets severely damaged by falling rocks (© Cantonal office for forestry Bern)

Fig. 7.18 (*Continued*) [J] Displaced wire rope clamps result in insufficient rope tension (© Margreth)

Table 7.18 Damage and causes for anchorage and micropiles (condition levels according to ONR 24807) [204] (Figure 7.19: K, L, M, N, O)

Index	Condition level	Damage and observations	Possible causes
K	5	Broken or bent anchor, loose screw connections, projection of anchor head	Overloading (pressure of snow, avalanche, rock fall), material defects, terrain displacements, assembly errors
L	5	Torn-off anchors with grout column, loosened anchors, strong oxidation	Anchor length too small, less favourable subsoil properties than assumed, damaged corrosion protection
M	5	Torn out anchor without grout column, position of the anchor head moved	Overload, unfavourable load direction (perpendicular to the anchor), insufficient bonding between anchor and grout
N	5	Micropile bent laterally or in the direction of the valley, destroyed grout column	Inadequate lateral ballast of micropile (irreversible process, structure collapses), critical if support is available only on a micropile, cracks in the ground are often visible
O	4	Projecting anchor head/ floor, exposed anchor, destroyed grout column	Assembly error, soil erosion, projecting by more than 20 cm critical because bending stress can occur

Fig. 7.19 Damage to anchors and micropiles: K, L, M, N, O. [K] Broken mountain-side anchor in a snow bridge with pressure bar (© Margreth). [L] Ripped out wire rope anchors with grout column (© Margreth). [M] Ripped out rock anchor without grout column (© Margreth)

Fig. 7.19 (*Continued*) [N] Laterally bent micropile for a snow net support (© Margreth). [O] Projecting anchor/soil caused by soil erosion (© Margreth)

Table 7.19 Damage and causes with ground plates and concrete foundations (condition levels according to ONR 24807) [204] (Figure 7.20: P, Q)

Index	Condition level	Damage and observations	Possible causes
P	5	Exposing or tilting of the ground plate, often with end structures	Erosion, inadequate connection depth, overloading, insufficient ground resistance
Q	2	Exposure or undermining of the foundation base, tilting or moving of the foundation, broken foundation	Erosion, overloading or eccentric load, insufficient foundation depth, dimensions that are too small, poor concrete quality, missing reinforcement

Fig. 7.20 Damage to ground plates and concrete foundations: P, Q. [P] Exposed ground plate (© Margreth)

Fig. 7.20 (*Continued*) [Q] Exposed foundation as a result of building settlements (© Margreth)

measures that restore the functioning of a defense structure in the event of impairment or restrictions – such as from damages, construction defects and other functional limitations or as a consequence of an extreme event. Small-scale construction measures without a significant increase in the use value, such as for example the exchange of beams in steel snow bridges after overloads. Along with measures on the structures themselves, this includes the stabilization of small ground movements, which are related to the stability of the protection structure.

Table 7.20 Damage and causes on earth dams and deflective walls of steel (condition levels according to ONR 24807) [204] (Figure 7.21: R, S, T, U)

Index	Condition level	Damages and observations	Possible causes
R	2	Surface erosion, geotechnical instabilities	Heavy rains, inadequate compaction
S	4	Bent steel elements, torn concrete elements, moved foundations	Excessive effects of avalanche, insufficient assessment
T	2	Reduced effect height due to material deposits, reduced retention volume as a result of filling the catchment room, danger of log jam in the area of the discharge opening (torrent)	Catchment area can be achieved by debris flows or falling stones, often in combined structures against avalanches and debris flows
U	2	Geotechnical instabilities, reduced effect height as a result of strong settlements	Insufficient compaction, bad bulk materials

Fig. 7.21 Damage to avalanche dams and deflecting walls in steel: R, S, T, U. [R] Erosion of the dam embankment (© Margreth). [S] Bent steel elements (© Margreth)

Fig. 7.21 (*Continued*) [T] Reduced effective level of the dam as a result of material deposition (© Margreth). [U] Geotechnical instabilities (© P. Mayoraz)

7.2.4.2 Methods of renovation

The restoration of avalanche protection structures also includes measures for the change or renewal of components in addition to the repair work. A pre-requisite for a sustainable renovation is the determination of the cause of the damage. Table 7.21 shows the regeneration methods in relation to possible causes of damage.

7.2.4.3 Urgency of maintenance measures

ONR 24807 [204] provides a possible approach for the assessment of the urgency of redevelopment measures. The *urgency* of the measures is defined in the measures grid based on the condition level (Table 7.22) which defines the minimum requirements. The measures grid describes the relationship between the condition level and the urgency of

(a)

(b)

Fig. 7.22 (a): In April 2009, a wet snow avalanche damaged the roof of the avalanche gallery Val Pischöt built in the 1920s (Samnaunerstraße). The avalanche fell almost vertically on the roof and broke into the gallery (© Margreth); (b): During an avalanche in February 1999, suction forces tore out two glass elements of the valley-side facade of Salezer Gallery at Davos. The glass elements were designed for pressure forces of 6 kN/m^2 (© Margreth)

redevelopment measures. The measures grid depends on the chosen preservation strategy (Section 7.2.2.2). Standard structures with a condition level 5 are to be placed immediately in the repair program or the expendability of the structure is to be determined (removal or controlled expiration). Instead of the measures in Table 7.22, at any time a redevelopment can be performed or a building check (stability, serviceability) can be arranged.

(a)

(b)

Fig. 7.23 (a): Snowdrift fence on the Hinterrugg (SG) with a height of 6 m. Due to strong winds, individual boards were ripped out; (b): snowdrift fence on the Hinterrugg (SG). The lateral diagonal braces (profile 70/70/5 mm FE 235) have bent under the settling forces of the snow cover (© R. Ehrbar)

Table 7.21 Regeneration methods based on the causes of damage

Cause of damage	Regeneration method	Measures
Ageing, construction defects, material defects	Corrective maintenance, renewal	Correction of defects, replacement of parts or renewal of the entire structure
Misjudgement of snow loads or the foundation soil, design error	Repair or renewal after new assessment	Raising, reinforcing, compacting of the structures
Miscalculation of the snow height, underestimation of the dangers from rock fall hazards, unrecognized slope instabilities	Repair with the supplementary arrangements	Repair, replacement of parts or renewal of the entire construction with additional measures (snowdrift structures, rock fall protection, slope stabilization, protection against avalanches from up-slope starting zones)

Table 7.22 Measures based on the conditional level in standard structures ONR 24807 [204]

Condition level for standard constructions		Measures and time points
0	Structure is not necessary	Controlled expiration, demolition, partial demolition
2	Good state of conservation	Long-term need for action: repair measures as required
4	Inadequate state of conservation	Mid-term need for action: Renovation, renewal period of three years and mandatory monitoring after larger events
5	Poor state of conservation	Need for immediate action: Renovation, modernization within a year, immediate inclusion in the maintenance program and ensuring the financing or determining the expendability of the structure

7.2.4.4 Renovation methods for avalanche walls

Avalanche walls are usually older than 80 years and constructed in dry-stone masonry. Such structures have reached their expected life from today's perspective. The decline of avalanche walls can lead to the rock fall danger downslope or other structure systems (steel support structures, snow nets) and also cause instabilities in the slope if the walls are high. For these reasons, avalanche walls must be preserved despite high costs. The most damage is in the crown of the wall, where the snowpack has high loads.

A permanent regeneration is the manufacture of a structurally reinforced concrete sheath on the wall crown and concrete stems at the base of the wall. Stabilization and preservation is possible through the concrete bond and the cover. The required concrete cubage is easily underestimated and causes very high transport costs when a helicopter is used. For extensive renovation projects and transport cubage, a cable crane can offer cost benefits.

When walls bulge (usually as a result of the earth pressure), there is the possibility of stabilization with heavyweight support ribs made of concrete. This restoration method also requires great amounts of concrete cubage and is therefore correspondingly expensive. Restoration can be less expensive with injection concrete and micropiles (Figure 7.24). In this method, in a first step, a construction grid is fastened and the entire wall area is sealed with injection concrete for stabilization. This establishes improved stability to prevent stones from breaking out during the drilling. As a second step, the injection anchor is drilled and additional drainage holes are made to allow outlets for seeping water.

After the installation of the injection anchor, anchor plates are screwed on and sealed with shotcrete.

Fig. 7.24 Renovation of an avalanche wall with micropiles and shotcrete (© WLV Bludenz)

In Switzerland, walls are mostly modernized by repackaging in wire gabions or with anchored network covers. In the future, walls are likely to be increasingly dismantled and replaced with support structures.

7.2.4.5 Renovation methods for snow supporting structures

Although defective crossbeams are not an immediate risk of damage for the protected zones and therefore no great urgency for renovation can be inferred, more damage can be prevented by a timely replacement of the crossbeams. Slightly deformed crossbeams have a reduced carrying capacity, so that when another impact event happens, larger deformations are expected. The bending of the beams causes warping of the girders by screwing with the girders, which lose load-carrying capacity especially in the upper area of the snow bridge. This results in considerable damage to the bearing construction. For this reason, it appears practical to replace slightly deformed beams and replace these, where appropriate, with stronger beams.

Structures filled with soil or debris can be emptied most efficiently with a climbing excavator. It is advisable to install the material in the form of terraces, in strong inclines in terraces with geo-grid reinforcement, in gabions or wooden cribs. If there is damage to the crossbeam due to rock fall or if this is to be expected, additional application of round wood can offer a certain degree of protection.

Damaged girders can be replaced relatively easily when the connection to the foundation is a jointed connection. For a rigid carrier connection, a separation of the carrier and

(a)

(b)

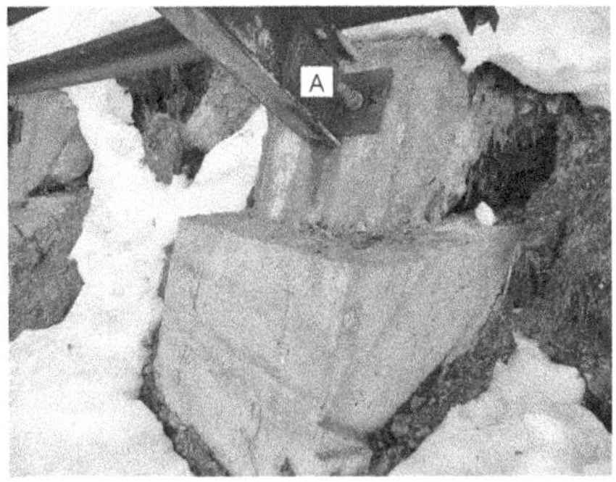

Fig. 7.25 Renovation of avalanche protection structures: (a): excavation for the underpinning of a carrier foundation. In addition, drill holes are made for tension and pressure piles and the load supports are added (A: Tension pile load support; B: Tension pile load support with stocking); (b): completed underpinning of a carrier foundation with concrete and additional anchoring with micropiles (A: fitting of tensile pile with the carrier) (© WLV Vorarlberg)

rewelding of a new part is possible if the foundation is intact. Defective supports can be replaced easily on construction types with micropile foundations or concrete foundation, since the support connection to the foundation is freely accessible. After reinforcing the grate or hanging this back on the upper series of structures, the support can be replaced. The replacement of supports for properly installed ground plate foundations is more complex, since the ground plates must be exposed in order to remove the connecting bolts in the ground plate.

The renewal of foundations is mainly required when foundations have been exposed by erosion or shallow ground-slope movements. Damage as a result of construction defects or overloads usually requires a renewal of the foundations.

Concrete foundations exposed and undercut by erosion can be renovated by concrete underpinning (Figure 7.25a, b). This renovation possibility can be used for beams as well as for supporting foundations. Ground plate foundations exposed by erosion, or those that were only set on the surface, can be repaired after making deeper excavations. To keep from changing the angle between the support and the girder, the support must either be extended or, if the permissible buckling length has been exceeded, replaced with a stronger one.

Concrete foundations, moving in loose material which is unstable in shallow ground can be stabilized in certain circumstances by the additional installation of micropiles. One variation is to place a micropile near the girder parallel to this in the concrete foundation and also a second vertically through the concrete foundation.

A micropile and a second holding pile can also be fitted nearly horizontally towards the support for the stabilization of the supporting foundation. In the case of larger-scale refurbishment, it is necessary to refurbish some foundations in a first stage and to

Fig. 7.26 Refurbishment of a bent micropile for the support of a snow net by a concrete foundation (© WLV Vorarlberg)

observe this for a corresponding amount of time to determine whether the desired restoration effect has been achieved.

Micropiles exposed by erosion, which have not yet suffered any deformation, can be refurbished with a concrete sheath. To do this, it is necessary to excavate and surround the piles with a structurally reinforced concrete. In addition to erosion damage, in this way construction defects in micropiles that were installed with an overly large projection can be restored. To renovate micropile foundations, it is necessary to dismantle the snow bridge with the helicopter, to place this in the immediate vicinity and to reinstall this after modernizing the micropiles.

7.2.4.6 Renovation methods for snow nets

For avalanche nets with a supporting structure made of ropes or grids, with the exception of the support, overloads will lead to tears primarily in the nets or edge ropes, generally requiring a replacement. Damage to the support foundations caused by bending stress in the form of bent micropiles can be repaired by installing concrete foundations or ground plates (Figure 7.26). Damaged anchors are to be replaced as a rule.

7.2.4.7 Costs of maintenance and renovation

There are practically no preservation costs for snow supporting structures in the starting zone – apart from the tightening of cable clamps for snow nets; the avalanche barrier protection is thus largely maintenance-free. If it is assumed that there should be an annual inspection of the built-up areas over an estimated life of 100 years, then about 2% are to be added to construction costs for the maintenance (operational support) [208].

The renovation costs for support structures can vary widely. Renovation costs can often be traced back to erroneous evaluation of the geological conditions, the snow loads, use of poorly proven models, a lack of compliance with the applicable directives and construction defects, in particular in the foundation. Examinations of 32 avalanche barriers in Switzerland showed an annual average value for preservation and restoration of 0.39% of the total cost. The values varied between 0.01% (no damage, only control costs) and >3% for problematic structures. In Austria, an investigation of approx. 100 km of steel snow bridges from 1996 to 2006 showed annual rehabilitation costs of 0.03% of the total costs.

8 Building protection (direct protection) measures

Jürgen Suda, Siegfried Sauermoser and Florian Rudolf-Miklau

8.1 Structural building protection measures

8.1.1 Principles of building protection against avalanches

The most effective measure of defense against avalanches is keeping endangered zones free of buildings. However, this is not always achievable in exposed mountain valleys, on the contrary, in these regions, areas that have already been built up are often threatened by extreme avalanches. In areas with a moderate risk of avalanches, the *technical building protection* (also called *direct protection* or *object protection*) presents an efficient protection concept and, in certain cases, an alternative to avalanche defense structures.

Different national regulations are in place for the construction of buildings in hazard zones. For example, the construction of buildings in Austria in yellow hazard zones and in Switzerland in blue hazard zones (Section 4.2.3) is legally permissible and technically possible with certain safeguards. This requires a risk-adapted design, sufficient dimensioning and structural design of the building suited to the actions to be expected by avalanches. Mountain huts or shelters, which are built in exposed areas that are frequently endangered by avalanches, are also to be assessed for avalanche effects – in the overall form as well as with regard to the resistance of the individual building parts. To plan building defense measures, it is important to know whether the building could potentially be affected by flowing avalanches only or by mixed and powder avalanches as well, and whether the flow will only go over or around the object.

The primary goal of building protection is to maintain the existing structures, but not to ensure the usability of the buildings during acute avalanche danger. Building protection measures are therefore often associated with temporary or seasonal use prohibitions, since the protective effect is limited to the building itself and its immediate surroundings; this however does not apply to the broader vicinity and the accessibility of the building. Even with efficient building protection measures there remains still a continued high-risk for people outside the building.

8.1.2 Avalanche action on buildings

Buildings are stressed by the flowing (fluidized) layer of an avalanche flowing around, flowing over or impacting – depending on the topography and shape of the building. The amount of stress depends on the density of the avalanche, the components (blocks, tree trunks) transported with it and the run-off height.

The dangers generated by the powder layer of an avalanche result from the powder pressure and suction of the aerosol. This is comparable to squalls. It produces pressure forces on the upwind side and suction forces on the lee walls, roofs and building ledges. The loads caused by powder avalanches are dependent on the density and velocity of the aerosol.

The Technical Avalanche Protection Handbook, First Edition. Edited by Florian Rudolf-Miklau, Siegfried Sauermoser and Arthur I. Mears.

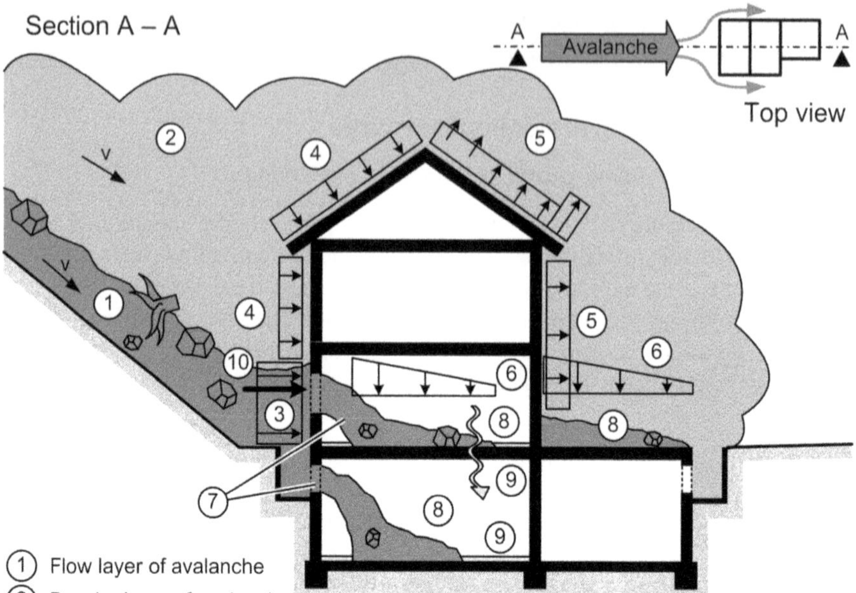

① Flow layer of avalanche
② Powder layer of avalanche
③ Dynamic pressure on outer wall by impact of flow layer
④ Dynamic pressure by impact of powder layer
⑤ Negative pressure (suction) by powder layer
⑥ Load by avalanche snow deposit
⑦ Penetration of avalanche snow into the building
⑧ Deposit of avalanche snow
⑨ Seeping water by melting of snow
⑩ Impulse pressure by impact of massive components

Fig. 8.1 Hazard figure Avalanche *flow around the building*: The avalanche acts the building and subsequently wraps around it. The powder layer creates a pressure or suction effect on the building, especially in the roof area (pursuant [265]; © Suda)

In most cases, buildings are surrounded by a flow or mixed-snow avalanche. The height of the flow layer is clearly below the eave height of the roof (Figure 8.1). After the avalanche hits the front side of the building, a compression of the snow takes place within milliseconds. At the same time, the avalanche snow backs up due to the flow obstruction by the building itself. The impact produces extensive dynamic pressure (similar to a mudflow). Individual components carried along (such as boulders, tree trunks) produce point-shaped action forces. The pressing of the snow masses upwards produces pressure on any roof eaves. This leads to snow deposits on the impact wall. Avalanche snow can penetrate into a building through building openings or knockout holes in the building construction. The remaining snow masses flow around the building or are deposited. The side walls and all walls with flow at an angle are stressed by reduced pressure and friction. The stress by the powder component is the same as with a pure powder avalanche.

This hazard situation results in the following effects on the building (Figure 8.1):

- Flow layer:
  - Dynamic flow avalanche pressure by the impact of the avalanche front
  - Impact of solid components
  - Friction of the avalanche snow on the building shell by flowing around the building
  - Snow penetrating the interior of the building
- Powder layer:
  - Pressure and suction stress from the powder layer
  - Friction on the surfaces with the overflow
- Loads from deposited avalanche snow

Low buildings can also be overflown by a flowing or mixed-snow avalanche. If a flowing avalanche hits a building and flows over it, only the forces from the flowing avalanche are relevant for the stress on the building (Figure 8.2).

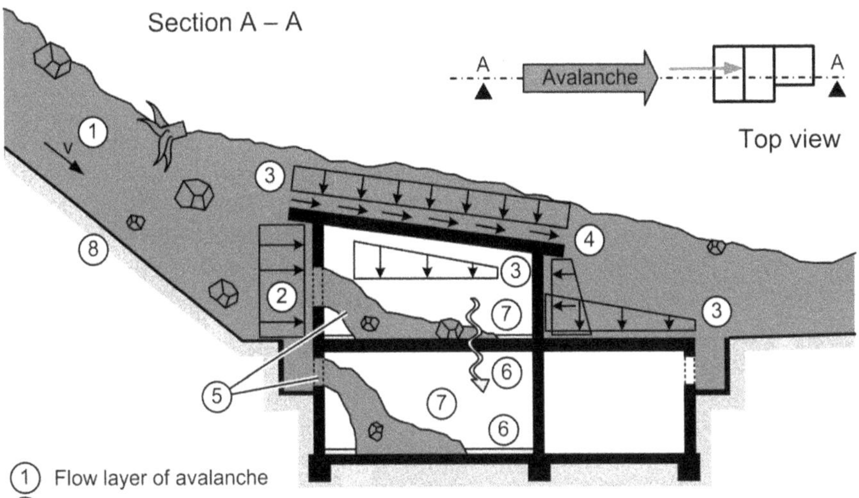

(1) Flow layer of avalanche
(2) Dynamic pressure on outer wall by impact of flow layer
(3) Load by avalanche flow over the building and avalanche snow deposit
(4) Friction by avalanche flow over the building
(5) Penetration of avalanche snow into the building
(6) Seeping water by melting of snow
(7) Deposit of avalanche snow
(8) Erosion

Fig. 8.2 Hazard figure Avalanche *flow over the building*: The avalanche impacts the building and subsequently flows over it. (pursuant [265]; © Suda)

The impacts and resulting damages in this hazard figure correspond to those from flows around the building. Additionally, the following occur

- Frictional forces on the roof skin from the overflow

In areas affected by powder avalanches, the freestanding structure can be stressed by the pressure and suction forces of the aerosol. In the assessment of powder avalanche effects and the effect of the saltation layer, consider that these two layers of the avalanche have a considerably lower density, on the other hand however a higher velocity and therefore flow ahead of the flowing components. The pressure distribution in the longitudinal and transverse direction of a powder avalanche can therefore be very irregular. Even just turbulence-induced pressure differences can cause turbulent peak pressures, which can be up to a factor of two over the averages. The overflow and surrounding flow of the powder avalanche results in the following effects on the building:

- Static pressure as the flow passes over (typical values between 0.5 and 2 kPa)
- Dynamic pressure on parts of the building orientated against the direction of flow
- Dynamic negative pressure (suction) on side walls and roofs, as well as inside the building

Approaches for dimensioning of avalanche action (impact) on objects are included in section 3.3.

8.1.3 Structural measures at the building

8.1.3.1 Shape and orientation of the building

By optimizing the *orientation and shape of the building*, the effects from an avalanche can be reduced. Generally, a building is to be situated so that the impact surface for the avalanche is minimized. This requirement is met largely by selecting a building site behind natural brow terrain or on the edge of a potential runout zone. In addition, the outer shape of the building is to be designed so that an avalanche can flow either over or around it (Figure 8.3a–d). Agricultural buildings which are located in avalanche areas and are not used in winter are often protected against overflowing avalanches with monopitch roofs (Figures 8.3d and 8.4a). Buildings which have avalanches flowing around them must have a building shape that is favourable for flow (narrow side facing the avalanche, favourably shaped ground plan shape for flows) (Figure 8.3a–c). An example of a wedge ground plan shape on the avalanche-exposed side in conjunction with an avalanche control splitting wedge integrated directly into the building is shown in Figures 8.6 and 8.4b.

8.1.3.2 Constructive building protection measures

In addition to reducing the impact surface, measures to counteract the penetration of the avalanche snow or the shockwave into the building interior are amongst the most important construction safeguards. Therefore, on the avalanche-exposed side of the building, as few niches or openings as possible are placed, whereby the utilization concept is to be taken into account. Ideally the avalanche side of the building should be free of openings and without roof protrusions. The latter form an impact surface for the vertical and upward moving components of avalanches; in particular with regard to potential powder avalanche effects, the roof overhang should be kept as small as

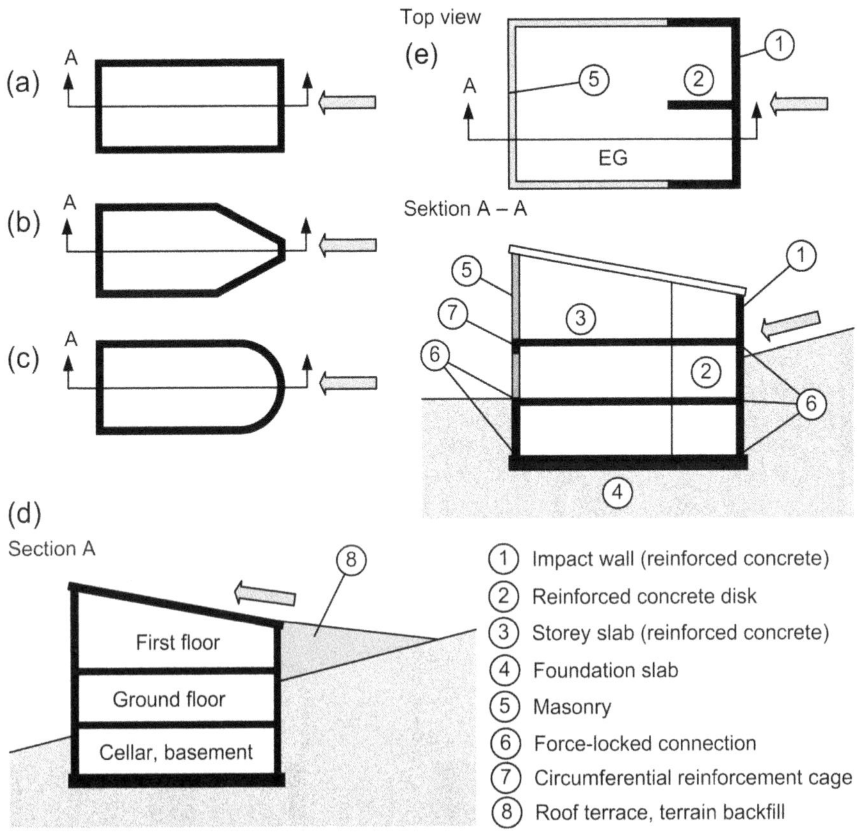

Fig. 8.3 Structural design for buildings in avalanche zones: (a,b,c): flow dynamic shaped ground plans; (d): favourably shaped roof and integration into the ground surface; (e): support structure reinforced with steel concrete plates (example) (© Suda)

possible (Figure 8.5a). Roof structures in powder avalanche zones are to be assessed for the vertical pressure and suction forces and to be anchored in the building accordingly. The anchoring of the rafters and purlins against the suction of powder avalanches is particularly important. Flat roofs offer a smaller impact surface than gable or hipped roofs. It is also possible to make the support structure for gable and hipped roofs of reinforced concrete. However, this is associated with high costs.

Directly exposed walls should generally be built from reinforced concrete (Figure 8.5a). Depending on the type of structure (solid construction, skeleton construction) this is to be braced. This is best done in building construction by placing shear walls on the outside and inside of the building (Figure 8.5e). The walls are to be arranged nearly parallel to the direction of the avalanche and connected in a friction-locked manner with the story ceilings and the basement. For reinforcement measures of the building shell, ensure that the actions from the avalanche can be deflected over the support structure of the building into the foundation and the substratum.

(a) (b)

Fig. 8.4 External shape of the building as avalanche protection: (a): Monopitch roof on an agricultural building (© Rudolf-Miklau); (b): brick avalanche splitting wedge integrated in the construction of a Swiss shelter (Lämmerenhütte SAC, Leukerbad, Valais) (© Sauermoser)

(a) (b)

Fig. 8.5 Design of building walls facing potential avalanches: (a): Vertical avalanche defense wall without a projecting roof on the impact side; (b): Perpendicular lamellae (supporting ribs in reinforced concrete) on the avalanche side for static reinforcement of the outer wall and limiting of lateral spread of the avalanche (© Sauermoser)

Figure 8.5b shows the construction of an avalanche-side building front (impact wall), that has been statically reinforced with vertical lamellae (steel concrete ribs). As an additional effect, these structures limit the lateral spreading of the avalanche. The side walls of buildings are to be secured with avalanche-side wing walls so that an avalanche meeting a lateral surface at an angle of 20° can no longer damage it and is sent past the building.

An additional construction protection measure is the arrangement of statically reinforced (assessed for avalanche pressure) impact walls on the avalanche flow side, which are to be designed free of openings (Figure 8.6a). Furthermore, solid parts of the building which exert a repellent, absorbing or deflecting effect on the avalanche can be set in front of the rest of the building (Figure 8.6b).

(a) (b)

Fig. 8.6 Reinforced building components: (a): solid design, opening-free impact wall on a single family dwelling, assessed for avalanche pressure action; (b): solid design building section located on the flow side of the main structure (© WLV Tyrol)

8.1.3.3 Building protection measures with temporary effect

In some cases, the utilization concept for the building requires placing windows or other openings on the avalanche side. Generally, the assessment of the window needs to take the avalanche pressure into consideration (Section 8.1.3.4). The installation of *avalanche windows* can only be waived in areas with rare avalanche occurrences and low risk (near the edge of the yellow hazard zone) or in buildings that are not used in the winter. In this case, the building can be secured by means of shutters that are closed in the event of avalanche danger (Figure 8.7a). This measure is also a proven method for seasonally securing buildings that are not used in the winter at all. Avalanche-secure shutters can be implemented in a variety of designs (for example, Figure 8.7b); in any case, they are to be assessed for the applicable avalanche pressure.

(a) (b)

Fig. 8.7 Protection of windows and building openings: (a): solid shutters on a shelter, which are to be closed when there is a risk of avalanches; (b): solid shutters to protect a seasonally used shelter against avalanches (© Sauermoser)

Fig. 8.8 Temporary protection of a cable car station: The deflection area for carriers is covered by steel elements in case of avalanche hazard (© Schilcher)

Due to the air pressure wave generated by avalanches, securing windows alone is not sufficient, all other building openings (basement ducts, ventilation ducts) must be sealed. To this end, for example, garage entrances or openings at cable car stations (Figure 8.8) can be secured by temporary securing measures – such as covers made of steel plates.

8.1.3.4 Design and commercial products for building protection against avalanches

As can be seen in Table 8.1, standard local constructions of wood or brick can resist a maximum avalanche pressure of $\leq 30\,kN/m^2$ as long as certain construction requirements are met. Avalanche pressures over 30 kN/m^2 can only be absorbed by solid reinforced concrete constructions. Therefore, in the yellow (blue) hazard zone, the avalanche side of a building is to be built with reinforced concrete (Figures 8.3e and 8.6a). This requirement may also apply to the side walls of a building if the flow direction of the avalanche is not precisely defined or if it can vary.

Table 8.1 Damage to constructions, according to [7]

Avalanche Pressure	Damages to Constructions
up to 1 kN/m^2	Windows are smashed in
up to 5 kN/m^2	Doors are pushed in
up to 30 kN/m^2	Wood or brick structures are damaged or destroyed
up to 100 kN/m^2	Trees will be uprooted
up to 1000 kN/m^2	Concrete structures are damaged or destroyed

Table 8.2 Load class (permissible surface load) in accordance with standard ÖNORM B 5301 in relation to the hazard zone

Avalanche Hazard Zone	Load Class	Load kN/m^2
Yellow zone	LP 5	5
	LP 10	10
Red zone	LP 15	15
	LP 20	20

In Austria, avalanche defense windows and doors are available in accordance with the standards ÖNORM B 5301 and ÖNORM B 5302, which applies to the 'avalanche secure' design of windows, French doors and exterior doors. The standards apply to single and multiple wing elements in all types of openings, as well as for non-moving parts (fixed glazing). Four load classes are defined in the context of the standard ÖNORM B 5301 (Table 8.2). The selection of the corresponding load class is based on the appraisal of a specialist for avalanche defense who will define the loads on the basis of the expected avalanche effects on the corresponding property in the scope of the official construction process or by the property owner. The requirements for this check are based on the corresponding load from the glass to the construction of an avalanche defense element on the building.

8.1.4 Structural measures in front of the building

8.1.4.1 General remarks

The protection of individual properties or a small group of buildings from avalanches using measures in the starting zone (Section 5.2), in the avalanche path or in the runout zone (Section 5.3) is normally not possible for economic reasons. The necessary protection can also be achieved by solid constructions that can be placed directly in front of the property to be protected (upslope building safeguards). These include avalanche splitting wedges, roof terraces and impact walls. The function of such defense measures is the direction of the flow over and around the building or in the avalanche deposits in front of the building. A disadvantage of the defense structures measures upslope from the building is that access to the property is generally still in the hazard zone since the protective structures is only effective for the building itself. Therefore, there is still a high risk to people outside of the protected building when these measures are implemented.

8.1.4.2 Avalanche splitting wedges

To protect individual properties in the avalanche path, *splitting wedges* can be created – directly on the avalanche-exposed side of the property or immediately in front of it (Figure 8.9a and b). This splits the avalanche and directs it around both sides of the property. The height of the splitting wedge walls and the stress due to avalanche forces

(a)

(b)

Fig. 8.9 Avalanche splitting wedges: (a): solid splitting wedges of reinforced concrete (with cyclopean masonry) to protect an agricultural property (Galtür, Tyrol) (© Rudolf-Miklau); (b): historic splitting wedge of dry stone masonry to protect an agricultural building in Galtür (Tyrol) (© WLV Tyrol)

are determined by the same method as for deflection dams (Section 6.4.2.2). The splitting wedge legs must have a sufficient length; the splitting wedge opening angle should be less than 60°.

In historic times, natural properties (such as boulders) were used as avalanche splitting wedges. Splitting wedges were subsequently built of dry or cement masonry mortar, then made of concrete or reinforced concrete. Figures 8.9a, b and 8.10a show various types and forms of splitting wedges.

Fig. 8.10 Splitting wedge and roof terrace: (a): earth splitting wedge to protect single family dwellings (© WLV Vorarlberg); (b): historic roof terrace for the protection of agricultural buildings (© Biarcha)

Today splitting wedges are mainly constructed to protect properties used agriculturally or seasonally, only in exceptional cases for residential buildings or higher-valued farm buildings. However, the measure is an effective single defense structure for high voltage pylons (Figure 8.10b) and cable car supports.

8.1.4.3 Roof terrace

A *roof terrace* is a building protection measure in the avalanche path often used in previous times. This had a solid, secured material filling situated on the side of the mountain directly above the house which caused the avalanche to flow over and around the building without any damage (Figure 8.10b).

The roof of the protected building seamlessly connects to the embankment. This causes the avalanche to flow around the side and over the building. Special attention must thus be given to the constructional design of the roof connection. The roof and sides of the building are to be assessed to compensate for the effects of the avalanche friction. Securing the fireplace is usually not possible, so that the use of roof terraces is limited to agricultural or seasonally used buildings.

8.1.4.4 Impact walls

Impact walls are solid catchment structures that are built at a certain distance from the property to be protected or built directly connecting to it (Figure 8.11a and b). The impact walls are to be assessed in the same way as avalanche catchment dams (Sections 6.4.1 and 6.4.2.2). Impact walls are designed for the full pressure action of an assessment event and built at a height that they cannot be overrun by the avalanche. Impact walls only have a very limited effect against powder avalanches. Powder avalanche effects are therefore to be assumed behind an impact wall as well.

8.1.5 Building defense measures for other structures

Below is a brief overview of the possibilities for direct avalanche defense (property protection) for other buildings located in the avalanche path.

(a) (b)

Fig. 8.11 Impact walls to protect buildings in the village of Galtür (Tyrol): (a): directly connected to the building or (b): separate from the building (© Rudolf-Miklau)

Power line masts or cable car supports are often located in the avalanche runout zones. As for buildings, the search for avalanche safe locations should be the primary goal. If construction cannot be avoided in avalanche-prone areas, then the buildings (constructions) are to be designed for avalanche forces (Figure 8.12a) or are to be secured by a defense measure located directly in front of them. For limited flow heights, it is recommended that a splitting wedge be erected to deflect the avalanche from the hindrance (Figure 8.12b). Depending on the placement of a pole in the avalanche path or the runout zone, this is to be assessed in accordance with Section 3.3.1.3 for the flow height of potential avalanches and the type of avalanche (powder, flowing components)

(a) (b)

Fig. 8.12 (a): steel mast for energy supply lines that is dimensioned for avalanche pressure on both sides (© Sauermoser); (b): splitting wedge to protect a high voltage power mast (© WLV Vorarlberg)

8.2 Safety concepts for buildings endangered by avalanches

The risk to people caused by avalanches penetrating buildings can be considerably reduced with a customized use of space. Areas directly along the outer wall exposed to the avalanche effects should have rooms that are mainly used for very limited times. Sleeping and living rooms should be located in the avalanche lee side of the building.

Intensively used areas in the outdoor area of a building should be in the area shielded by the building. Balconies are to be placed exclusively on the avalanche lee side of the building. The drive-up and access areas to the building should also be arranged in the protected area according to feasibility.

9 Artificial release and monitoring technology for avalanches

Lukas Stoffel, Patrick Nairz, Karl Kleemayr, Emily Procter, Arnold Kogelnig, Roderich Urschler, Marco Larghi and Siegfried Sauermoser

9.1 Methods of temporary avalanche defense

Temporary avalanche defense measures are methods to mitigate acute avalanche danger for a specific duration of time (i.e. temporary) and can be defined as active or passive [254]. Active temporary measures require assessment of the avalanche danger (for example by an avalanche commission) immediately prior to taking action, whereas passive temporary measures are done as a preventive measure regardless of the prevailing avalanche situation. A classification of temporary avalanche defense measures is shown in Table 9.1.

Temporary avalanche defense measures are used in areas where permanent measures are not possible or not reasonable, especially if there are economic constraints. Temporary measures usually require human intervention, but this brings an additional potential source of error. The most important application of active temporary avalanche protective measures is securing traffic routes and ski areas (cable cars, lifts, pistes).

The combination of permanent and temporary avalanche protective measures is part of a comprehensive (i.e. holistic) approach to prevention (see Sections 4.3.3.3, 4.3.3.4) and is gaining importance for safety plans in settlement areas. Permanent protective measures are dimensioned for a specific design event (avalanche event with specific probability of recurrence); an avalanche event greater than the design event can be mitigated with additional temporary protective measures. It is important to define the threshold value at which the effectiveness of permanent protective measures is compromised and thus when additional temporary measures have to be used.

9.2 Artificial release of avalanches

9.2.1 General remarks

Artificial release of avalanches is the use of technical methods to artificially trigger an avalanche at a specified time with the aim of temporarily securing potential release areas, avalanche tracks and deposition areas. Temporary protection in this context means that the likelihood is low that an avalanche event will affect secured zones, at least for some period after blasting. By triggering several small avalanches with little or no damage potential, larger spontaneous avalanches and more severe damage can be avoided. *Negative blasting* is a blasting attempt that fails to release an avalanche. Assessments after negative blasting are important, but usually it can be assumed that the probability of spontaneous avalanche release is low. Use of artificial release methods requires familiarity with the area and practical experience, as well as an understanding of the blasting and avalanche sciences.

The Technical Avalanche Protection Handbook, First Edition. Edited by Florian Rudolf-Miklau, Siegfried Sauermoser and Arthur I. Mears.

Table 9.1 Classification of temporary avalanche protective measures, modified from [254]

Active temporary avalanche protection	Passive temporary avalanche protection
Artificial release of avalanches, acute closure of roads and infrastructure, evacuation	Avalanche warning, avalanche forecasting, seasonal closure of roads or infrastructure, organizational measures

9.2.2 Fundamentals of artificial release of avalanches

Artificial release in a catchment area should be done with periodical blasting that is appropriate for the situation. Planning an operation requires consideration of the characteristics of the release area, possible blasting sites (including blasting order), blasting method, time and additional safety precautions (e.g. closures).

Artificial release of avalanches is the standard method for temporary protection of ski slopes. In Switzerland it is also used for traffic routes and in exceptional cases for settlement areas; this method is not used in Austria for settlement areas. The difference between these two applications is that only small avalanches with frequent recurrence need to be triggered in ski areas, whereas settlement areas and traffic routes are often located in runout areas of large avalanches (*valley avalanches*) that have a medium to high return period.

Advantages of artificial avalanche release include [262]:

- blasting gives information about the stability of the snow cover in the release area and the probability of further avalanche activity
- the time of avalanche release is controlled
- the size of avalanche events are minimized, if frequent and situation-appropriate blasting is used and avalanche release occurs (no negative blasting)
- the probability of large, wet snow avalanches is reduced if release occurs during dry snow conditions
- investment costs are less and construction time is shorter compared to structural measures.

Disadvantages of artificial avalanche release include [262]:

- assessment of the current avalanche situation can be difficult; blasting my cause release of avalanches that are larger than expected or in some cases secondary avalanches
- avalanches may cause smoothing of the surface in the track and thus better flow conditions (glide) for subsequent avalanches
- existing avalanche deposits may deflect subsequent avalanche flow and increase risk to other areas
- high personnel costs, in particular for controlling road and railway closures
- any induced damage may have legal consequences (civil and criminal).

A prerequisite for use of artificial release methods is a well-defined *safety plan*. The safety plan should refer to the release area, blasting method, (approximate) location of the blasting sites and planned safety precautions including closures of potentially affected areas such as ski pistes, roads, railways, and so on.

Operations should be supported by continual observation of the current snow and avalanche conditions by the avalanche warning services. Pre-blasting assessment requires estimating both the magnitude of an avalanche (including possible secondary avalanches) and the damage potential. The effectiveness of the operation needs to be assessed after blasting, before further measures are considered. Any substantial change in the avalanche situation always requires a comprehensive reassessment of the situation. During heavy snowfall lower blasting sites should be chosen first to attain at least partial avalanche release, though this depends on the situation and blasting method. If the probability of release is low, a larger number of explosive charges or larger loads can be applied. Additionally, it is recommended that areas with less snow cover are blasted.

Spontaneous avalanche release and effective blasting are more common in areas with a grade >35°. In contrast, avalanche release is less frequent if the grade is <35°, which can lead to accumulation of snow and potentially to formation of large avalanches. During extensive periods of snowfall, blasting should be carried out at an early stage (already during snowfall) to trigger avalanches that are as small as possible. This requires good visibility and release methods adapted to all weather conditions. Blasting is not recommended if the likelihood of releasing a large avalanche or causing severe damage is high.

Safety precautions are of paramount importance. Persons must not be endangered by artificial release of avalanches. It is important to document all data (snow, weather), observations, decisions and actions.

9.2.3 Effects of artificial release

Detonation of an explosive charge creates a pressure wave that spreads over the snow cover and exerts an additional load, which can ultimately lead to failure of the snow pack. Detonation of explosives or gas is possible and a distinction is made between blasting above, on or inside the snow cover.

Blasting above or on the snow surface creates a pressure wave over a wide area. In dry new snow, the additional tension created by the detonation declines rapidly within the top half metre of the snow cover (this does not limit the effect to the top half metre and release of a thicker slab is possible). In comparison, the effect of blasting inside the snow cover is mainly limited to the area of the depression caused by the explosion, though shocks into the ground may also result. The most influential factors for an effective blast (apart from the position relative to the snow cover) are the location of the blasting site, size of the explosive charge and type of explosive.

The topography of the blasting site also has to be considered. Areas around the explosive charge that cannot be seen from the blasting site are said to be in the *lee side* of the pressure wave. In the lee side no direct additional load is generated on the snow cover. The effects of blasting are increased by reflection off rock faces. In dry new snow blasting is most effective with detonation above the snow cover, ideally at a height between 1 and 4 m. Up to 90% success (i.e. avalanche release) can be achieved by blasting above the snow cover with large explosive charges, 60% by blasting on the snow cover and <50% by blasting inside the snow cover. Artificial release can be attempted during wet snow conditions, but blasting should be on or inside the snow cover and done in the late afternoon.

For most blasting methods the size of the explosive charge ranges from 1 to 5 kg. There are several explosives specific for avalanche blasting (e.g. Slurry, Gelatin, Lawinit) and effectiveness is comparable. Effectiveness is primarily determined by the measured explosion heat. In addition, handling safety and functionality (e.g. influence of low ambient temperature) are important. Detonation of gas has a similar effect as detonation of explosives, depending on the gas volume.

Detonation can cause either release of an avalanche (*positive blasting*), failure of release (*negative blasting*), settling of the snow cover (i.e. a slab) or partial release. The width of a slab avalanche may vary between a few meters and several hundred meters depending on the structure of the snow cover and terrain.

If no avalanche is triggered the effect of the explosion(s) has to be evaluated, commonly through so-called *areas of impact*. For a 2 kg explosive charge detonated on the snow cover an impact radius of 35 m is expected; for a 5 kg explosive charge detonated above the snow cover the impact radius is approx. 120 m. Whether further spontaneous avalanche release is likely after negative blasting, provided no significant change in the situation, depends on the explosive effect (e.g. number of blasts), structure of the snow cover and inclination.

9.2.4 Methods of artificial avalanche release: overview

Rapid technological developments in recent years have led to sophisticated artificial release systems. The most common blasting methods are shown in Table 9.2 and include hand charging, helicopter blasting, blasting ropeway (Figure 9.1a), Gazex® (Figure 9.1c), Inauen-Schätti Avalanche Tower and Avalanche Guard® (Figure 9.1e) (originally Doppelmayr Avalanche Guard®), Wyssen Avalanche Tower® (Figure 9.1d) and Avalancheur® (Figure 9.1f). Other methods include lowering or release of explosives from cable cars, Daisybell®, O'Bellx®, avalanche pipe®, a charge fixed to at beam, a cantilever at a ridge, Avalhex® and the blasting sledge. In Switzerland military projectiles or mortars are used in some cases and in Austria the Avalanche Locker® (pyrotechnic system) is also used. However, the development of fixed systems (Gazex®, O'Bellx®, Wyssen Avalanche Tower®, Inauen-Schätti Avalanche Tower®) are of particular interest since they are independent from weather conditions, provide reliable effectiveness and enable blasting at any time remotely from a computer. Details of these systems are given in Chapter 9.2.7.

9.2.5 Comparison of methods: effects and efficiency

The most important criteria for artificial avalanche release systems are safety of the operators, explosive effect, reach, applicability (visibility and weather conditions), execution time and costs. Table 9.3 compares the most common avalanche release

Table 9.2 Commonly used methods of artificial avalanches release (selection)

Method	Description
Hand charge	A single charge is thrown from a safe location, either as a secured throw (the charge is attached to a rope and can be withdrawn over the snow surface) or an unsecured throw (without a rope). Common charge size ranges from 1.5–2 kg.
Helicopter blasting	Explosive charges of 4–5 kg are thrown by hand out of the open helicopter door into the avalanche release area while flying at low altitude (approximately 2–5 m). This method requires good weather. In order to avoid failures the charges are equipped with dual ignition (two detonators, two fuses).
Blasting ropeway	The explosive charges are transported by ropeway from the station (deflexion pulley) to the release area and are detonated above the snow cover. Lowering devices can also be used. Blasting ropeways spanning >400 m are usually operated by an engine. The risk of icing can be reduced with slow, continuous operation (Figure 9.1a).
Gazex®	The first fixed blasting system for permanent installation in the release zone was built in 1988. From a central gas shelter, a propane-oxygen mixture is piped underground to the pipe *exploder*, which is located aboveground with a shape that directs the explosion towards the surface of the snow cover. The detonation is triggered remotely by radio signal. Exploders are available with capacities of 0.8, 1.5, 3.0 and 4.5 m^3. The gas supplies are usually sufficient for a whole winter season. For the 0.8, 1.5 and 3.0 m^3 exploders an autonomously operating version is also available, that is the gas tank is situated directly at the exploder (Figure 9.1c).
Daisybell®	The Daisybell® is a bell-shaped structure attached to a helicopter; ignition of the gas occurs in the bell and is controlled from the helicopter. The explosive power is slightly higher than that of a 0.8 m^3 Gazex® exploder, though the effect is also dependant on the blasting height above the snow cover.
O'Bellx®	The O'Bellx® consists of a cladded Daisybell® on a support. Refilling the gas requires helicopter removal and transport of the O'Bellx®.
Avalhex®	This system uses a set of balloons (17 or 26) filled with gas that are individually detonated above the snow cover.

(continued)

Table 9.2 *(Continued)*

Method	Description
Inauen-Schätti Avalanche Guard®	This system was developed in 1996 and consists of a magazine box fixed to a supporting pillar, from which explosive charges (2.7 or 5.4 kg) are shot into the release area up to 150 m. The box contains 10 explosive charges and 2 boxes can be installed per pillar. Reloading must be done manually on site. Adaptations to the system in 2002 allowed charges to be ejected on a string and detonated on the snow cover (Figure 9.1e).
Avalanche pipe®	The avalanche pipe® is similar to the avalanche guard but with a single defined direction and shot distance up to 400 m.
Wyssen Avalanche Tower®	This system was developed in 1999 and consists of a magazine box fixed to a pillar, from which 5 kg explosive charges are dropped on a string and detonated above the snow cover. Each box contains 12 explosive charges. Reloading the magazine requires removal by helicopter. Only the pillar remains in the release area during the off-season (Figure 9.1d).
Avalancheur®	This system uses a gas pressure gun to shoot arrows loaded with explosives up to 2 km into the release area. The arrows are 1.8 m long and are loaded with 2.2 kg liquid explosive. In case of a failure, the explosive is no longer capable of reaction 48 h after the mixture of the two component liquids (Figure 9.1f).

methods/systems. Gas systems have the advantage of producing no waste and avoiding that non-detonated charges remain in the field. An advantage of the avalanche tower over the Gazex® is that no supply pipes are necessary from the storage to the exploder. The effect of blasting from avalanche towers is generally very good. Fixed systems such as Gazex® or avalanche towers are limited to the same location, which may be a disadvantage in some cases.

9.2.6 Safety requirements and risks of artificial avalanche release

Stated simply, risk is inherent when handling explosives and damage is possible when using artificial release methods. Accidents have occurred while approaching the blasting site, carrying out a blast (especially with hand charges) and controlling closures/restricted areas. Proper execution of closures or other safety precautions may be difficult depending on the situation and terrain. Further avalanche activity after positive and negative blasting operations cannot be excluded. Working with helicopters also represents a general risk (e.g. helicopter blasting and transportation). However, with proper execution the risk of an accident is low or negligible. Correct handling of

(a)

(b)

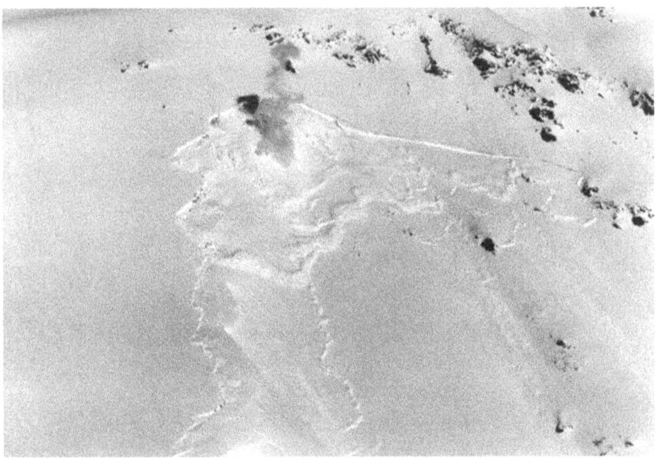

Fig. 9.1 Artificial avalanche release (avalanche blasting): (a) a blasting ropeway and the snow cloud after detonation; (b) propagation of a fracture in the snow cover after blasting (© L. Stoffel)

explosives is taught in specific blasting courses and instruction on the use of gas and in general on the use of different systems is done by the manufacturers.

Artificial avalanche release can cause unexpectedly large avalanches. Catastrophic avalanches cannot be excluded if blasting is not done repeatedly during extensive snowfall periods or if blasting fails and is re-attempted later. Extensive safety precautions and closures are essential, but assessment of the actual avalanche situation in a specific area may be difficult, especially predicting the potential runout distance. There

(c)

(d)

Fig. 9.1 (*Continued*) (c) exploder of a Gazex® system; (d) Wyssen Avalanche Tower® (© L. Stoffel)

is also risk of releasing secondary avalanches (avalanches in adjacent catchment areas). Therefore, with high avalanche danger it is recommended to enlarge the closure area to the deposition areas of secondary avalanches. Involved personnel often have to stay in endangered terrain during blasting operations. Coordination of the blasting time and extent are essential for safe controlling of closures/restricted areas.

(e)

(f)

Fig. 9.1 (*Continued*) (e) Inauen-Schätti Avalanche Tower®; (f) Avalancheur® (© L. Stoffel)

Residual risk cannot be avoided, even after blasting. Spontaneous avalanche activity can occur at any time after negative blasting and in rare cases even shortly after. It is therefore advisable to delay the reopening of areas at least 15 minutes, especially in low temperature conditions. It should be noted that blasting only gives information on those

Table 9.3 Comparison of the most common methods/systems of artificial avalanche release

Method	Evaluation criteria							
	Safety of operators	Blasting effect: Position relative to snow cover	Blasting effect: Load size	Range	Dependence on weather/ security	Duration of operation	Investment costs	Operating costs
Hand charge	–/+	–/+	+	–/+	+	–/+	++	++
Helicopter blasting	++	–/+	++	++	–	++	(++)	+/++
Blasting ropeway	+/++	++	+/++	+/++	+/++	+	– to ++	+/++
Gazex®	++	++	++	+/++	++	++	–	++
Inauen-Schätti Avalanche Guard®	++	–/+	+/+	+/++	++	++	–	+/++
Inauen-Schätti Avalanche Tower®	++	++	+/+	+/++	++	++	–	+/++
Wyssen Avalanche Tower®	++	++	++	+/++	++	++	–	+/++
Avalancheur®	+/++	+	+	+/++	+/++	+	+	–/+
Avalanche pipe	+/++	–/+ (–)	–	+	+/++	+	++	+/++

Table 9.3 (*Continued*)

Method	Evaluation criteria							
	Safety of operators	Blasting effect: Position relative to snow cover	Blasting effect: Load size	Range	Dependence on weather/ security	Duration of operation	Investment costs	Operating costs
Release from cable car	++	–	+/++	–/+	+/++	+/++	(++)	++
Lowering from cable car	++	++	+/++	–/+	+/++	+/++	(++)	++
Legend:			+	good				
–	insufficient		+/++	good to very good				
–/+	sufficient		++	very good				

avalanche areas that were blasted and at the given load and time, that is it may not be representative of other areas or conditions.

9.2.7 Construction and operation of selected artificial release systems

Editor's note: In this chapter some of the most prevalent and best established artificial release systems on the European market, all protected by patent, are presented. The content of the following sections, especially concerning the technical description and properties of the systems as well as application examples, was provided and elaborated by the manufacturing companies. The editors lay emphasis on the fact that this chapter does not contain any rating or benchmarking of the systems, furthermore in no respect advertisement was intended.

9.2.7.1 Gazex®

Gazex® was invented in 1988 by Jacob Schippers, CEO of Alpine Security Technology SA (Technologie Alpine de Sécurité, TAS) based in France, who wanted to overcome the legal restrictions on using explosives for artificial release of avalanches. The principle of the Gazex® system is to create an explosion using an oxygen-propane mixture to trigger an avalanche. The system is installed in autumn and stays operational throughout the winter without needing human intervention. Gas reserves are located in a central gas shelter in the field and the system is remotely controlled by radio or GSM from an office or vehicle. The explosion produces only water vapour and carbonic gas in a small quantity but no harmful waste.

With more than 2100 installations to date, today, the system is used to protect ski areas, villages, roads, tunnels, railway lines and open air mines located above 5000 m. It is also suitable for other strategic points and areas with difficult access.

A gas power shelter located in a safe area is connected by pipeline to one or more pipe *exploders* in the avalanche starting zone. The exploders can be placed up to 500 m in a radial form from the shelter. The gas power shelter is made of composite material. It houses the gas reserves and electronic remote control system that is supplied by a solar panel-battery. Each shelter is also equipped with a seismic sensor that records vibrations from the explosion and any subsequent avalanche activity. The seismic data are monitored in real-time on the computer to assess the quality of the explosion.

The *exploder* is a steel tube that is resistant to explosions. The tube is closed on the uphill side and open on the front end with a bent shape to direct the explosion towards the snow surface. Oxygen and propane are transported in two separate polyethylene piping systems that feed into the exploder; thus the gases mix only in the machine. An autonomous system is used to ignite the mixture. Exploders are available in different capacities (0.8, 1.5, 3.0 and 4.5 m^3) to suit local configurations on site.

The original Gazex product had *standard exploders* that were anchored in the ground on a concrete base on the uphill side and supported by bars fixed into the ground on the front side. A newer model, the *inertia exploder*, was developed in 1996/97 for sites with low ground resistance. Unlike the standard exploder that is fixed to the ground, the inertia exploder is entirely articulated. The fixed bars were replaced by a sliding

Fig. 9.2 The exploder pipe of a Gazflex® system (© Interfab)

counterweight and the base of the device is articulated, which greatly reduces the force on the anchors during the explosion. The most recent product was introduced in 2011; the exploder Gazflex® (Figure 9.2) is fixed on a concrete block on the uphill side and rests on steel torsion bars with a high elastic limit (spring support) on the front side. During the explosion, the force is dissipated by the torsion bars before being transmitted to the anchors. This system is suitable for sites with stable ground and incurs lower installation costs.

The system is operated remotely from an office using software that controls the valves in the gas shelter. A specified quantity of oxygen and propane is released into the pipes and transported to the exploder where they mix. A branch of the oxygen pipe near the exploder activates an autonomous system that ignites the gas mixture. The average duration of a detonation is 30 seconds, not including the recuperation of data from the seismic sensor; the average waiting time between two detonations from two exploders connected to the same shelter is 3 minutes. There is no waiting time if the operator changes shelters. If the explosion fails, the operator can retry in 15 minutes, that is the time needed to refill the propane tank and cool the exploder in case the mixture burnt without exploding. The explosion involves successive phases:

1. A progressive explosion of the gas mixture from the base of the exploder towards the opening creates overpressure (during a few milliseconds). The chemical reaction of the explosion produces CO_2 and water vapour.
2. The overpressure created as the mixture is consumed results in ejection of most of the gas through the opening of the exploder (after a few hundredth seconds).
3. The resulting shockwave collides with the snow cover at an angle of approximately 30°. This wave generates overpressure that acts on the snow surface; this is followed by a depression phase that has a tendency to weaken and facilitate movement in the snowpack.

The explosion creates both a direct force on the snow under the mouth of the exploder and a shockwave that creates an overpressure then a depression on the snow cover, which can destabilize the adjacent layers. The major advantages of the system Gazex® are:

- Security for the user
  - No handling of explosives or detonators
  - No storage and no handling of explosive gas mixtures, as the gases are mixed just prior to explosion and without human intervention
  - No operators in a dangerous zone during the explosion.
- Flexibility
  - The system is operational at any time (also without daylight) and independent of weather conditions
  - Fast start up (7 seconds per detonation and 10 seconds between each detonation)
  - The system avoids constraints imposed by legal regulations for use of explosive charges
  - Location of the explosion is controlled (no throwing, no hazardous sliding of explosives).
- Low consumable product costs
- No pollution
- Autonomy for a whole winter season.

Other avalanche release systems based on gas explosions are Daisybell®, which is a bell-shaped structure attached to a helicopter with ignition of the gas occurring in the bell, and O'Bellx®, which consists of a cladded Daisybell® on a support (Figure 9.3).

(a)

Fig. 9.3 Avalanche release systems Daisybell®: (a) transported by helicopter (© Interfab)

(b)

Fig. 9.3 (*Continued*) (b) O'Bellx® (© Interfab)

9.2.7.2 Wyssen Avalanche Tower LS12-5®

The Wyssen Avalanche Tower LS12-5® (Figure 9.4) was developed in 1998 from the principle that detonation above the snow cover is most effective to reduce residual risk of unforeseen avalanches. The goal of the development team was to reduce risk for the avalanche control team, optimize positioning of the explosive charge, and enable rapid operation and control work independent of the actual weather conditions.

The system consists of a simple tower that supports a finger-type docking system for a compact magazine that is remotely controlled by computer and houses twelve explosive charges. For recharging, maintenance and off-season storage the magazine is removed by helicopter and brought to the operation base station. The release of an explosive charge is executed via coded radio command sent from a laptop or PC. The command is

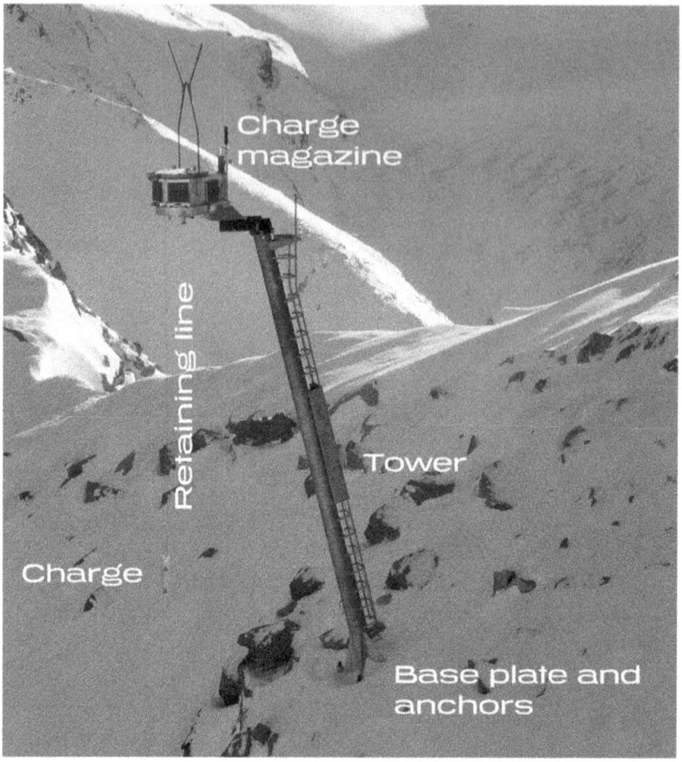

Fig. 9.4 Wyssen Avalanche Tower LS12-5 ®: The magazine holds twelve explosive charges and is placed on top of a steel tower, which is fixed with four anchors. Through remote control the charges can be dropped and remain hanging on a retaining line (© Wyssen)

received at the avalanche tower and an explosive charge is dropped. The magazine contains twelve charges of up to 5 kg (11 lbs) that are activated sequentially. After the charge is dropped it remains hanging on a retaining line above the snow. During the drop, dual ignition is pyrotechnically activated and the detonation takes place after a time delay.

The main features of this system are:

- Detonation takes place above the snow, which is known to produce the maximum effective range (260 m diameter with 5 kg charges; 850 ft with 11 lbs charges)
- Independent from weather and rapid release
- Remote control allows fast intervention by the avalanche control headquarters; avalanche release can be triggered with powerful charges and without delay
- No personnel in the avalanche danger zone
- Secure magazine locations (on the towers during winter and in storage in the off-season)
- Low environmental impact
- Energy supply through solar cells; no external supply lines.

The Wyssen Avalanche Tower has two components – a removable charge magazine and a supporting tower. Further information can be found in [86].

The magazine consists of twelve charges, a mechanical system to deploy charges, a microcontroller, a battery, solar panels and a radio or GSM transceiver. The magazine is designed to be removed and placed on the docking system of the tower by a helicopter equipped with a special hook. These operations do not require any on-site assistance. The weight of a fully charged magazine is approximately 650 kg (1400 lbs); an alternative lighter version for extreme high altitude operations weighs 350 kg (770 lbs). After placing the magazine on the docking mechanism its direction is automatically adjusted. The magazine activates the electronic control only after it is correctly docked on the supporting structure. The charge container is made from 5 mm steel (optionally aluminium) and is secured by three independent locks. The container holds up to twelve charges from 2 kg to 5 kg (4.5 lbs to 11 lbs).

The explosive charges are placed in vertical tubes in the magazine. The tubes are mounted to a turntable that moves the charges over a deployment hole in the bottom plate of the cylindrical magazine. In the home position of the turntable, a piston closes the deployment hole. At initialization, this piston first moves downward to remove rime in front of the hole and then moves upward to its operating position. If the piston has reached its operating position, the turntable starts to spin until the first loaded tube is detected close to the hole.

Each step of the process is checked twice by independent sensors and is indicated in the application window of the control computer at the base station. The process can be stopped anytime by a special command. The turntable moves until the charge reaches the position over the deployment hole in the bottom plate. The charge falls through the deployment hole attached to a cord. As the charge falls two pull igniters are activated and fire two slow fuses attached to the detonators within the charge. The detonation is detected by a geophone in the magazine and is displayed in real-time on the control screen. After detonation the cord is detached from the magazine and the system returns to its home position.

The Wyssen Avalanche Tower performs continuous self-testing, and additional testing can be activated by remote control from the computer. The results of the automatic self-tests are downloaded and displayed on the control computer at regular intervals.

The system complies with standard safety requirements and has been approved by the Swiss Federal Police responsible for the handling of explosives, Austrian TÜV, Norwegian DSB and Natural Resources in Canada.

The carrier is a simple tower (or a cantilever extending from a rock wall) designed to withstand snow creep, snow settlement, small rock falls, heavy riming, wind gusts, and so on. The installation can be modified to each location. For example, the tower is available in 8, 10 or 12 m to suit the local topography and counteract shadowing of the air pressure wave as much as possible. Depending on the ground conditions, the towers are anchored with 4 to 5 rock anchors or micropiles.

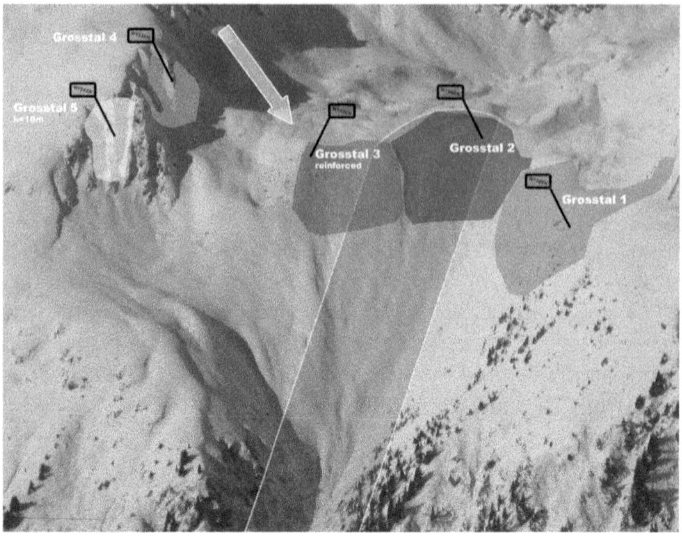

Fig. 9.5 Artificial release system in the Grosstal avalanche (Kappl, Tyrol) (© Wyssen)

The Wyssen Avalanche Tower LS12-5 is effective for large-area avalanche staring zones because of its wide effective range (i.e. use of powerful charges and detonation above the snow). The ability to control detonations remotely makes this system rapidly operable for fast interventions, independent of weather or time of day and safer for personnel. The Wyssen Avalanche Tower was designed specifically for the inaccessible and remote areas that are characteristic of avalanche terrain near roads, railroads or power lines. Additionally, completely integrated adjunct systems are available, including remote snow and weather stations and avalanche detection via radar and infrasound technology, which can be used to support decision-making for avalanche control work.

In Austria *temporary measures* (artificial release, detection by Doppler radar) were implemented for road protection for the first time in 2011. Several sections of the state road B 188 in the Paznaun Valley had to be closed regularly due to high avalanche risk. As the main economic activity of the is winter tourism and the Silvretta Strasse B 188 is a very important transport link, closing times of the B188 are very expensive for the region since the villages of See, Kappl, Ischgl and Galtür and their ski resorts are dependent on it. Therefore, seven Wyssen avalanche towers and a radar unit were installed in the Grosstal (Figure 9.5) and Hoher-Zug avalanche catchments [144].

9.2.7.3 Avalanche protection system Innauen-Schätti in the Scuol-Ftan-Sent ski area, Switzerland

This section describes a project of Inauen-Schätti AG to implement avalanche release systems in the Motta Naluns Scuol-Ftan-Sent ski area in Switzerland. The purpose of these installations was to avoid use of helicopter blasting in this area. An initial investigation of four areas of interest (Figure 9.6) was done by local avalanche experts, a blasting officer and the manager of the ski area. Follow-up investigations were done to

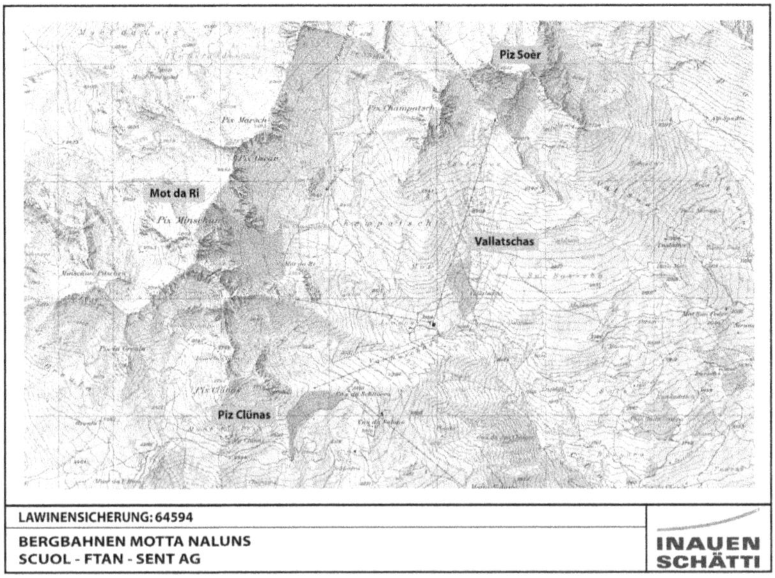

Fig. 9.6 Target avalanche sectors (chaded) in the Scuol-Ftan-Sent ski area, Switzerland (© Inauen, Schätti)

examine the avalanche-prone slopes and collect information on past avalanche events. In two areas (Mot da Ri and Piz Soèr) helicopter blasting was previously used, in one area (Vallatschas) hand-thrown charges were used and in one area (Piz Clünas) a new chair lift and ski slopes were in the planning phase and protection measures were yet to be implemented.

The Piz Clünas sector is basin-shaped with a width of approximately 270 m, length of 170 m and average grade between 30° and 40°. The location of the blasting point is crucial for wide release areas and a strong explosive effect is required for large and moderately steep release areas (<35°), preferably using detonation above the snow surface [264]. The Avalanche Master 5400® with 5 kg explosives was chosen for this sector.

The *Avalanche Master®* is an automated system with explosive charges that are triggered remotely from a computer. The magazine houses up to 10 charges of 5 kg. Charges are released on a rope and detonated 2 m above the snow surface to ensure positioning (i.e. charges cannot slide) and guarantee maximum explosive effect. A seismic sensor is used to detect the detonation and the signal is monitored in real-time in the control centre.

The Mot da Ri sector is very steep and rocky with several parallel couloirs. Using the *Avalanche Guard®* system several couloirs can be targeted from a single location; the *Avalanche Guard 5400®* uses 5 kg explosives and has a throw distance up to 180 m (possible deployment over 360 m). This system produces good results despite that detonation occurs in the snow. Each installation can hold up to four magazines, each with 10 charges. The charges are secured in the magazine and triggered individually. Three locations with a total of eight explosives boxes or eight blasting locations were planned for this avalanche sector (Figure 9.7). A seismic sensor is used to detect the detonation and the signal is monitored in real-time in the control centre.

The Piz Soèr sector is located above the Salaniva summit station at 2917 m. This area covers two large couloirs with an average grade of 35°. The *Avalanche Guard*® with 5 kg charges can target both couloirs. Rapid and reliable blasting is important in this area because of the proximity to the chair lift.

The Vallatschas sector is located in close proximity to infrastructure and ski pistes where the main hazard is wet-slab avalanches. For wet-slab avalanches, large explosive charges and careful timing is required. The optimal timing is early to late afternoon, or when there is no longer direct sunlight in the area [264]. The *Avalanche Pipe Mörser 5400*® can target several areas in a very short time using 5 kg explosives (max. distance up to 180 m). The *Avalanche Pipe*® is a device for throwing single charges that can be rotated 360° and fired at three inclinations. It serves as an aid to the blasting office and is installed on a foundation or as a mobile unit on snow groomers. There are three models (Figure 9.8); the *Avalanche Pipe Booster 500*® has a range up to 450 m.

(a)

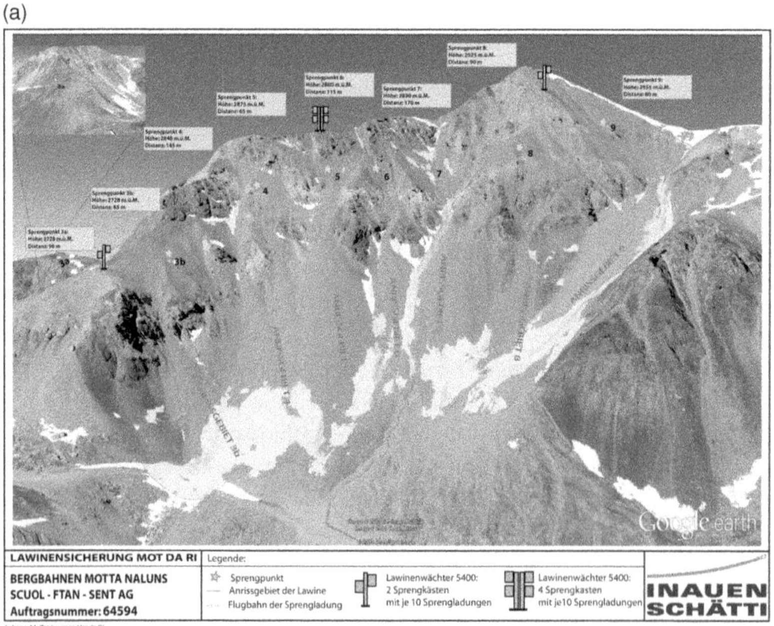

Fig. 9.7 Overview of the Mot da Ri sector (© Inauen-Schätti)

(b)

Fig. 9.7 (*Continued*) Avalanche Guard system (© Inauen, Schätti)

9.3 Avalanche monitoring technology

9.3.1 Principles of avalanche monitoring

Avalanche monitoring includes all methods of direct systematic recording, observing or monitoring of an avalanche event using technological devices or other monitoring systems. Depending on the purpose of an investigation it may consist of (i) observing climatic and nivologic factors that lead to a specific avalanche situation, or (ii) characterising the actual process, which includes documenting data at the time of occurrence and/or performing a comprehensive analysis of the situation and consequences (effects) after an event. Avalanche monitoring mainly serves scientific purposes and its application for early warning is restricted because events are rapid and warning time is negligibly short. Nevertheless avalanche monitoring gains growing importance related to artificial

(a)

(b)

Fig. 9.8 Avalanche Pipe models: (a) Booster 250®; (b) Booster 500® (© Inauen-Schätti)

avalanche release (Section 9.2) and calibration of numerical avalanche models (Section 3.2) as well as for avalanche forecasting.

Systematic documentation of avalanches dates to the end of the nineteenth century when natural hazard management gained importance during construction of trans-alpine railway lines. Documentation of avalanche events in databases for scientific purposes, however, dates only to a few decades ago.

(c)

Fig. 9.8 (*Continued*) (c) Mörser 5400® (© Inauen-Schätti)

In the Alpine countries, data on extreme snowfall events is collected from both a national network of partially automated meteorological stations and a regional network of fully automated meteorological stations and volunteer observers from the weather services. Newer technology such as weather radar is also used for real-time monitoring and analysing extreme weather situations.

9.3.2 Meteorological monitoring

9.3.2.1 Fundamentals

Meteorological monitoring refers to general monitoring of relevant climatic factors (atmospheric pressure, temperature, wind speed, wind direction, humidity and precipitation) as well as using recorded data to understand specific conditions that precipitate avalanche activity. Measurements from automatic weather stations are used for continuous documentation of weather patterns and direct interpretation of risk, and for input to avalanche forecast software to evaluate avalanche hazard.

9.3.2.2 Automatic weather stations

A weather station is a configuration of instruments for collecting meteorological data and observing weather or climate in a particular area. *Fully automatic weather stations* used in avalanche catchments are optimized for snow related measurements (Figure 9.9) and generally consist of a wind and a snow station. The wind station should be located on a ridge and the snow station in a calm, wind-free depression. These stations are connected via radio link to a central data storage system.

The most relevant snow related parameters include snow depth [cm], surface temperature of the snow cover [°C], temperature within the snow cover at 4 levels [°C], snow water equivalent (SWE) [mm/m^2], air temperature [°C], air humidity [% rh], global radiation [W/m^2] and radiation balance [W/m^2]. The most relevant wind related parameters include air temperature [°C], air humidity [% rh], wind direction [degrees], wind speed [m/s] und wind gusts [m/s].

Data can be transmitted from the station via interface (RS 232) or remotely via radio or telephone (analogue, ISDN, GSM) or GPRS. Solar panels are usually sufficient for the power supply.

9.3.2.3 Weather radar

Weather radar refers to any radar device used for recording weather data, especially precipitation. In the Alps weather radars are widely used for monitoring and predicting precipitation, both for hydrological applications and risk/catastrophe management in case of extreme events as well as for tourism.

Real-time monitoring of precipitation with high spatial resolution for large areas in the Alps necessitates installation of radars on an isolated height of land, for example on a mountain peak (Figure 9.10) that offers both optimal visibilities but also reliable measurement of precipitation close to the ground. Further considerations are that the radar beam may be partially or totally shaded by surrounding mountains (precipitation behind the mountains is hidden) or echoed off surrounding mountains (filtering ground echoes out of the data is time-consuming and challenging). Operation and maintenance of radar stations in such terrain is very cost-intensive. Simple rain gauges, however, are also not optimal in the mountains as they are influenced by wind.

For meteorological purposes the spatial and temporal course of precipitation is of interest and should be detectable in a range from light precipitation (0.1 mm/h) to heavy rainfall (>100 mm/h); for hydrological purposes high precision of measurement is important. Radar devices emit directional short microwave pulses with a beam width of 1 degree (at a distance of 60 km the radar beam is 1 km wide). Depending on the number and size of particles (rain, snow, sleet, hail) a part of the emitted radiation is reflected back. The intensity of the precipitation can be derived from the strength of the reflected signal and the position of a precipitation area can be determined from the signal runtime

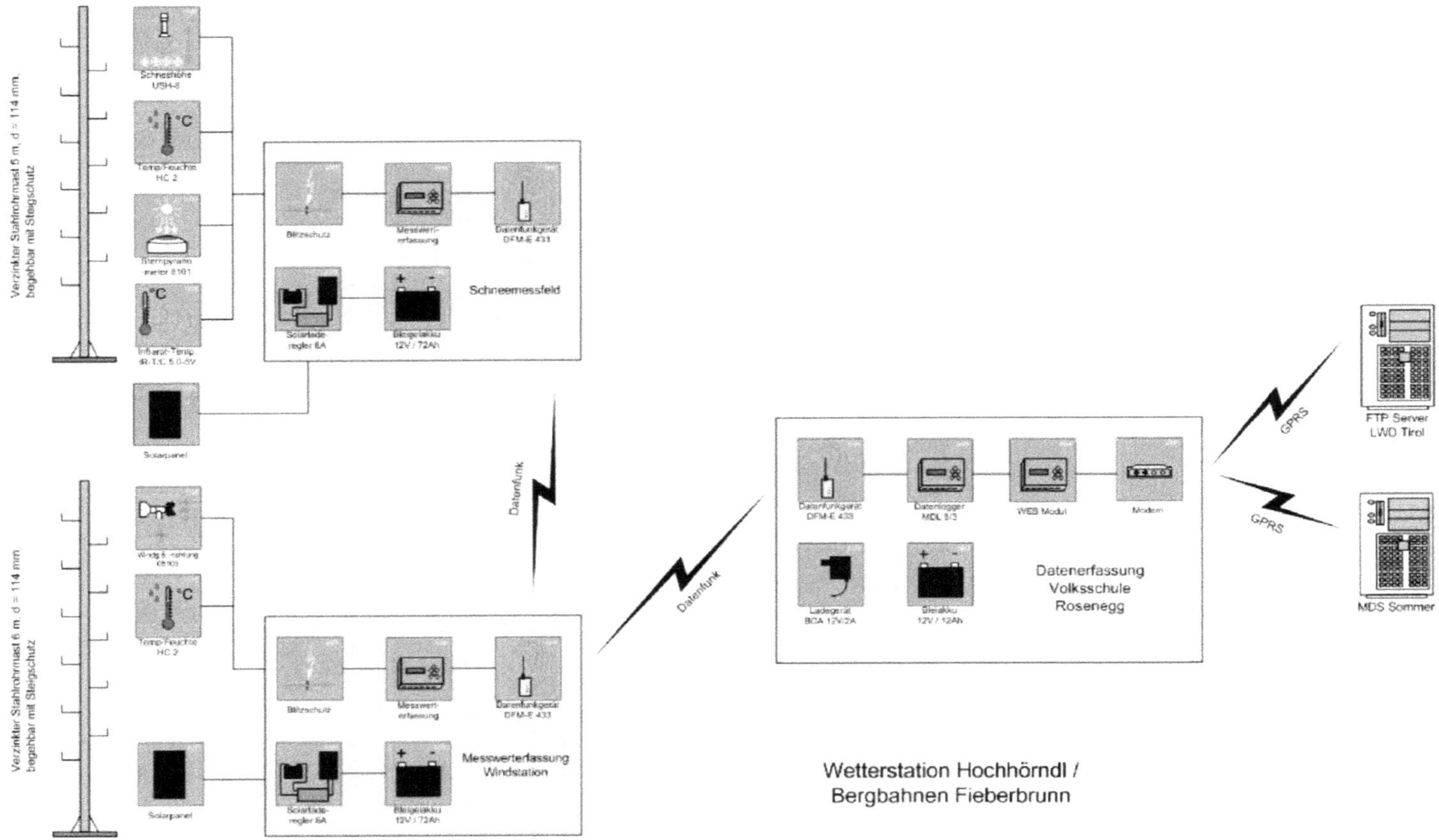

Fig. 9.9 Scheme of an automatic weather (snow) station located at the Hochhörndl (Mountain Cableways Fieberbrunn, Tyrol) (© Sommer)

Fig. 9.10 Weather radar station on the Valluga (Arlberg, Tyrol) (© Schaffhauser)

(Figure 9.11). Weather radars are very sensitive and can detect the reflected signal though it is much weaker (10–8 to 10–13 watts) than the radiated peak power (250 kilowatts). The antenna is rotated for 5 minutes in the horizontal and vertical direction and covers an altitude of 15 km within a radius of 220 km. Theoretically, weather radar can provide quantitative data on precipitation up to 100 km, but only qualitative information at further distances.

9.3.3 Monitoring snow cover

The seasonal snow cover can be monitored with numerous instruments (Table 9.4). Some are integral components of existing automatic weather stations (Section 9.3.2.2) that are installed near avalanche release areas and operated by local avalanche warning centres.

Terrestrial laser scanners (LIDAR) have become more common in recent years for generating distributed snow cover surface models. The scanner emits a laser beam (20–200 nm wavelength) that is reflected off the surface (backscattering); the signal runtime gives the distance between the surface and LIDAR.

(a)

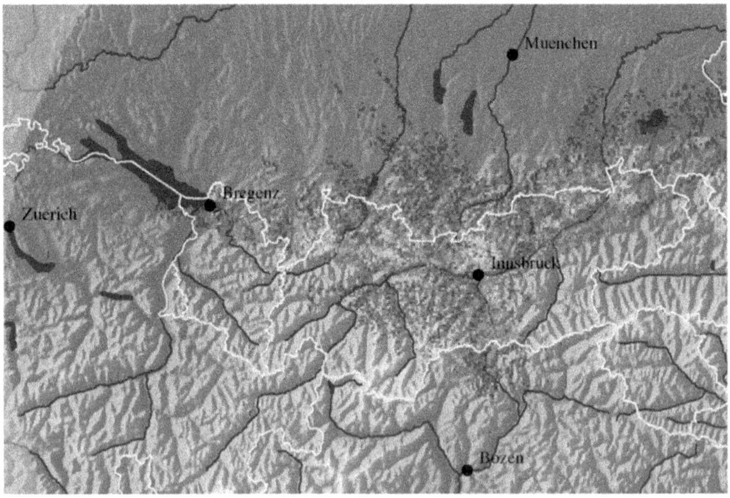

(b)

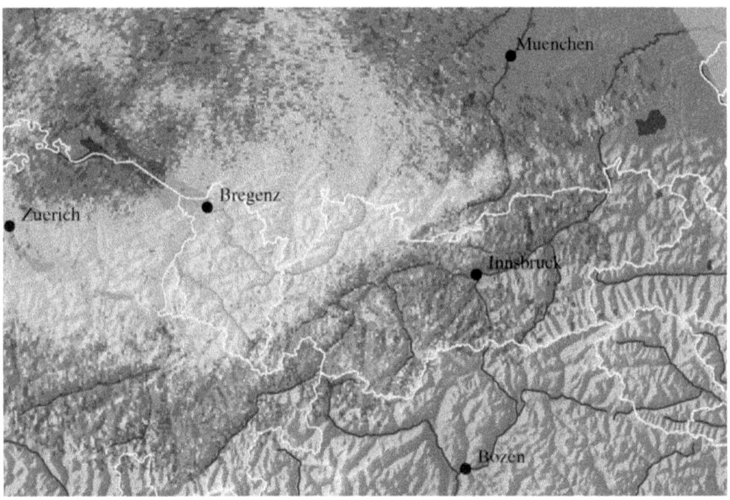

Fig. 9.11 Weather radar data from (a) Patscherkofel (Innsbruck, Tyrol) showing extensive snowfall in the western part of Austria and (b) Valluga (Arlberg, Tyrol) (© ZAMG)

Back triangulation from reflective reference targets gives the exact position of the LIDAR, which is critical if taking serial images. The spatial extension of the seasonal snow cover can be derived by subtracting the snow-free terrain model from a snow surface scan. Multiple scans give information on snow depth changes and can be used for optimizing technical control measures (construction height) or determining artificial release points.

Table 9.4 Overview of instruments for monitoring snow cover

Parameters	Technology
Air temperature and humidity	These parameters are generally measured with coupled sensors (thermometer, hygrometer). The instruments are encased in a light-coloured, slat-siding housing to protect the sensors against direct solar radiation and to facilitate natural ventilation. If there is a power supply for artificial ventilation, the measurement quality can be increased significantly.
Snow depth	The standard method for snow depth measurement uses ultrasonic sensors mounted on a mast. The runtime of the reflected short ultrasound signal (which also depends on air temperature) is used to calculate snow height by subtracting the reference sensor height.
Short wave radiation	There are two types of short wave radiation – incident and reflected. Measurements are typically carried out with a pair of *pyranometers*, one positioned upward and one downward. The incoming (incident) short wave radiation depends on the time of day and season, cloudiness and topography. The reflected radiation depends on these factors as well as the surface conditions and shape of the superficial layers. The ratio between incident and reflected radiation gives the albedo of the snow surface.
Long wave radiation	Measurements are done either in combination with short wave radiation (*pyrradiometer*) or directly (*pyrgeometer*). The instrument positioned upward measures the atmospheric counterradiation, which depends mainly on cloudiness. The pyrgeometer positioned downward measures the temperature of the snow surface, which can also be measured with infrared thermometers (though these have a narrower monitoring angle).
Wind velocity and direction	Propeller anemometers and cup anemometers are used to measure wind velocity (in exposed locations propeller anemometers are favoured). Problems are mainly caused by ice and solid precipitation particles. It is uncommon to have a power supply for heating the device. For wind profiles – important for estimating turbulent and latent heat flux – at least three sensors are mounted in a vertical array.
Snow water equivalent	The snow water equivalent (SWE) is typically monitored by snow pillows. The pressure of the snow cover is detected with a pillow-shaped gauge that is filled with anti-frost liquid and dug into the ground. The liquid pressure gives the depth-averaged SWE. The average snow density can be derived if the snow depth is measured.

(continued)

Table 9.4 (*Continued*)

Parameters	Technology
Snow moisture	Continuous monitoring of snow moisture is possible with capacity sensors. A flat ribbon cable placed diagonally through the snow cover measures the dielectric properties for multiple frequencies, which determines the volume fracture of liquid water, ice and pore volume.
Snow cover temperature	Snow temperature is typically measured with *Pt-100 sensors*. In order to avoid bias the sensors are often mounted on small white plastic tubes of varying lengths (*snow candles*). Continuous measurement enables analysis of metamorphosis or stability changes and detection of melting periods.

In contrast to LIDAR, the GB-SAR signal (Ground Based Interferometric Synthetic Aperture Radar) (Figure 9.12b) penetrates the dry snow cover and is reflected by the boundary layer between the snow and ground. If snow density is known the snow depth and changes in depth can be calculated [158, 238].

(a) (b)

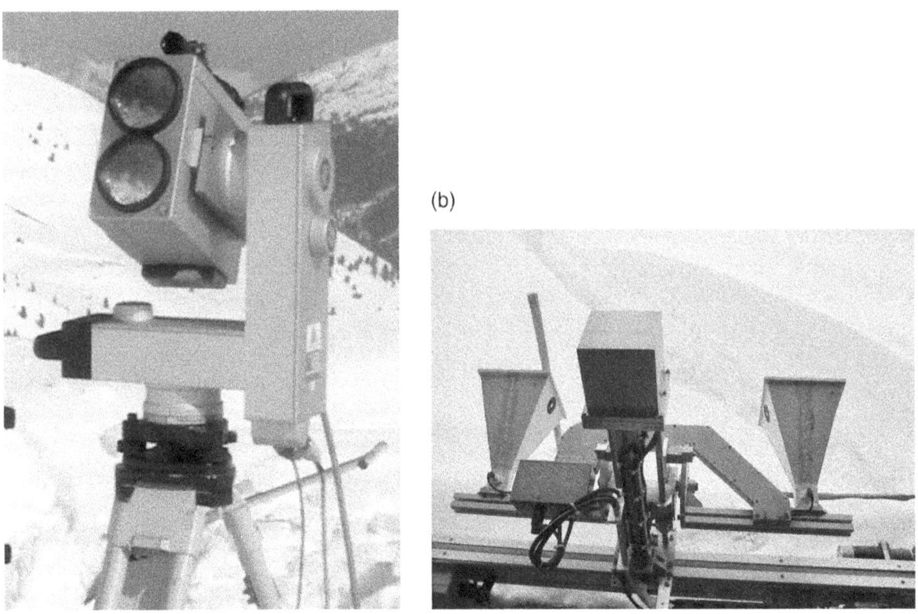

Fig. 9.12 Instruments to determine snow cover surface models (a) LIDAR und (b) GB SAR (© Kleemayr)

9.3.4 Monitoring snow mechanics

The most common parameters used for monitoring the mechanical properties of snow are (i) snow density, measured with cylindrical sample rings, (ii) snow strength with a *Rammsonde* or *snow penetrometer* (electro-mechanical thrust), (iii) shear strength with *shear frame* or shear ring and (iv) snow gliding (movement of the snow cover on the ground) with *gliding shoes* (small metal plates covered by the snow). Height marked profiles (typically filled with sawdust) are used to monitor snow cover deformation (snow creeping) and estimate snow viscosity. In contrast to automatic weather stations (Section 9.3.2.2) mechanical properties of the snow cover are rarely measured continuously. The technological efforts necessary for continuous measurement are usually only suitable for specific scientific purposes.

9.3.5 Monitoring with remote sensing technology

In the last decades, remote sensing (aerial or air-based monitoring) has become a keystone for engineers to assess events and mass movements or improve planning and dimensioning. Satellite data provide information on land use and land use change, vegetation cover and snow distribution. Aerial images are used to derive surface and terrain models. For aerial images the camera system is mounted in an aeroplane and flown some hundred meters above the ground. Aerial images also can be generated with RPAS (remotely piloted aircraft systems; drone), which have become more important in the last years after significant technical, economic and software advances. The low operational flying altitude of RPAS (80–150 m above the ground) has inherent advantages (high resolution even with standard cameras) and disadvantages (high relative velocity).

Remote sensing systems in general can be described as:

1. *Passive*
   - aerial images in black/white, true colour or near infrared (NIR)
   - thermal images for surface temperature
   - passive microwave radiometry (e.g. ground humidity).
2. *Active*
   - RADAR (e.g. microwave range for detecting the ground surface)
   - LIDAR (distance, terrain or object surface).

9.3.6 Monitoring snow forces on avalanche defense measures

To construct, maintain or improve technical constructions (e.g. snow nets, steel snow bridges) it is important to understand the forces acting on the construction and the reactions (stress, strain). Various instruments are available to monitor impact pressure, anchor stress and flux of forces in the construction. Such data can validate load and dimensioning assumptions from the design phase. Continuous monitoring of pole and anchor forces is also possible (Figure 9.13) [160, 164].

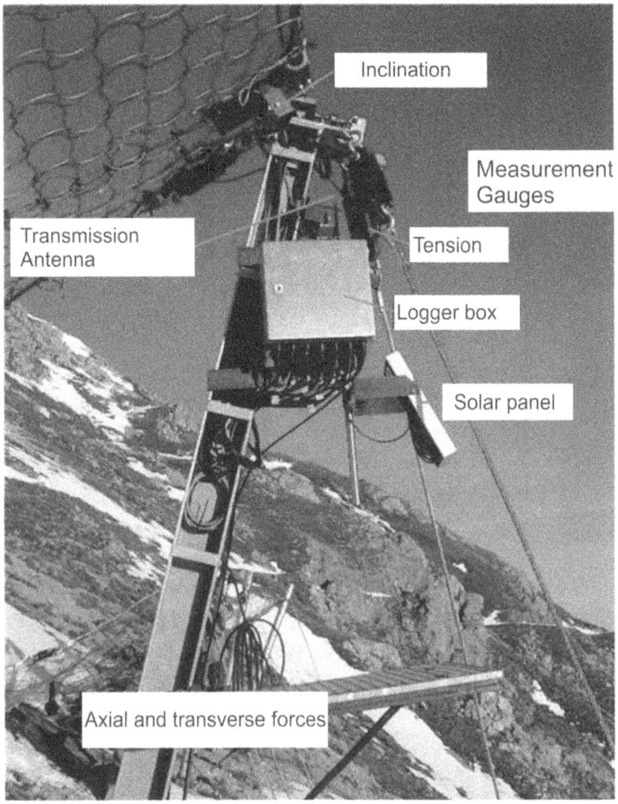

Fig. 9.13 Snow net monitoring site at Hafelekar (Tyrol, Austria) [214] (© WLV Tyrol/Austrian Research Centre for Forests)

9.3.7 Monitoring avalanche dynamics

9.3.7.1 Systems for monitoring avalanche motion

Some monitoring sites are installed specifically to study the dynamic process of avalanches (e.g. velocity, flow depth, pressure) or to serve as warning stations (or sometimes both). Field experiments with artificially released avalanches give essential information on the underlying physics, which form the basis for improving and calibrating models.

In 1997 the WSL Institute for Snow and Avalanche Research (SLF) in Davos established an avalanche monitoring site in Vallée de la Sionne (Kanton Wallis) [6] that has since become the most extensive site worldwide. Avalanches are triggered artificially with a volume up to 100 000 m^3. The release area spans ~30 ha and has an inclination of 23 to 45°. The maximum elevation difference of the track is 1200 m and maximum length is 2500 m.

The monitoring area is equipped with numerous instruments. Doppler radar and video cameras located in a bunker on the opposite slope are used to measure avalanche front

(a)

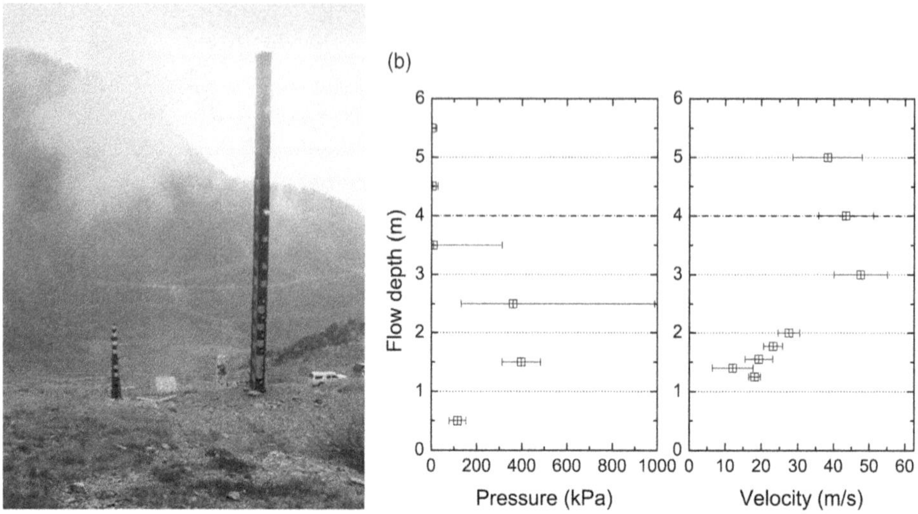

Fig. 9.14 Load measurement in Vallée de la Sionne (Wallis, Switzerland); (a) instrumented pylons in the track (height: 20 m) and the avalanche cone (© SLF); (b) the avalanche pressure and related velocity profile of avalanche Nr. 7226 (pole height: 20 m; width: 0.6 m) with a flow depth of 4 m (source [258, 259], © B. Sovilla)

velocity. Additionally, three FMCW (frequency-modulated continuous-wave) radars are located along the track to measure flow depth, snow erosion and snow entrainment. In the lower part of the track permanent structures are equipped with load cells and optical sensors to measure flow velocity at different heights, flow height, and air and particle pressure (Figure 9.14a, Figure 9.15). Photo- or videogrammetry are used to determine the snow/avalanche volume before and after the event (i.e. mass balance) [258, 259].

During the catastrophic winter of 1999 three avalanches hit these constructions (total volume >1.2^6 m^3) and destroyed most of them. The maximum recorded velocity was 80 m/s and maximum pressure was 400 kN/m^2. Similar instrumentation can also be found in the Norwegian test area at Rygfonn (test site of the Norwegian Geotechnical Institution, NGI) (Figure 9.16).

9.3.7.2 Measuring avalanche impact forces with load cells

Direct measurement of dynamic forces in an avalanche is usually done with *load cells* that are integrated into pressure plates. In Vallée de la Sionne the 1 × 1 m pressure plates are mounted on a concrete wedge (height: 4.5 m; width: 1 m) (Figure 9.15c). There are four special load cells (Magtrol, model LB 217) in the pressure plate that measure normal and shear stress simultaneously up to 600 kN (normal) and 300 kN (shear). The pressure plate is a welded, joint steel construction with four basic elements: base

(a)

(b)

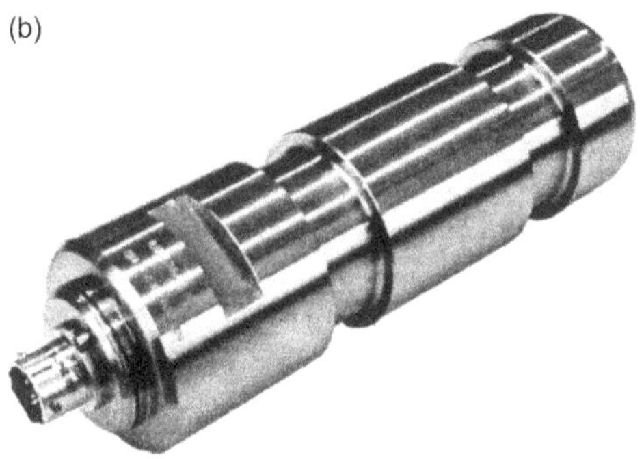

(c)

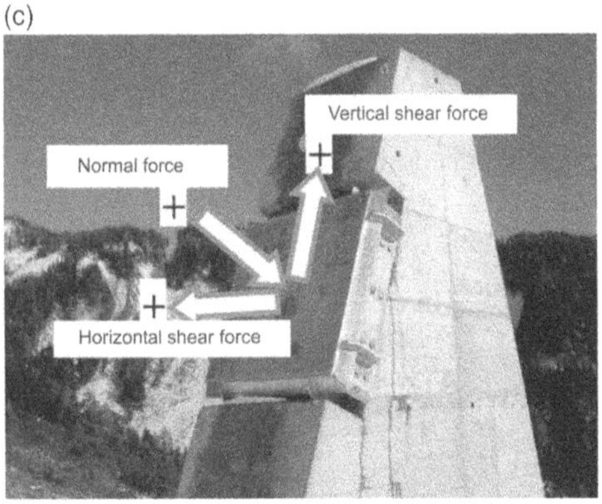

Fig. 9.15 Avalanche pressure instruments: (a) pressure plate (BFW); (b) load cell (Magtrol LB 237) and (c) installation of the plate on the concrete wedge (Vallée de la Sionne, Switzerland) (© Kleemayr)

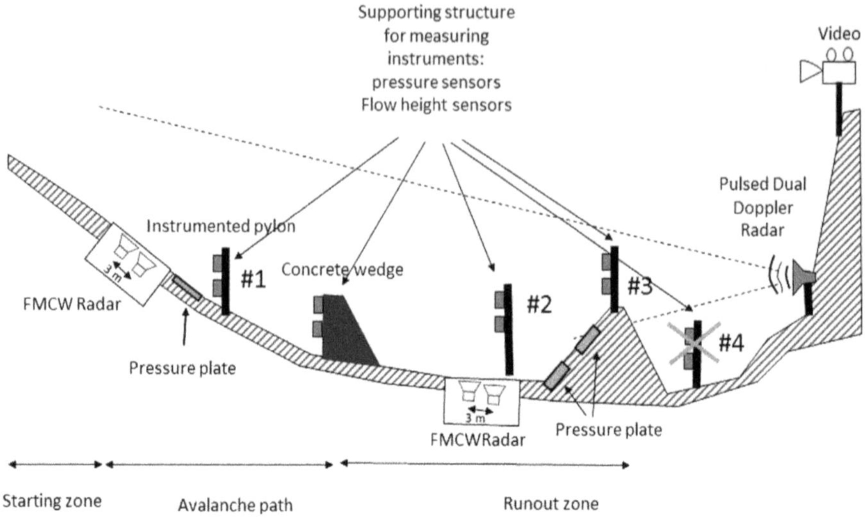

Fig. 9.16 Schematic of the planned measurement configuration at the Norwegian avalanche test site Rygfonn (© NGI, 2009)

plate, a force redirection element with the load cell and a side element to avoid force impact from the side (Figure 9.15a). Six similar pressure plates have been installed in Tyrol (Austria) by the Austrian Research Centre for Forests.

Measurements at these test sites have improved the understanding of the processes and forces behind avalanches, including three noteworthy aspects:

1. Avalanche velocities in the track have been widely underestimated up to now.
2. In most major avalanche events both the new snow and settled old snow cover have been eroded and entrained, leading to a significant change in the mass balance.
3. The resulting pressure depends largely on the flow regime (Figure 9.14b) [258]. In particular for dense-flow avalanches (<10 m/s) measurements yielded considerably higher pressure values than would be expected from common calculation methods.

9.3.7.3 Measuring avalanche flow depth

In Vallée de la Sionne, flow depth is measured with sensors mounted on the 20 m pylon (Figure 9.14a) [258]. There are two configurations:

- toggle switches mounted laterally on the pylon at 25-cm intervals are activated by the passing snow masses; the highest activated switch represents the flow height of the dense flow part;
- pressure sensors distributed over the front of the pylon and the corresponding time series also indicate the flow height – as soon as the value permanently exceeds 20 kPa it is assumed that the sensor was hit by the dense flow part.

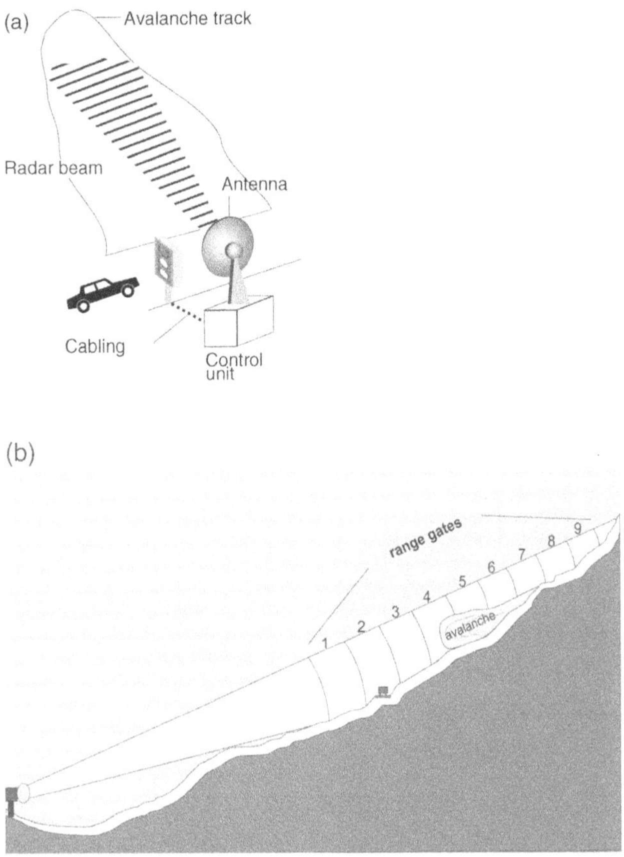

Fig. 9.17 A radar-coupled system (a) to protect the main road at the Multereck avalanche (Styria, Austria) (© Kleemayr); Doppler radar offers the possibility of dividing the monitored area into different sections, so called *Range Gates* (b), which as a result only allows some of the sections to be used for signal analysis [144] (© Kogelnig)

9.3.7.4 Measuring velocity with optical sensors

In Vallée de la Sionne, optical sensors are installed on the pylon at 1-m intervals between 3 and 6 m. The sensors point parallel to the flow direction and the ground surface. The signal correlation of two infrared sensors measuring the displacement in the flow direction gives the velocity. Additionally, video signal processing is used to derive avalanche front velocity.

9.3.7.5 Measuring velocity with pulsed dual doppler radar

Pulsed Doppler radar is used to measure velocity and turbulence parameters. The radar system emits short microwave pulses that are reflected off surrounding objects and received by the antennae. Interpretation of the signal response gives the velocity and distance (between object and antenna).

Fig. 9.18 The avalanche radar device installed down in the valley facing the Grosstal avalanche track (Kappl, Tyrol) [144] (© Kogelnig)

By interpreting the reflected Doppler frequencies of various sections in the avalanche track (*range-gating*) the measured intensity, velocity and turbulence parameters can be related spatially. The data give spectra for each range-gate for the different time slots. With offline data processing software the time dependent velocity for the range-gates can be determined.

Technical or economic reasons may prevent use of active technical control measures in many endangered areas in the Alps, especially traffic routes. In these areas Doppler radar may be a valuable alternative to monitor or detect avalanche activity. One potential application couples radar to a road closure system – if avalanche movements are detected a signal is transmitted directly or indirectly to a traffic light or gate on the road. Such a radar-coupled system has been installed at the Multereck avalanche (Styria, Austria) (Figure 9.17a).

Similar systems are also used to document artificial avalanche release in ski resorts. Radar measurements are better than common optical and sonic methods to verify release and can also be done in unfavourable (weather) conditions. Such a system, for example was installed to monitor the success of artificial avalanche release in Grosstal avalanche (Kappl, Tyrol, see Figs. 9.17b and 9.18) (Section 9.2.7.2) [144].

10 Technical avalanche protection international: facts and figures

Florian Rudolf-Miklau, Stefan Margreth, Francois Rapin, Bernhard Zenke, Yasuo Ishii, Rudolf Pollinger, Krister Kristensen, Tomas Johannesson, Pere Oller, Art Mears, Chris Stetham and Emily Procter

Technical avalanche protection is a topic of growing importance in many mountain regions worldwide. This section provides an overview of the state of status of development in the countries (regions) most affected by avalanche risks. The following table is the result of an expert survey based on a questionnaire, which was carried out in the course of the publication of the German edition of this handbook [223]. Avalanche expert from 11 countries have contributed to this survey. The comparative table shall provide a rough overview of the similarities and differences of technical avalanche protection in Austria, Switzerland, France, Italy, Germany, Norway, Iceland, Spain, Canada, USA and Japan.

The following institutions and experts have contributed to the survey:

Austria (A)	Florian Rudolf-Miklau (Federal Ministry for Agriculture, Forestry, Environment and Water Management)
Switzerland (CH)	Stefan Margreth (WSL Institute for Snow and Avalanche Research SLF)
France (F)	Francois Rapin (IRSTEA, ETNA St. Martin D'Herès)
Germany (Bavaria) (D)	Bernhard Zenke (Bavarian Federal Office for Environment, Avalanche Warning Service)
Italy (South Tyrol) (IT)	Rudolf Pollinger (Autonomous Region of Trentino-Alto Adige, Bolzano)
Norway (N)	Krister Kristensen (Norges geotekniske institutt NGI, Oslo)
Iceland (IS)	Tomas Johannesson (Veðurstofa Islands, Reykiavik)
Spain (Catalonia) (E)	Pere Oller (Institut Geológic de Catalunya, Barcelona)
Canada (CND)	Chris Stethem (Chris Stethem & Associates Ltd., Canmore)
United States of America (USA)	Art Mears (Arthur I. Mears, P.E., Inc., Gunnison/CO)
Japan (J)	Yasuo Ishii (Snow Avalanche and Landslide Research Centre, Myoko-City, Niigata Precture)

The Technical Avalanche Protection Handbook, First Edition. Edited by Florian Rudolf-Miklau, Siegfried Sauermoser and Arthur I. Mears.

Technical avalanche protection: International overview	Country										
	A	CH	F	IT	D	N	IS	E	CND	USA	J
Degree of importance of technical avalanche protection	very high	very high	regionally high	regionally high	middle – low	regionally high	regionally high	low, locally high	regionally high	middle – low	regionally high
Documented avalanche types											
Dense flowing avalanche	•	•	•	•	•	•	•	•	•	•	•
Powder avalanche	•	•	•	•	•	•	—	•	•	•	•
Mixed avalanche	•	•	•	•	•	•	—	•	•	•	•
Wet-snow avalanche	•	•	•	•	•	•	•	•	•	•	•
Slush flows	—	—	—	—	—	•	•	—	•	•	•
Combination debris flow-snow avalanche	•	•	•	•	—	•	•	—	•	•	•
Other: glacier avalanche	•	•	•	•	—	•	—	—	—	—	—
Number of known destructive avalanches paths	~18 000	20 000	4000	15 000	650	several hundred	several hundred	500	30 000	several tens	20 500
Avalanche damage (people, property)											
Fatality by avalanche (per year)	30	25	28	N/A	N/A	N/A	2	1.5	15	N/A	N/A
Property damage by avalanches (per year, in millions of Euros)	N/A	58	N/A	N/A	N/A	N/A	~14	N/A	N/A	N/A	N/A
Hazard maps, hazard zone maps											
In existence	yes	yes	yes	yes	no	yes	no	yes	yes	locally	yes

Technical avalanche protection: International overview	Country										
	A	CH	F	IT	D	N	IS	E	CND	USA	J
Scale 1:	2000	5000	25 000 (10,000)	25 000 (10,000)	—	50 000 2000	—	25 000	5000 to 10 000	1000 to 24 000	25 000
Hazard level red (considerable danger)	•	•	•	•	—	•	—	—	•	—	N/A
Hazard level blue (moderate danger)	yellow = moderate	•	•	•	—	—	—	—	•	—	—
Hazard level yellow (low danger)		•	—	•	—	•	—	—	—	—	—
Hazard level yellow/white (residual danger)	—	•	—	•	—	—	—	—	—	—	—
Recurrent design event (return period)	150	300/ 30	300/100	300/30	—	5000/ 1000/ 100	—	—	300/30	—	N/A
Criterion for zoning, avalanche pressure (kPa)	10/1	30/3	30/1	30/1	—	/1	—	—	30/1/ 0.1	—	N/A
Legal basis to avalanche protection	yes	yes	yes	yes	—	yes	yes	no	yes	—	no
Legally binding for land use planning and construction engineering	yes	yes	yes	yes	—	yes	yes	no	yes	—	no
Guidelines for safety and emergency planning in existence	yes	yes	partially	yes	—	yes	yes	no	partially	—	no
Risk maps											
In existence	no	no	no	partially	no	partially	yes	no	no	no	no

Technical avalanche protection: International overview	Country										
	A	CH	F	IT	D	N	IS	E	CND	USA	J
Scale 1:	—	—	—	varies	—	—	5000	—	—	—	—
Recurrent design event (return period)	—	—	—	varies	—	—	3000/300/ 100	—	—	—	—
Permanent technical avalanche defence											
Practiced technique (y/n)	yes	yes	yes	yes	yes	yes	yes	locally	locally	locally	yes
Protection target (relevant for avalanche protection mearsures)											
Settlements	•	•	•	•	•	•	•	•	•	•	•
Streets and highways	•	•	•	•	•	•	•	•	•	•	•
Railways	•	•	•	•	—	•	—	•	•	•	•
Power lines, pipelines	•	•	•	•	•	•	•	—	•	•	—
Ski areas	•	•	•	•	—	•	—	•	•	•	•
Industrial sites	•	•	•	•	•	•	—	—	•	•	—
Military sites	•	•	—	—	—	•	—	—	—	—	—
Structural avalanche protection measures (constructions), other measures											
Avalanche defence structures in the starting zone	•	•	•	•	•	•	•	•	•	•	•
Snow drift control structures	•	•	•	•	—	•	—	•	•	•	•
Avalanche catching dams	•	•	•	•	—	•	•	—	•	•	—
Avalanche breakers	•	—	•	•	—	•	•	—	•	•	—

Technical avalanche protection: International overview	Country										
	A	CH	F	IT	D	N	IS	E	CND	USA	J
Avalanche retarding structures	•	•	•	•	—	•	•	—	•	•	—
Avalanche galleries (or tunnels)	•	•	•	•	•	•	•	•	•	•	—
Avalanche deflection dams	•	•	•	•	—	•	•	•	•	•	•
Roof terraces/Splitting Wedges	•	•	•	•	—	—	—	—	—	•	—
Snow glide protection	•	•	—	•	—	•	—	—	—	•	—
Existing national technical norms	yes	yes	yes	(Swiss Guidelines)							yes
Existing legal basis to protection measures	yes	yes	yes	yes	yes	yes	yes	no	no	no	yes
Avalanche protection forest (maintenance, high elevation afforestation)	•	•	•	•	•	(•)	—	—	(•)	(•)	(•)
Public investments in technical avalanche protection (in millions of € per year):	30	35.0	N/A	10.0–12.0	1.0	N/A	3.0	1.0	—	0.5	N/A
Private contributions required	yes	yes	no	no	no	no	no	no	yes	yes	no
Implementation of measures by public (governmental) institutions	yes	yes	yes	yes	yes	no	partially	no	no	no	yes

Technical avalanche protection: International overview	Country										
	A	CH	F	IT	D	N	IS	E	CND	USA	J
Technical protection of buildings											
Practiced technique (y/n)	yes	yes	yes	yes	yes	yes	yes	yes	yes	yes	no
Obligatory for adherence to legal norms	yes	yes	yes	no	yes	yes	yes	no	yes	yes	no
Official expert's report required	yes	yes	yes	yes	yes	partially	partially	no	N/A	N/A	no
Temporary avalanche defence											
Temporary measures											
Avalanche warning	•	•	•	•	•	•	•	•	•	•	—
Artificial avalanche release	•	•	•	•	•	•	tests only	•	•	•	•
Closures (transit routes)	•	•	•	•	•	•	•	•	•	•	•
Evacuation	•	•	•	•	•	•	•	•	•	•	•
Catastrophe plans (avalanches)	•	•	•	•	—	•	•	•	•	•	—
Warning and alarm systems	•	•	•	•	•	•	•	•	•	•	•
Artificial avalanche release (blasting): systems											
Hand charg	•	•	•	•	—	—	only tests	•	•	•	•
Helicopter blasting	•	•	•	—	•	—		—	•	•	—
Blasting ropeway	•	•	•	•	•	•		—	•	•	—

Technical avalanche protection: International overview	Country										
	A	CH	F	IT	D	N	IS	E	CND	USA	J
Gaz.Ex®	•	•	•	•	•	—	only tests	—	•	•	•
DaisyBell®	—	•	•	•	—	•		•	—	—	—
Avalhex®	—	—	•	—	—	—		•	—	—	—
Lawinenwächter Inauen-Schätti®	•	•	—	—	—	—		—	•	•	—
Sprengmast Wyssen®	•	•	—	—	—	—		—	—	—	—
Avalancheur®	—	•	—	—	—	—		•	•	•	•
Military arms: mortar (mine launchers), long range artillery	—	•	—	—	—	—		—	•	•	—
blasting of cornices	—	•	—	—	—	•		—	—	—	—
Avalanche warning services											
In existence	yes	yes	yes	yes	yes	yes	yes	regionally	yes	regionally	no
Services mandatory upon a legal basis	partially	no	N/A	partially	yes	no	partially	no	no	no	—
Existing observation networks (meteo. stations, in-field observations)	yes	yes	yes	yes	yes	partially	yes	partially	yes	partially	partially
Avalanche report/bulletin available on the internet	•	•	•	•	•	•	—	•	•	(•)	—

Technical avalanche protection: International overview	Country										
	A	CH	F	IT	D	N	IS	E	CND	USA	J
Avalanche commission											
In existence	yes	yes	yes	yes	yes	locally	yes	no	no	locally	no
Legal foundation to activities and/or decisions	partially	partially	no	yes	yes	partially	yes	no	no	yes	no
Disaster (catastrophe) management											
Organizational structure at the community or municipality level	yes	yes	yes	yes	yes	yes	yes	yes	yes	yes	yes
Disaster management, emergency, and/or evacuation plans	partially	yes	yes	yes	yes	partially	yes	yes	partially	partially	yes
Regional or national early warning and crisis intervention system	yes	yes	yes	yes	yes	yes	yes	yes	yes	no	yes
Avalanche monitoring and event documentation											
Avalanche monitoring system	•	•	•	•	•	•	•	—	•	—	•
Avalanche warning sites (locally)	•	•	•	•	—	•	N/A	•	•	—	•
Event register/avalanche register (i.e. database)	yes	yes	yes	yes	yes	partially	partially	yes	N/A	no	N/A

Technical avalanche protection: International overview	Country										
	A	CH	F	IT	D	N	IS	E	CND	USA	J
Research, standardizations and training											
Specialized research facilities	•	•	•	•	—	•	•	—	—	—	•
University and non-university (e.g. private institution) based research	•	•	•	•	•	•	•	(•)	•	•	•
Test sites, laboratory facilities	•	•	•	—	—	•	•	—	—	—	•
Academic training in the field of technical avalanche protection	•	—	—	—	—	—	—	—	—	—	—
Academic training within other technical or college programs	•	•	•	•	—	•	•	—	•	—	•
Advanced training or professional development available for practitioners	•	•	•	•	•	•	•	—	•	•	•

Literature

1. Agerer, H. and Rudolf-Miklau, F. (2007) Lebenszyklus-basiertes Management von Wildbach- und lawinenverbauung: Strategische und praktische Dimension. *Journal for Torrent, Avalanche and Erosion Control*, **155**, 30–42 (in German).
2. AUVA/General Accident Insurance Institution (2004) Arbeiten im Freien: Risiken und Schutzmaßnahmen, Vienna (in German).
3. AUVA/General Accident Insurance Institution (2005) Handzeichen für Einweiser – Sicherheit kompakt (M 203), Vienna (in German).
4. AUVA/General Accident Insurance Institution (2007) Höhenarbeiten am Seil und Arbeitsplatzpositionierung – Leitlinien, Vienna (in German).
5. Ammann, W.J., Buser, O. and Vollenwyder, U. (1997) *Lawinen*, Birkhäuser, Basel Boston Berlin (in German).
6. Ammann, W.J. (1999) A new Swiss test-site for avalanche experiments in the Vallée de la Sionne/Valais. *Cold Regions Science and Technology*, **30** (1–3), 2–11.
7. Amt der Tiroler Landesregierung/Office of the Tyrolean Government (1996) *Lawinenhandbuch*, Tyrolia, Innsbruck (in German).
8. Armstrong, B.R. and Williams, K. (1992) *The Avalanche Book*, Fulcrum Publishing, Golden/Colorado.
9. Arnalds, P., Jónasson, K. and Sigurðsson, S. (2004) Avalanche hazard zoning in Iceland based on individual risk. *Annals of Glaciologie*, **38**, Davos: 285–290.
10. ASTRA/Federal Roads Office (2007) Einwirkungen infolge Lawinen auf Schutzgalerien, Bern (in German).
11. Aulitzky, H., Fiebiger, G., Kastner, F., Kemmerling, W. and Kaupa, H. (1977) *Wirtschaftlichkeitsuntersuchungen in der Wasserwirtschaft, Teil 1*, Federal Ministry for Finances, Vienna (in German).
12. Aulitzky, H. (1992) Die Sprache der "Stummen Zeugen". Proceedings of the International Symposium INTERPRAEVENT Vol. 6, Bern, 139–174 (in German).
13. Bader, H., Haefeli, R., Bucher, E., Neher, J., Eckel, O., Thams, C. and Niggli, P. (1939) Der Schnee und seine Metamorphose. Beiträge zur Geologie der Schweiz. Geotechnische Serie: Hydrologie, Lieferung 3 (in German).
14. Baillifard, M. (2007) Interaction between snow avalanches and catching dams. Federal Technical University of Zurich, Phd.-thesis 17301.
15. Baillifard, M., Kern, M. and Margreth, S. (2007) Anleitung zur Dimensionierung von Lawinenauffangdämmen. WSL Institute for Snow and Avalanche Research SLF, Davos (in German).

The Technical Avalanche Protection Handbook, First Edition. Edited by Florian Rudolf-Miklau, Siegfried Sauermoser and Arthur I. Mears.

16. Bartelt, P. and Stöckli, V. (2001) The influence of tree and branch fracture, overturning and debris entrainment on snow avalanche flow. *Annals of Glaciology*, **32**, 209–216.

17. Bartelt, P. and Buser, O. (2009) Production and decay of random kinetic energy in granular snow avalanches. *Journal of Glaciology*, **55**, 3–12.

18. Bebi, P., Kienast, F. and Schönenberger, W. (2001) Assessing structures in mountain forests as a basis for investigating the forests dynamics and protective function. *Forest Ecology and Management*, **145**, 3–14.

19. Bebi, P., Kulakowski, D. and Rixen, C. (2009) Snow avalanche disturbances in forest ecosystems – state of research and implications for management. *Forest Ecology and Management*, **257**, 1883–1892.

20. Bergmeister, K., Suda, J., Hübl, J. and Rudolf-Miklau, F. (2009) *Schutzbauwerke gegen Wildbachgefahren: Grundlagen, Entwurf und Bemessung, Beispiele*, Ernst & Sohn, Berlin (in German).

21. Bommer, C., Phillips, M., Keusen, H.-R. and Teysseire, P. (2009) Bauen im Permafrost – Ein Leitfaden für die Praxis. WSL Institute for Snow and Avalanche Research SLF, Davos (in German).

22. Borter, P. (1999) *Risikoanalyse bei gravitativen Naturgefahren*, Swiss Agency for the Environment, Forests and Landscape (SAEFL), Bern, Publication 107/I+ II (in German).

23. Brang, P., Schönenberger, W., Frehner, M., Schwitter, R. and Thormann, J.J. (2006) Management of protection forests in the European Alps: an overview. *Forest Snow and Landscape Research*, **80**, 23–44.

24. Brücker, W.J. (1981) Vegetationsuntersuchungen in Lawinenablagerungsgebieten des Kantons Uri. *Berichte der Naturforschenden Gesellschaft Uri*, **9**, 1–254 (in German).

25. Bründl, M., Schneebel, M. and Flühler, H. (1999) Routing of canopy drip in the snowpack below a spruce crown. *Hydrological Processes*, **13**, 49–58.

26. Bründl, M., Romang, H., Holthausen, N., Merz, H. and Bischof, N. (2009) Risikokonzept für Naturgefahren – Leitfaden, Teil A: Allgemeine Darstellung des Risikokonzepts. National Platform for Natural Hazards PLANAT Bern (preliminary version) (in German).

27. Bründl, M. and Romang, H. (2009) Risikokonzept für Naturgefahren – Leitfaden, Teil B: Anwendung des Risikokonzepts: Prozess Lawine. National Platform for Natural Hazards PLANAT Bern (preliminary version) (in German).

28. Bucher, E. (1948) Beitrag zu den theoretischen Grundlagen des Lawinenverbaus. Beiträge zur Geologie der Schweiz Bern. *Geotechnische Serie: Hydrologie, Lieferung*, **6**, 113 (in German).

29. Federal Agency for Forestry/Swiss Federal Institute for Snow and Avalanche Research (1984) Richtlinien zur Berücksichtigung der Lawinengefahr bei raumwirksamen Tätigkeiten. Davos/Bern (in German).

30. Swiss Agency for the Environment, Forests and Landscape (SAEFL)/Federal Office for Spatial Planning; (Federal Agency for Water Management (1997) Berücksichtigung von Hochwassergefahren bei der raumwirksamen Tätigkeit. Biel (in German).

31. Federal Transport Office (2006) Berücksichtigung der Lawinen- und Schneedruckgefährdung bei touristischen Anlagen: Guideline. Bern (in German).

32. Federal Forest Law. SR 921.0 form 4th October 1991 (as of: 1st January 2008) (Switzerland).

33. Federal Ministry for Agriculture and Forestry (1984) 100 Jahre Wildbachverbauung in Österreich. In self-publishing. Vienna (in German).

34. Federal Ministry of Agriculture, Forestry, Environment and Water Management (BMLFUW) (2006) Technische Richtlinie für die Wildbach- und Lawinenverbauung, gemäß § 3 Abs 1 Z 1 und Abs 2 WBFG 1985, Vienna (in German).

35. Federal Ministry of Agriculture, Forestry, Environment and Water Management (BMLFUW) Facts and Figures 2013, Vienna.

36. Federal Ministry of Agriculture, Forestry, Environment and Water Management (BMLFUW) (2010) Richtlinie für die Gefahrenzonenplanung. Vienna (internal guideline) (in German).

37. Federal Ministry of Transport, Innovation and Technology (BMVIT) (2004) Erlass über den "Lawinenschutz im Bereich von Seilbahnen", GZ 238961/3-II/Sch3-2004. Vienna (in German).

38. Federal Ministry of Transport, Innovation and Technology (BMVIT) (1975) Erlass der Obersten Seilbahnbehörde für Seilbahnen und Schlepplifte, die vor dem 22.1.1975 errichtet worden sind, GZ EB 6403/31-II/3-1975. Vienna (in German).

39. Butler, D.R. (1979) Snow avalanche path terrain and vegetation. Glacier National Park, Montana. *Arctic and Alpine Research*, **11**, 17–32.

40. Canadian Avalanche Association CCA (2002) Guidelines for Snow Avalanche Risk Determination and Mapping in Canada. Revelstoke (B.C.).

41. Canadian Avalanche Association CCA (2007) Observation Guidelines and Recording Standards for Weather, Snowpack and Avalanches. Revelstoke (B.C.).

42. Coaz, J. (1881) Die Lawinen der Schweizeralpen. Eidgenössisches Handels- und Landwirtschaftsdepartement. Bern, 147 (in German).

43. Christen, M., Bartelt, P. and Gruber, U. (2002) AVAL-1D: Numerische Berechnung von Fließ- und Staublawinen. WSL Institute for Snow and Avalanche Research SLF. Davos (in German).

44. Christen, M., Bartelt, P., Kowalski, J. and Stoffel, L. (2008) Calculation of Dense Snow Avalanches in three-dimensional Terrain with the Numerical Simulation Program RAMMS. International Snow Science Workshop, Whistler (Canada), 709–716.

45. CRESTAGEO AG (2008) Rempar Grischun– Lawinenschutzsystem: Montagehandbuch. Chur (in German).

46. Decree of the govenor of South Tyrol/Alto Adige 42 (2008) Durchführungsverordnung betreffend die Gefahrenzonenpläne, Bolzano (in German).

47. De Quervain, M. (1978) Wald und Lawinen. Proceedings of the IUFRO Seminar "Mountain Forests and Avalanches". Davos, 219–231 (in German).

48. Dietz, P., Knigge, W. and Löffler, H. (1984) Wegerschließung. Paul Parey Hamburg Berlin (in German).

49. Dragosits, F. (1987) Die Praxis der Sprengfundierung in der Lawinenverbauung. Internationales Symposium über die Bildung, Bewegung und Wirkung von Lawinen Davos. *Mitteilungen des Eidgenössischen Instituts für Schnee- und Lawinenforschung EISLF*, **43**, 29–35.

50. WSL Institute for Snow and Avalanche Research SLF (1961) Lawinenverbau im Anbruchgebiet. Richtlinien für den permanenten Stützverbau. Mitteilungen des EISLF 15. Davos (in German).

51. WSL Institute for Snow and Avalanche Research SLF (1968) Lawinenverbau im Anbruchgebiet. Richtlinien des Eidgenössischen Oberforstinspektorats für den Stützverbau. Mitteilungen des EISLF, Davos (in German).

52. WSL Institute for Snow and Avalanche Research SLF (1990) Richtlinien für den Lawinenverbau im Anbruchgebiet. Swiss Agency for the Environment, Forests and Landscape (SAEFL) Bern (in German).

53. Erschbamer, B. (1989) Vegetation on avalanche paths in the Alps. *Vegetatio*, **80**, 139–146.

54. Florineth, F. (2004) Pflanzen statt Beton – Handbuch zur Ingenieurbiologie und Vegetationstechnik. Patzer Berlin Hannover (in German).

55. Forest Act BGBl Nr 440/1975 idgF (ForstG) (Austria) (in German).

56. Feix, J., Walkner, R. and Schustereder, C. (2008) Einfluss der Boden-Bauwerk-Interaktion auf die Bemessung von Lawinengalerien. *Bauingenieur*, **83**, 158–166.

57. Fierz, C., Armstrong, R.L., Durand, Y., Etchevers, P., Greene, E., McClung, D.M., Nishimura, K., Satyawali, P.K. and Sokratov, S.A. (2009) The International Classification for Seasonal Snow on the Ground. IHP-VII Technical Documents in Hydrology No 83, IACS Contribution No 1, UNESCO-IHP, Paris.

58. Flaig, W. (1935) Lawinen. Brockhaus Leipzig (in German).

59. Fraccaro, S., Bednarz, R. and Poscher, G. (2002) Sicherung alpiner Verkehrswege vor Steinschlag, Muren und Lawinen mittels Tunnellösungen. *Felsbau*, **20/5**, 50–56 (in German).

60. Fredston, J. (2005) *Snowstruck – In the Grip of Avalanches*, A Harvest Book Harcourt.

61. Frehner, M., Schwitter, P. and Wasser, B. (2005) Nachhaltigkeit und Erfolgskontrolle im Schutzwald. Wegleitung für Pflegemaßnahmen in Wäldern mit Schutzfunktion. Swiss Agency for the Environment, Forests and Landscape (SAEFL) Bern (in German).

62. Frey, W. (1977) Wechselseitige Beziehung zwischen Schnee und Pflanze – eine Zusammenstellung anhand von Literatur. *Mitteilungen des Eidgenössischen Instituts für Schnee- und Lawinenforschung EISLF*, **34**, 1–223 (in German).

63. Friedl, H. (1966) Verlauf der alpinen Waldgrenze im Rahmen anliegender Gebirgsgelände. Ökologie der Alpinen Waldgrenze. Symposium Innsbruck. Mitteilungen der Forstlichen Bundesversuchsanstalt FBVA, Vienna (in German).

64. Frutiger, H. (1972) Zur Geschichte des schweizerischen Lawinenverbaus. *Bündnerwald*, (Suppl 9), 57–66.

65. Fuchs, S., Bründl, M. and Stötter, J. (2004) Development of avalanche risk between 1950 and 2000 in the municipality of Davos, Switzerland. *Natural Hazards and Earth System Sciences*, **4/2**, 263–275.

66. Fuchs, S. (2004) Development of Avalanche Risk in Settlements – Comparative Studies in Davos, Grisons (Switzerland). University of Innsbruck, doctorial thesis.

67. Fusek, G. (1992) Mikroverpfählung im Stützverbau. *Journal for Torrent, Avalanche and Erosion Control*, **120**, 167–180 (in German).

68. Gabl, K. (2000) Der Schnee im Februar 1999 im Westen Österreichs aus meteorologischer und klimatologischer Sicht. *Journal for Torrent, Avalanche and Erosion Control*, **141**, 69–80 (in German).

69. Gauer, P., Kern, M., Kristensen, K., Lied, K., Rammer, L. and Schreiber, H. (2007) On pulsed Doppler radar measurements of avalanches and their implication to avalanche dynamics. *Cold Regions Science and Technology*, **50**, 55–71.

70. Gauer, P. (2003) SIP8 offshore geohazards: Feasibility study for a slushflow model within CFX4. NGI-Report 20021023˜35. Norwegian Geotechnical Institute NGI Oslo.

71. Gauer, P., Issler, D., Lied, K., Kristensen, K. and Sandersen, F. (2008) On snow avalanche flow regimes: Inferences from observations and measurements. Proceedings of the International Snow Science Workshop. Whistler (British Columbia), 717–723.

72. Gauer, P., Lied, K. and Kristensen, K. (2009) Analysis of avalanche measurements out of the runout area of NGI's full-scale test-site Ryggfonn. *Cold Regions Science and Technology*, **57**, 1–6.

73. Building insurance of canton Grison (1994) Vorschriften für bauliche Maßnahmen an Bauten in der blauen Lawinenzone (in German).

74. Egli, T. (2005) Wegleitung Objektschutz gegen gravitative Naturgefahren. Association of Canton Fire Insurances (ed.): Bern (in German).

75. Gefahrenzonenplanverordnung (Ordinance of the Federal Ministry of Agriculture, Forestry, Environment and Water Management). BGBl. Nr. 436/1976 (Austria) (as of 2010) (in German).

76. Gislason, E. (2009) 2D-modelling of Icelandic snow avalanches for hazard zoning. International Snow Science Workshop, Davos, 529–532.

77. Glade, T. and Stötter, J. (2008) Gravitative Massenbewegungen und Schneelawinen, in *Naturrisiken und Sozialkatastrophen* (eds C. Flegentreff and T. Glade), Springer Spektrum, Berlin, Heidelberg, pp. 151–162 (in German).

78. Glazovskaya, T.J. (1987) Global distribution of snow avalanches. University of Moscow, doctorial thesis.

79. Glossar der europäischen Lawinenwarndienste (URL: www.lawine.at) (in German).

80. Gorton, F. (1985) Österreichischer Weg der landschaftsangepassten Trassierung von Forstwegen. *Internationaler Holzmarkt*, **76/12**, 1–6 (in German).

81. Gorton, F. (1985) Praxis und Kosten einer landschaftschonenden Bauführung von Forststraßen. Allgemeine Forstzeitung, 241–243 (in German).

82. Granig, M. and Luxner, M.H. (2006) Handbuch zum erweiterten α-β-Modell-06. Austrian Service for Torrent and Avalanche Control, Schwaz (in German).

83. Granig, M. and Luxner, M.H. (2009) Handbuch zum erweiterten α-β-Modell-09. Austrian Service for Torrent and Avalanche Control, Schwaz (in German).

84. Granig, M., Fischer, J.T. and Jörg, P. (2013) Systematic practical evaluation of snow entrainment with the avalanche simulation model SamosAT. International Snow Science Workshop, Grenoble, 599–603.

85. Gruber, U., Bartelt, P. and Margreth, S. (1999) Anleitung zur Berechnung von Fließlawinen: Neue Berechnungsmethoden in der Lawinengefahrenkartierung. WSL Institute for Snow and Avalanche Research SLF Davos.

86. Gubler, H. and Wyssen, S. (2002) Artificial release of avalanches using the remote controlled Wyssen Avalanche Tower. Proceedings of the International Snow Science Workshop (ISSW), Pentiction, Canada, 688–695.

87. Gürer, I., Tuncel, H., Murat Yavas, Ö., Erenbilge, T. and Sayin, A. (1995) Snow avalanche incidents in North-Western Anatolia, Turkey during December 1992. *Natural Hazards*, **11**, 1–16.

88. Haid, H. (2007) Mythos Lawinen – Eine Kulturgeschichte. Studien Verlag Wien, Innsbruck, Bozen (in German).

89. Haefeli, R. (1951) Neuere Entwicklungstendenzen und Probleme des Lawinenverbaus im Anbruchgebiet. *Mitteilungen des Eidgenössischen Institutes für Schnee- und Lawinenforschung EISLF*, **9**, 28–56.

90. Haefeli, R. (1954) Lawinenverbau im Anbruchgebiet – Vorschläge und Berechnung von Netzwerken, Zürich (in German).

91. Hagen, F. and Heumader, J. (1998) Die in Österreich entwickelte gasdynamische Lawinensimulation (Staublawinensimulation) – Beschreibung, Anwendung, Grenzen. *Journal for Torrent, Avalanche and Erosion Control*, **135**, 7–26 (in German).

92. Hákonardóttir, K.M. and Hogg, A.J. (2005) Oblique shocks in rapid granular flows. *Physics of Fluids*, **17** (7), 77–101.

93. Hákonardóttir, K.M., Jóhannesson, T., Tiefenbacher, F. and Kern, M. (2001) A laboratory study of the retarding effect of breaking mounds in 3, 6 and 9 m long chutes. Reykjavík, Veðurstofa (Íslands). (Technical Report V01007, 2001).

94. Hákonardóttir, K.M., Hogg, A.J., Jóhannesson, T., Kern, M. and Tiefenbacher, F. (2003) Large-scale avalanche braking mound and catching dam experiments with snow: a study of the airborne jet. *Survey of Geophysics*, **24** (5/6), 543–554.

95. Harbitz, C. (1998) SAME – Snow Avalanche Modelling, Mapping and Warning in Europe. A survey of computational models for snow avalanche motion. Fourth European Framework Programme: Environment and Climate. Contract ENV4-CT96-0258, Oslo.

96. Heierli, J., Gumbsch, P. and Zaiser, M. (2008) Anticrack nucleation as triggering mechanism for snow slab avalanches. *Science*, **321**, 240.

97. Heierli, J. and Zaiser, M. (2008) Failure initiation in snow stratifications containing weak layers: Nucleation of whumpfs and slab avalanches. *Cold Regions Science and Technology*, **52**, 385–400.

98. Heimgartner, M. (1986) Fundation im Lawinenverbau. Internationales Symposium über die Bildung, Bewegung und Wirkung von Lawinen Davos 1986. *Mitteilungen des Eidgenössischen Instituts für Schnee- und Lawinenforschung EISLF*, **43**, 19–27.

99. Heinimann, H., Hollenstein, K., Kienholz, H., Krummenacher, B. and Mani, P. (1998) Methoden zur Analyse und Bewertung von Naturgefahren. Swiss Agency for the Environment, Forests and Landscape (SAEFL), Environmental publications 85, Bern.

100. Hess, J. (2008) Schutzziel im Umgang mit Naturrisiken: Wandeln auf dem Pfad der Gerechtigkeit. INTERPRAEVENT Conference Proceedings Dornbirn, vol. 2, 347–358.

101. Hestnes, E. and Sanderson, F. (1998) Slushflow acitivity in the Rana district (North Norway). *IHAS- Publ.*, **162**, 317–330.

102. Hestnes, E. and Kristersen, K. (2010) The diversities of large slushflows illustrated by eight case studies. Proceeding of the International Snow Science Workshop, Lake Tahoe (USA).

103. Heumader, J. (1987) Pflugbermen als Gleitschneeschutz und Verjüngungshilfe. *Journal for Torrent, Avalanche and Erosion Control*, 51 (1987), **105**, 105–110 (in German).

104. Heumader, J. (1987) Schutz vor Gleitschneeschäden durch Verpfählung mit Drahtverspannung. *Journal for Torrent, Avalanche and Erosion Control*, **105**, 111–117 (in German).

105. Hofmann, R. (2005) Baugrunderkundung und Nachweis von Flächengründungen nach ÖNORM B 4402, ÖNORM EN 1997-1 und ÖNORM B 4435-1/2. Gründungstechnik im Hochbau: Planung und Ausführung von Gründungskörpern. VÖBU Bauakademie Innsbruck (in German).

106. Hofmann, R. (2005) Versagenswahrscheinlichkeit von Böschungen, Dämmen und Hängen im Lockergestein. Institute for Engineering Geology, Technical University of Vienna: Sicherheit, Versagenswahrscheinlichkeit und Risikomanagement in der Geotechnik (in German).

107. Hofmann, R. (2006) Globale Sicherheit, Konzept der Teilsicherheitsbeiwerte (EN 1997-1), Sicherheitsfaktoren in Abhängigkeit der Nutzungsdauer und der Ereigniswiederkehr. Workshop "Bemessungssicherheit von Schutzbauwerken und gefährdeten Objekten im Zusammenhang mit gravitativen Naturgefahren", Institute for Engineering Geology, Technical University of Vienna (in German).

108. Hofmann, R. (2009) Bemessung von Stützbauwerken nach ÖNORM EN 1997-1 und ÖNORM B 1997-1-1. Austrian Standards Institute, Vienna (in German).

109. Hofmann, R. (2009) ON AG 023.07-Technischer Lawinenschutz: Systemstütze – Grundbruchberechnungen nach ÖNORMB 4435-2 und ÖNORM B 4433, Vienna (in German).

110. Hofmann, R. and Sauermoser, S. (2008) Schutzbauten gegen Naturgefahren in Verbindung mit dem Eurocode 7. 6. Kolloquium Bauen in Boden und Fels, Technische Akademie Esslingen (in German).

111. Höller, P. and Fromm, R. (2008) Temporäre Schutzmaßnahmen und Lawinenprognose. *Bundesforschungs- und Ausbildungszentrum für Wald, Naturgefahren und Landschaft. BFW-Praxisinformation*, **15**, 11–12.

112. Horvat, A. (2005) Bestimmung erosionsgefährdeter Flächen – Slowenische Erfahrungen. *Journal for Torrent, Avalanche and Erosion Control*, **152**, 63–76 (in German).

113. Huber, T., Sailer, R. and Jörg, P. (2005) Risikostudie Lebensraum "Hinteres Pitztal". *Journal for Torrent, Avalanche and Erosion Control*, **153**, 75–113.

114. Hübl, J., Fuchs, S. and Agner, P. (2007) Optimierung der Gefahrenzonenplanung. Weiterentwicklung der Methoden der Gefahrenzonenplanung. IAN-Report Nr. 90. University of Natural Ressources and Applied Life Sciences Vienna (in German).

115. Hübl, J. and Kraus, D. (2004) Erweiterungsvorschläge zur Kosten-Nutzen-Untersuchung der Wildbach- und Lawinenverbauung. WLS Report 94, University of Natural Ressources and Applied Life Sciences Vienna (in German).

116. In der Gand, H. (1968) Aufforstungsversuche an einem Gleitschneehang – Ergebnisse der Winteruntersuchungen 1955/56 bis 1961/62. *Schweizerische Anstalt für das forstliche Versuchswesen, Mitteilungen*, **44** (3), 163–174 (in German).

117. INTERPRAEVENT (Eds.) (2009) *Alpine Naturkatastrophen*, Leopold Stocker Graz.

118. Issler, D. (1999) Berücksichtigung der Staublawinen in der Gefahrenkartierung – Neue Berechnungsmethoden in der Lawinengefahrenkartierung. WSL Institute for Snow and Avalanche Research SLF, Davos (in German).

119. Issler, D. (1999) European avalanche test sites. Overview and analysis in view of coordinated experiments. WSL Institute for Snow and Avalanche Research SLF Davos 59 (in German).

120. Issler, D. (2003) Experimental information on the dynamics of dry snow avalanches, in *Dynamic Response of Granular and Porous Materials Under Large and Catastrophic Deformations* (eds K. Hutter and N. Kirchner), Springer, 109–160.

121. Issler, D. and Gauer, P. (2008) Exploring the significance of the fluidised flow regime for avalanche hazard mapping. *Annals of Glaciology*, **49**, 193–198.

122. Jäger, G. (2005) Lawinenschutz bei Bergbauernhöfen. *Der Alm- und Bergbauer*, **12** (5), 25–29 (in German).

123. Jaedicke, C., Kern, M.A., Gauer, P., Baillifard, M.A. and Platzer, K. (2008) Chute experiments on slushflow dynamics. *Cold Regions Science and Technology*, **5**, 156–167.

124. Jecklin, R. (1986) Die Sprengankerfundierung. *Bündnerwald*, **39** (5), 46–49 (in German).

125. Jóhannesson, T. and Margreth, S. (1999) Adaptation of the Swiss Guidelines for supporting structures for Icelandic conditions. Veðurstofa Íslands Report VÍ-G99013-ÚR07, Reykjavík.

126. Jóhannesson, T. and Hopf, J. (2008) Loading of supporting structures under Icelandic conditions: The type of structures and structural requirements in future projects (Results of a field experiment in Siglufjördur). Proceedings of International Symposium on Mitigative Measures against Snow Avalanches, Egilstadir, 143–150.

127. Jóhannesson, T., Gauer, P., Issler, D. and Lied, K. (eds) (2009) *The Design of Avalanche Protection Dams. Recent Practical and Theoretical Developments European Commission*, Directorate General for Research.

128. Jóhasson, K., Sigurðsonn, S. and Arnalds, P. (1999) Estimation of avalanche risk. Veðurstofa Íslands Report VÍ-G99001-ÚR01, Reykjavík.

129. Johnson, E.A. (1987) The relative importance of snow avalanche disturbance and thinning on canopy plant populations. *Ecology*, **68**, 45–53.

130. Jörg, P., Sampl, P. and Granig, M. (2010) Implementation of a new suspension model and calibration of the powder snow layer in SamosAT. European Geophysical Union General Assembly, Vienna.

131. Kajimoto, T., Daimaru, H., Okamoto, T., Otani, T. and Onodera, H. (2004) Effects of snow avalanche disturbance on regeneration of subalpine Abies mariesii forest, northern Japan. *Arctic, Antarctic and Alpine Research*, **36**, 436–445.

132. Kapeller, G., Kleemayr, K. and Fellin, W. (2008) Untersuchung der Effektivität eines aufgelösten Lawinenbrechers. *Journal for Torrent, Avalanche and Erosion Control*, **158**, 90–97.

133. Keiler, M., Fuchs, S., Zischg, A. and Stötter, J. (2004) The adaptation of technical risk analysis on natural hazards on a regional scale. *Zeitschrift für Geomorphologie N. F.*, (Suppl 135), 95–110.

134. Keiler, M. and Stötter, J. (2004) Aspekte der Entwicklung des Schadenspotenzials bei Lawinenrisiken am Beispiel Galtür. Internationales Symposion INTERPRAEVENT Riva/Trient, 2004, Tagungspublikation VI, 139–150 (in German).

135. Kern, M., Tiefenbacher, F. and McElwaine, J. (2004) The rheology of snow in large chute flows. *Cold Regions Science and Technology*, **39** (2/3), 181–192.

136. Kern, M., Bartelt, P., Sovilla, B. and Buser, O. (2009) Measured shear rates in large dry and wet snow avalanches. *Journal of Glaciology*, **190**, 327–338.

137. Kern, M., Bartelt, P. and Sovilla, B. (2010) Velocity profile inversion in dense avalanche flow. *Annals of Glaciology*, **54**, 27–31.

138. Kienholz, H. (2005) Gefahrenzonenplanung im Alpenraum – Ansprüche und Grenzen. *Journal for Torrent, Avalanche and Erosion Control*, **152**, 135–151 (in German).

139. Klaus, W. (1997) IST-Zustandserhebung von Stahlschneebrücken in Tirol. University for Natural Ressources and Applied Life Sciences Vienna, master thesis (in German).

140. Klenkhart, C. (1985) Derzeit übliche Fundationsarten in der Lawinenstützverbauung in Vorarlberg, Tirol und der Schweiz. *Journal for Torrent, Avalanche and Erosion Control*, **101**, 145–156.

141. Krapfenbauer, R. and Krapfenbauer, T. (2004) Bautabellen. Jugend und Volk, Vienna (in German).

142. Krist, G. (2007) The White Cascade, the Great Northern Railway disaster and America's Deadliest Avalanche, Holt Paperbacks.

143. Kristensen, K. (1998) A survey of snow avalanche accidents in Norway. Norwegian Geotechnical Institute NGI, Publications 203, Oslo.

144. Kogelnig, A. and Sauermoser, S. (2013) Sicherung von Straßen mit künstlicher Lawinenauslösung (Pilotprojekt Großtallawine). *Journal for Torrent, Avalanche and Erosion Control*, **171**, 158–170 (in German).

145. Kolymbas, D. and Kirsch, A. (2009) Zum Erddruck auf Lawinenschutzgalerien. *Bautechnik*, **86**, 774–779.

146. Körner, H. (1976) Reichweite und Geschwindigkeit von Bergstürzen und Fließlawinen. *Rock Mechanics*, **8**, S.225–256.

147. Kulakowski, D., Rixen, C. and Bebi, P. (2006) Changes in forest structure and in the relative importance of climatic stress as a result of suppression of avalanche disturbance. *Forest Ecology and Management*, **223**, 66–74.

148. Legge 3 agosto 1998, n. 267: Conversione in legge, con modificazioni, del D.L. 11 giugno 1998, n. 180, recanti misure urgenti per la prevenzione del rischio idrogeologico ed a favore delle zone colpite da disastri franosi nella regione Campania (pubblicata nella GU n. 183 del 7 agosto 1998) (Italy) (in Italian).

149. Leichtfried, A. (2010) Bestimmung der extremen Schneehöhen für die Stützverbauung in Anbruchgebieten von Lawinen für Tirol und Vorarlberg. Austrian Central Institute for Meteorology and Geodynamics ZAMG Innsbruck (in German).

150. Léon, Gallardo. and René, E. (2003) Nieve y avalanchas. Una blanca historia de riesgos en montana. Ediciones de Santa Rosa de Los Andes, Chile (in Spanish).

151. Leuenberger, F. (1992) Bewährung von technischen Maßnahmen zum Lawinen- und Gleitschneeschutz im Aufforstungs- und Verjüngungsgebiet. *Internationales Symposion INTERPRAEVENT Bern, Tagungspublikation Bd*, **5**, 9–18 (in German).

152. Lied, K. and Bakkehoi, S. (1980) Empirical calculations of snow avalanche run-out distance based on topographic parameters. *Journal of Glaciology*, **94**, 165–177.

153. Lied, K., Hopf, J., Bakkehoi, S. and Weiler, C. (1995) Calculation Methods for Avalanche Run-Out Distance for the Austrian Alps, Norwegian Geotechnical Institute NGI Publication, 581210.3, Oslo.

154. Loat, R. (2005) Die Gefahrenzonenplanung in der Schweiz. *Journal for Torrent, Avalanche and Erosion Control*, **152**, 77–92.

155. Loi n 1995-101 du 2 février 1995 relative au renforcement de la protection de l'environne- ment. (loi BARNIER) et décret no 95-1089 du octobre 1995 relative aux plans de prévention des risques naturels prévisibles (France).

156. Loi n 2003-699 du 30 juillet 2003 relative à la prévention des risques technologiques et na- turels. (loi BACHELOT) (France).

157. LOV 1985-06-14 nr 77 Plan- og bygningsloven (Norwegian Planning and Building Act).

158. Luzi, G., Noferini, L., Mecatti, D., Macaluso, G., Pieraccini, M., Atzeni, C., Schaffhauser, A., Fromm, R. and Nagler, T. (2009) Using ground-based SAR interferometer and a terrestrial laser scanner to monitor a snow-covered slope: results from an experimental data collection in Tyrol (Austria). *IEEE Transactions on Geoscience and Remote Sensing*, **47** (2), 382–393.

159. Maierhofer, F. (1988) Erkenntnisse und Erfahrungen über die Wasserableitung nach 4 Jahrzehnten Forstwegebau. *Internationales Symposium INTERPRAEVENT Graz, Tagungspublikation Bd.*, **2**, 61–71 (in German).

160. Margreth, S. (1995) Snow pressure measurements on snow net systems. Proceedings of the ANENA Symposium May 30-June 03 Chamonix, 241–248.

161. Margreth, S. (2004) Technische Lawinenschutzmaßnahmen. Federal office for the Environment FOEN, Bern (in German).

162. Margreth, S. (2004) Die Wirkung des Waldes bei Lawinen. Forum für Wissen, 21–26 (in German).

163. Margreth, S. (2007) Defense structures in avalanche starting zones. Technical guideline as an aid to enforcement. The environment in practice 0704. Federal office for the Environment FOEN, Bern/WSL, Institute for Snow and Avalanche Research SLF, Davos.

164. Margreth, S. (2007) Snow pressure on cableway masts. Analysis of damages and design approach. *Cold Regions Science and Technology*, **47**, 4–15.

165. Margreth, S. and Platzer, K. (2006) Baulicher Lawinenschutz. tec 21, **8**, 8–10 (in German).

166. Margreth, S. (2008) PROTECT – Beurteilung der Wirkung von Schutzmaßnahmen gegen Naturgefahren: Arbeitsanleitung Lawinen. WSL Institute for Snow and Avalanche Research SLF, Davos (in German).

167. Mayer, H. (1974) Wälder des Ostalpenraums, Gustav Fischer (in German).

168. McClung, D. (1984) Comparison of snow pressure measurements and theoretical predictions. *Canadian Geotechnical Journal*, **21** (2), 250–258.

169. McClung, D. (1999) Return period of the Galtür avalanche 1999, Expert opinion on order of the Austrian Service for Torrent and Avalanche Control.

170. McClung, D. (2001) Characteristics of terrain, snow supply and forest cover for avalanche initiation caused by logging. *Annals of Glaciology*, **32**, 223–229.

171. McClung, D. (2003) Magnitude and frequency of avalanches in relation to terrain and forest cover. *Arctic, Antarctic and Alpine Research*, **35**, 82–90.

172. McClung, D. and Shearer, P. (2006) *The Avalanche Handbook*, The Mountaineers Books, Seattle.

173. Mears, A.I. (1975) Dynamics of dense snow avalanches interpreted from broken trees. *Geology*, **3**, 521–523.

174. Ménégoz, M., Gallée, H. and Jacobi, H.W. (2013) Precipitation and snow cover in the Himalaya: from reanalysis to regional climate simulations. *Hydrology and Earth System Sciences*, **17**, 3921–3936.

175. Mock, C.J. and Birkeland, K.W. (2000) Snow avalanche climatology on the Western United States Mountain Ranges. *Bulletin of the American Meteorological Societey*, **81** (10), 2367–2390.

176. Naaim, M. and Ancey, C. (1992) *Dense Avalanche Model*, European Summer University Chamonix, Cemagref Publications, 173–181.

177. Naaim, M. (1995) Modélisation numérique des avalanches aérosols. *La Houille Blanche*, **5/6**, 56–62.

178. Naaim, M. (1998) Dense avalanche numerical modeling: Interaction between avalanche and structures. Norwegian Geotechnical Institute, NGI-Publications 203, Oslo. (25 Years of Snow Avalanche Research at NGI, Proceedings Anniversary Conference Voss 12–16 May, 1998).

179. Naaim-Bouvet, F., Naaim, M. and Michaus, J.L. (2002) Snow fences on slopes at high wind speed: physical modeling in the CSTB cold wind tunnel. *Natural Hazards and Earth System Sciences*, **3/4**, 137–145.

180. Mair, R. and Nairz, P. (2010) Lawinen – Die 10 entscheidenden Gefahrenmuster erkennen. Tyrolia Innsbruck (in German).

181. Neuschmid, K. (1996) Methoden der Böschungssicherung von Hangquerungen im steilen Gelände. Self edition (in German).

182. Newesely, C., Tasser, E., Spadinger, P. and Cernusca, A. (2000) Effects of land-use changes on snow gliding processes in alpine ecosystems. *Basic and Applied Ecology*, **1**, 61–67.

183. Norem, H., Irgens, F. and Schieldrop, B. (1987) A continuum model for calculating snow avalanche velocities. Proceedings of Avalanche Formation, Movements and Effects, IAHS publication 162, 363–378 (also in: Norwegian Geotechnical Institute Report 58120-9).

184. Norem, H., Irgens, F. and Schieldrop, B. (1989) Simulation of snow-avalanche flow in run-out zones. *Annals of Glaciology*, **13**, 218–225.

185. Norme, Française (1992) Equipements de protection contre les avalanches. Filets paravalanches. Paris: NF P 95-304 (in French).

186. Nussbaumer, J. (1996) Die Gewalt der Natur: Eine Chronik der Naturkatastrophen von 1500 bis heute. Grünbach Edition Sandkorn (in German).

187. Oberndorfer, S. and Granig, M. (2007) Modellkalibrierung des Lawinensimulationsmodells SamosAT. Austrian Service for Torrent and Avalanche Control, Schwaz 2007.

188. Onesti, L.J. (1985) Meteorological conditions that initiate slushflows in Brooks Range, Alaska. *Annuals of Glaciology*, **6**, 23–25.

189. ÖNORM B 4433 (1987) Erd- und Grundbau – Böschungsbruchberechnung. Ausgabe: 1987-12-01.

190. ÖNORM B 1991-1-4 Eurocode 1 (2006) Einwirkungen auf Tragwerke – Teil 1-4: Allgemeine Einwirkungen – Windlasten. Nationale Festlegungen zu ÖNORM EN 1991-1-4 und nationale Ergänzungen. Ausgabe: 2006-12-01 (in German).

191. EN 1991-1-4 Eurocode 1 (2005) Einwirkungen auf Tragwerke – Teil 1-4: Allgemeine Einwirkungen – Windlasten. Ausgabe: 2005-01-04 (in German).

192. ÖNORM EN 1991-2 (2004) Eurocode 1 – Einwirkungen auf Tragwerke – Teil 2: Verkehrslasten auf Brücken. Ausgabe: 2004-08-01 (in German).

193. ÖNORM EN 1997-1 (2009) Eurocode 7: Entwurf, Berechnung und Bemessung in der Geotechnik – Teil 1: Allgemeine Regeln (konsolidierte Fassung): Ausgabe: 2009-05-15 (in German).

194. ÖNORM B 1991-2 (2004) Eurocode 1: Einwirkungen auf Tragwerke – Teil 2: Verkehrslasten auf Brücken – Nationale Festlegungen zu ÖNORM EN 1991-2 und nationale Ergänzungen. Ausgabe: 2004-08-01 (in German).

195. ÖNORM B 1997-1-1 (2007) Eurocode 7: Entwurf, Berechnung und Bemessung in der Geotechnik – Teil 1: Allgemeine Regeln – Nationale Festlegungen zu ÖNORM EN 1997-1 und nationale Ergänzungen. Ausgabe: 2007-11-01 (in German).

196. ÖNORM B 1997-2 (2012) Eurocode 7: Entwurf, Berechnung und Bemessung in der Geotechnik – Teil 2: Erkundung und Untersuchung des Baugrunds – Nationale Festlegungen in Zusammenhang mit ÖNORM EN 1997-2 und nationale Ergänzungen. Ausgabe: 2012-06-15 (in German).

197. ÖNORM B 1997-1-3 (2007) Eurocode 7: Entwurf, Berechnung und Bemessung in der Geotechnik – Pfahlgründungen. Ermittlung der Tragfähigkeit. Entwurf: 2007-09-07 (in German).

198. ÖNORM B 1997-1-4 (2010) Eurocode 7: Entwurf, Berechnung und Bemessung in der Geotechnik – Erddruckberechnung und Stützbauwerke. Ausgabe: 2010-03-15 (in German).

199. ÖNORM EN 14199 (2005) Ausführung von besonderen geotechnischen Arbeiten (Spezialtiefbau)– Pfähle mit kleinen Durchmessern (Mikropfähle) (in German).

200. ÖNORM-Regel 24800 (2009) Schutzbauwerke der Wildbachverbauung – Begriffe und ihre Definitionen sowie Klassifizierung. Ausgabe: 2009-02-15 (in German).

201. ÖNORM-Regel 24802 (2011) Schutzbauwerke der Wildbachverbauung – Projektierung, Bemessung und konstruktive Durchbildung. Ausgabe: 2011-01-15 (in German).

202. ÖNORM-Regel 24805 (2010) Permanenter technischer Lawinenschutz – Benennungen und Definiti- onen sowie statische und dynamische Einwirkungen. Ausgabe: 2010-06-01 (in German).

203. ÖNORM-Regel 24806 (2011) Permanenter technischer Lawinenschutz – Bemessung und konstruktive Ausgestaltung. Ausgabe: 2011-12-15 (in German).

204. ÖNORM-Regel 24807 (2010) Permanenter technischer Lawinenschutz – Überwachung und Instandhaltung. Ausgabe: 2010-06-01 (in German).

205. Perla, R.I. and Martinelli, M. Jr (1976) *Avalanche Handbook. Agriculture Handbook*, US Department of Agriculture, (489).

206. Perla, R., Cheng, T.T. and McClung, D. (1980) A two parameter model of snow avalanche motion. *Journal of Glaciology*, **94**, 197–202.

207. Pestal, E. (1961) *Seilbahnen und Seilkrane für Holz- und Materialtransport*, Georg Fromme & Co., Vienna (in German).

208. Pfammatter, C. (2006) Abschätzung des Finanzbedarfs zur Erhaltung der Funktionstauglichkeit von Lawinenstützverbauungen. Confederate Technical High School Zurich ETH, master thesis.

209. Plate, E. and Merz, B. (eds) (2001) Naturkatastrophen. Ursachen – Auswirkungen – Vorsorge. Schweizerbart'sche Verlagsbuchhandlung Stuttgart (in German).

210. National Platform for Natural Hazards PLANAT (2008) Strategie Naturgefahren Schweiz – Umsetzung des Aktionsplans PLANAT 2005-2008, Projekt B 2.2: Schutzziele. Schlussbericht, Federal office for the Environment FOEN Bern (in German).

211. National Platform for Natural Hazards PLANAT (2008) Strategie Naturgefahren Schweiz – Umsetzung des Aktionsplans PLANAT 2005–2008, Projekt A 1.1, Risikokonzept für Naturgefahren – Leitfaden. Schlussbericht/Testversion, Bundesamt für Umwelt BAFU. Bern 2009 (in German).

212. Prinz, H. (1993) Felsbau durch Sprengung oder Hydromeisel? *Österreichische Forstzeitung*, **7**, 15–17 (in German).

213. Rammer, L., Kern, M., Gruber, U. and Tiefenbache, F. (2007) Comparison of avalanche-velocity measurements by means of pulsed Doppler radar, continuous wave radar and optical methods. *Cold Regions Science and Technology*, **50**, 35–54.

214. Rammer, L. and Granig, M. (2009) Three years snow net project Hafelekar/ Innsbruck (eds J. Schweizer and A. Van Herwijnen), Proceedings of the International Snow Science Workshop Davos, 582–586.

215. Rapin, F. (2002) A new scale for avalanche intensity. International Snow Science Workshop Penticton, British Columbia.

216. Rapin, F. and Queffélean, Y. (1997) La pathologie des ouvrages de défense active contre les avalanches. Ingénieries – EAT No 9: 35–42 (in French).

217. Redaktionsartikel, Baublatt (2008) Resistent gegen Lawinen aller Art. *Baublatt*, **119**, 20–21.

218. DIRECTIVE 2007/60/EC OF THE EUROPEAN PARLIAMENT AND OF THE COUNCIL of 23 October 2007 on the assessment and management of flood risks.

219. Richtlinien zur Erstellung der Gefahrenzonenpläne. gemäß Landesraumordnungsgesetz 13, Artikel 22/bis vom 26, August 2008 (Autonomous Province of Bolzano/Italy) (in German).

220. Romang, H. (2004) Wirksamkeit und Kosten von Wildbach-Schutzmaßnahmen. GEOGRAPHICA BERNENSIA Reihe G 73. University of Bern, Institute for Geography (in German).

221. Rudolf-Miklau, F., Schilcher, W., Kessler, K. and Suda, J. (2008) Life cycle management for technical avalanche protection systems. International Symposium on Mitigative Measures against Snow Avalanches Egilstadir, Iceland.

222. Rudolf-Miklau, F. (2009) Naturgefahren-Management in Österreich: Vorsorge – Bewältigung – Information, Lexis Nexis ORAC Vienna (in German).

223. Rudolf-Miklau, F. and Sauermoser, S. (eds) (2001) *Handbuch Technischer Lawinenschutz*, Ernst & Sohn, Berlin (in German).

224. Sailer, R., Fellin, W., Fromm, R., Jörg, P., Rammer, L., Sampl, P. and Schaffhauser, A. (2008) Snow avalanche mass balance calculation and simulation model verification. *Annals of Glaciology*, **48**, 183–192.

225. Salm, B. (1977) Snow forces. *Journal of Glaciology*, **19** (80), 67–100.

226. Salm, B. (1978) Snow forces on forest plants. Proceedings of the IUFRO Seminar "Mountain Forests and Avalanches". Davos, 157–182.

227. Salm, B. (2004) A short and personal history of snow avalanche dynamics. *Cold Regions Science and Technology*, **39** (2/3), 83–92.

228. Salm, B., Burkard, A. and Gubler, H. (1990) Berechnung von Fließlawinen, eine Anleitung für Praktiker mit Beispielen. Mitteilung des Eidgenössischen Institutes für Schnee und Lawinenforschung EISLF 47. Davos (in German).

229. Sampl, P. and Zwinger, T. (2004) Avalanche simulation with SAMOS. *Annals of Glaciology*, **38**, 393–398.

230. Sampl, P. (2007) SamosAT – Modelltheorie und Numerik. AVL List GmbH Graz (in German).

231. Sampl, P. and Granig, M. (2009) Avalanche Simulation with SamosAT. International Snow Science Workshop Davos.

232. Sauermoser, S. (1996) Die Aufforstung von Gleitschneehängen. *Österreichische Forstzeitung*, **3**, 18–20 (in German).

233. Sauermoser, S. (1998) Erfahrungen mit Gleitschutzböcken. *Journal for Torrent, Avalanche and Erosion Control*, **135**, 93–103 (in German).

234. Sauermoser, S., Sailer, R., Herbert, A. and Hagen, F. (2004) Recalculation of the Galtür avalanche with different simulation models – a comparison of the results (ed. P. Bartelt) Snow Engeneering. Proceedings of the 5th International Conference on Snow Engineering Davos, 335–342.

235. Sauermoser, S. (2006) Avalanche Hazard Mapping – 30 years of experience in Austria. Proceedings of the International Snow Science Workshop (ISSW) Telluride, 314–321.

236. Sauermoser, S., Huber, T., Jenni, M. and Domaas, U. (2007) Hazard mapping in Norway and Austria – A comparision. *Journal for Torrent, Avalanche and Erosion Control*, **157**, 22–31.

237. Schach, R., Otto, J., Häupel, H. and Fritzsche, M. (2006) Lebenszykluskosten von Brückenbauwerken. *Bauingenieur*, **81**, 343–350 (in German).

238. Schaffhauser, A., Adams, M., Fromm, R., Jörg, P., Luzi, G., Noferini, L. and Sailer, R. (2008) Remote sensing based retrieval of snow cover properties. *Cold Regions Science and Technology*, **54**, 164–175.

239. Scheidl, C., Kurz, D. and Wieshofer, S. (2003) Handbuch zur Bestimmung des Risikos in Schigebieten aufgrund von Lawinengefahren. Federal Ministry for Agriculture, Forestry, Environment and Water Management Vienna (in German).

240. Schiechtl, H.M. and Stern, R. (1992) Handbuch für naturnahen Erdbau. Eine Anleitung für ingenieurbiologische Bauweisen. Österreichischer Agrarverlag Vienna (in German).

241. Schilcher, W. (2002) Technischer Lawinenschutz im Anbruchgebiet. *Journal for Torrent, Avalanche and Erosion Control*, **148**, 20–38.

242. Schönenberger, W. and Brang, P. (2004) Silviculture in mountain forests, in *Encyclopedia of Forest Sciences* (eds J. Burley, J. Evans and J. Youngquist), Elsevier, 1085–1094.

243. Schönenberger, W. and Noack, A. (2005) Effect of timber removal from wind throw slopes on the risk of snow avalanches and rockfall. *Forest Ecology and Management*, **213**, 197–208.

244. Schneebeli, M. and Bebi, P. (2004) Snow and avalanche control, in *Encyclopedia of Forest Sciences* (eds J. Burley, J. Evans and J. Youngquist), Elsevier, 397–402.

245. Schneebeli, M. (2009) Schneemetamorphose: Schnee im ständigen Wandel. Newsletter Naturgefahren (www.slf.ch) 6–7 (in German).

246. WSL Institute for Snow and Avalanche Research SLF (2000) Der Lawinenwinter 1999. Ereignisanalyse (in German).

247. WSL Institute for Snow and Avalanche Research SLF (1999) Neue Berechnungsmethoden in der Lawinengefahrenkartierung. Kursunterlagen zur Einführung des Programms AVAL-1D Davos (in German).

248. Schneider, K.-J. (Hrsg.) (2006) Bautabellen für Ingenieure. Werner Neuwied (in German).

249. Schröder, M. (2005) DIN German Institute for Standardisation e.V (Eds.): Der Wartungsvertrag, Vertragsgestaltung der Inspektion – Wartung – Instandsetzung von baulichen Anlagen und Rechtsfolgen. Beuth Berlin Vienna Zürich (in German).

250. Schweizer, J., Jamieson, J.B. and Schneebeli, M. (2003) Snow avalanche formation. *Reviews of Geophysics*, **41** (4), 1016.

251. Sedlak, O. (1993) Walderschießung und Naturschutz. *Österreichische Forstzeitung*, **7**, 10–12 (in German).

252. Swiss Ropeway Association (SBS) (2002) Die Verkehrssicherungspflicht für Schneesportabfahrten. Richtlinien mit Erläuterungen, 5th edtion (in German).

253. Sereinig, N. and Schober, S. (2009) Gefahrenzonenausweisung Szenarien. The Provincial Government of Carinthia Klagenfurt (in German).

254. Schaerer, P. and McClung, D. (2006) *The Avalanche Handbook*, 3rd edn, Mountaineers Books Seattle.

255. Sigrist, C. and Schweizer, J. (2007) Critical energy release rates of weak snowpack layers determined in field experiments. *Geophysical Research Letters*, **34**, L03502.

256. Skolaut, H. (1998) Erfahrungen mit Lawinenbrechern – Ist das Brechen von Lawinen möglich? *Journal for Torrent, Avalanche and Erosion Control*, **135**, 69–76 (in German).

257. Sommerhalder, E. (1972) Ablenkverbau, in *Lawinenschutz in der Schweiz* (eds F. Casteleberg, H.R. In der Gand, F. Pfister and B. Rageth), Bündnerwald, (Suppl. 9), pp. 155–169 (in German).

258. Sovilla, B. and Bartelt, P. (2002) Observations and modeling of snow avalanche entrainment. *Natural Hazard and Earth Systems Sciences*, **2**, 169–179.

259. Sovilla, B., Schaer, M., Kern, M. and Bartelt, P. (2008) Impact pressures and flow regimes in dense snow avalanches observed at the Vallée de la Sionne test site. *Journal of Geophysical Research*, **113**, F01010.

260. Sovilla, B., Schaer, M. and Rammer, L. (2008) Measurements and analysis of full-scale avalanche impact pressure at the Vallée de la Sionne test site. *Cold Regions Science and Technolology*, **51** (2/3), 122–137.

261. Statham, G. (2008) Avalanche hazard, danger and risk: A practical explanation. Proceedings International Snow Science Workshop, Whistler (British Columbia), 224–227.

262. Stoffel, L. (2001) Künstliche Lawinenauslösung. Praxishilfe. Mitteilungen des Eidgenössischen Instituts für Schnee- und Lawinenforschung 53 Davos (in German).

263. Stoffel, L. (2006) Künstliche Lawinenauslösung. tec21, 4–7 (in German).

264. Stoffel, L. (2013) Comparison of blasting methods: Gazex, Avalanche Guard/ Avalanche Master Inauen-Schätti, Wyssen Avalanche Tower, Avalancheur. WSL Institute for Snow and Avalanche Research SLF Davos/Swiss Association for Avalanche Warning Systems Münster.

265. Suda, J. and Rudolf-Miklau, F. (eds) (2012) *Bauen und Naturgefahren*, Verlag Springer, Vienna, (in German).

266. Tabler, R.D. (1991) *Snow Fence Guide*, Strategic Highway Research Council, Washington DC.

267. Takeuchi, Y., Kobajashi, S., Sato, T., Izumi, K., Kosugi, K., Xin, W., Jiapin, Z. and Yongheng, P. (2001) The effect of wind direction of drift control by snow fences. *Annals of Glaciology*, **32**, 159–162.

268. Teich, M. and Bebi, P. (2009) Evaluating the benefit of avalanche protection forest with GIS-based risk analyses – a case study in Switzerland. *Forest Ecology and Management*, **257**, 1910–1919.

269. Tiefenbacher, F. and Kern, M. (2004) Experimental devices to determine snow avalanche basal friction and velocity profiles. *Cold Regions Science and Technology*, **38** (1), 17–30.

270. Tollinger, C. and Jörg, P. (2008) DGM – Analyse anhand von Lawinenauslauflänge. Austrian Service for Torrent and Avalanche Control, Schwaz.

271. Turnbull, B. and McElwaine, J.N. (2007) A comparison of powder snow avalanches at Vallée de la Sionne with plume theories. *Journal of Glaciology*, **180**, 30–40.

272. Turnbull, B., McElwaine, J.N. and Ancey, C. (2007) The Kulikovsky-Sveshnikova-Beghin model of powder snow avalanches: Development and application. *Journal of Geophysical Research*, **112**, F01004.

273. UNESCO (1981) Lawinen-Atlas. Bebilderte Internationale Lawinenklassifikation. United Nations Educational, Scientific and Cultural Organization (UNESCO) Paris.

274. Vallet, J., Turnbull, B., Joly, S. and Dufour, F. (2004) Observations on powder snow avalanches using videogrammetry. *Cold Regions Science and Technolology*, **39**, 153–159.

275. Van Herwijnen, A. (2009) Experimente zur Bruchausbreitung und Lawinenauslösung: Die Auslösung von Schneebrettlawinen. WSL-Swiss Federal Institute for Snow & Avalanche Research (SLF) Davos (in German).

276. Veblen, T.T., Hadley, K.S., Nel, E.M., Kitzberger, T., Reid, M. and Villalba, R. (1994) Disturbance regime and disturbance interactions in a Rocky Mountain subalpine forest. *Ecology*, **82**, 125–135.

277. Voellmy, S. (1964) Über die Zerstörungskraft von Lawinen. *Schweizerische Bauzeitung*, **73** (12), 159–165, 212–217, 246–249, 280–285 (in German).

278. Volk, G. and Kleemayr, K. (1999) Lawinensimulationsmodell ELBA. *Journal for Torrent, Avalanche and Erosion Control*, **138**, 23–32 (in German).

279. Volk, G. (2005) *Handbuch ELBA+*, NiT Technisches Büro GmbH, Pressbaum, (in German).

280. Weiler, C., Hopf, J., Lied, K. and Bakkehoi, S. (1995) Methoden zur Berechnung der Auslauflänge von Fließlawinen im Österreichischen Alpenraum. Austrian Service for Torrent and Avalanche Control Tyrol (in German).

281. Wilhelm, C. (1997) Wirtschaftlichkeit im Lawinenschutz – Methodik und Erhebungen zur Beurteilung von Schutzmaßnahmen mittels quantitativer Risikoanalyse und ökonomischer Bewertung. WSL Swiss Federal Institute for Snow and Avalanche Research SLF Davos (in German).

282. Winkler, C. and Burkard, A. (2006) Korridorplanung. tec21 41, 4–6 (in German).

283. Witt, K.J. (2009) Grundbau-Taschenbuch. Tl. 3. Gründungen und geotechnische Bauwerke, Ernst und Sohn/Wiley VCH, Berlin (in German).

284. Wopfner, H. and Hopf, J. (1963) Versuche mit Kolktafeln an der Schneeforschungsstelle Wattener Lizum (Tirol) in den Jahren 1950–1955. Forstliche Bundesversuchsanstalt Mariabrunn Vienna (in German).

285. Yoshimatsu, H. (1991) Study of mechanics of debris flow and its simulation model. *Journal of Japan Landslide Society*, **28**, 2.

286. Zsigmondy, E. and Paulke, W. (1933) *Gefahren der Alpen*, 6th edn, Rother.

Index

The Technical Avalanche Protection Handbook, First Edition. Edited by Florian Rudolf-Miklau, Siegfried Sauermoser and Arthur I. Mears.

D

www.ingramcontent.com/pod-product-compliance
Ingram Content Group UK Ltd.
Pitfield, Milton Keynes, MK11 3LW, UK
UKHW051117290926
15471UKWH00010B/1731

9 783433 030349